Hilbert Spaces of Analytic Functions

Volume 51

CRM PROCEEDINGS & LECTURE NOTES

Centre de Recherches Mathématiques
Montréal

Hilbert Spaces of Analytic Functions

Javad Mashreghi
Thomas Ransford
Kristian Seip
Editors

The Centre de Recherches Mathématiques (CRM) of the Université de Montréal was created in 1968 to promote research in pure and applied mathematics and related disciplines. Among its activities are special theme years, summer schools, workshops, postdoctoral programs, and publishing. The CRM is supported by the Université de Montréal, the Province of Québec (FQRNT), and the Natural Sciences and Engineering Research Council of Canada. It is affiliated with the Institut des Sciences Mathématiques (ISM) of Montréal. The CRM may be reached on the Web at www.crm.math.ca.

American Mathematical Society
Providence, Rhode Island USA

The production of this volume was supported in part by the Fonds Québécois de la Recherche sur la Nature et les Technologies (FQRNT) and the Natural Sciences and Engineering Research Council of Canada (NSERC).

2000 *Mathematics Subject Classification.* Primary 46E20, 46E22, 47B32, 31C25.

Library of Congress Cataloging-in-Publication Data

Hilbert spaces of analytic functions / Javad Mashreghi, Thomas Ransford, Kristian Seip, editors.
p. cm. — (CRM proceedings & lecture notes ; v. 51)
Includes bibliographical references.
ISBN 978-0-8218-4879-1 (alk. paper)
1. Hilbert space—Congresses. 2. Analytic functions—Congresses. 3. Potential theory (Mathematics)—Congresses. I. Mashreghi, Javad. II. Ransford, Thomas. III. Seip, Kristian, 1962–

QA322.4.H55 2010
515′.733—dc22

2010003337

This volume was submitted to the American Mathematical Society
in camera ready form by the Centre de Recherches Mathématiques.
Visit the AMS home page at `http://www.ams.org/`

10 9 8 7 6 5 4 3 2 1   15 14 13 12 11 10

Contents

List of Participants

Evgueni Abakoumov
Université Paris-Est

Fatma Z. Abdessameud
Université Saad Dahlab Blida

Jim Agler
UC San Diego

Tayeb Aïssiou
McGill University

Nadya Askaripour
University of Western Ontario

Hirbod Assa
Université de Montréal

Amoah G. Atingabunor
Khomanani Business College

Joseph A. Ball
Virginia Tech

Ferenc Balogh
Concordia University

Laurent Baratchart
INRIA Sophia Antipolis – Méditerranée

André Boivin
University of Western Ontario

Alexander Borichev
Université de Provence

Abdellatif Bourhim
Université Laval

Marcus Carlsson
Purdue University

Nicolas Chevrot
Université Laval

Yemon Choi
Université Laval

Joseph A. Cima
UNC Chapel Hill

Constantin Costara
Universitatea Ovidius

Galia Dafni
Concordia University

Ronald G. Douglas
Texas A&M University

Omar El-Fallah
Université Mohammed V

Yasser Farhat
Université Laval

Tatyana Foth
University of Western Ontario

Richard Fournier
Université de Montréal

Emmanuel Fricain
Université Claude Bernard Lyon 1

Frédéric Gaunard
Université Bordeaux 1

Paul M. Gauthier
Université de Montréal

Kohur Gowrisankaran
McGill University

Dominique Guillot
Université Laval

Andreas Hartmann
Université Bordeaux 1

Tarik Jari
Université Laval

Dmitry Khavinson
University of South Florida

Daniela Kraus
Universität Würzburg

Alex Louis
University of Swaziland

Yura Lyubarskii
NTNU

Mostafa Mache
Université Laval

Davood Malekzadeh
Université Laval

Jordi Marzo
NTNU

Javad Mashreghi
Université Laval

Tesfa Mengestie
NTNU

Putinar Mihai
UC Santa Barbara

Joules Nahas
UC Santa Barbara

Mostafa Nasri
Université Laval

Joaquim Ortega-Cerdà
Universitat de Barcelona

Vladimir Peller
Michigan State University

Mihai Putinar
UC Santa Barbara

Quentin Rajon
Université Laval

Thomas J. Ransford
Université Laval

Richard Rochberg
Washington University in St. Louis

William T. Ross
University of Richmond

Stephan Ruscheweyh
Universität Würzburg

Sergey Sadov
Memorial University of Newfoundland

Kristian Seip
NTNU

Alexis Selezneff
Université Laval

Mahmood Shabankhah
Université Laval

Richard Spjut
UC Santa Barbara

Qingyun Wang
Washington University in St. Louis

Hasi Wulan
Shantou University

Nicolas Young
Leeds University

Hong Yue
Concordia University

Jaroslav Zemánek
Polish Academy of Sciences

Nina Zorboska
University of Manitoba

List of Speakers

Evgueni Abakumov
Université Paris-Est

Fatma Z. Abdessameud
Université Saad Dahlab Blida

Jim Agler
UC San Diego

Joseph A. Ball
Virginia Tech

Ferenc Balogh
Concordia University

Laurent Baratchart
INRIA Sophia Antipolis – Méditarranée

André Boivin
University of Western Ontario

Alexander Borichev
Université de Provence

Abdelatif Bourhim
Université Laval

Nicolas Chevrot
Université Laval

Joseph Cima
UNC Chapel Hill

Constantin Costara
Universitatea Ovidius

Galia Dafni
Concordia University

Ronald G. Douglas
Texas A&M University

Omar El-Fallah
Université Mohammed V

Tatyana Foth
University of Western Ontario

Richard Fournier
Université de Montréal

Emmanuel Fricain
Université Lyon 1

Paul Gauthier
Université de Montréal

Kohur GowriSankaran
McGill University

Andreas Hartmann
Université Bordeaux 1

Dmitry Khavinson
University of South Florida

Danieal Kraus
Universität Würzburg

Yura Lyubarskii
NTNU

Jordi Marzo
NTNU

Joules Nahas
UC Santa Barbara

Joaquim Ortega-Cerdà
Universitat de Barcelona

Vladimir Peller
Michigan State University

Mihai Putinar
UC Santa Barbara

Thomas Ransford
Université Laval

Richard Rochberg
Washington University in St. Louis

William T. Ross
University of Richmond

Stephan Ruscheweyh
Universität Würzburg

Mahmood Shabankhah
Université Laval

Hassi Wulan
Shantou University

Nicolas Young
University of Leeds

Hong Yue
Concordia University

Jaroslav Zemánek
Polish Academy of Science

Preface

The workshop entitled Hilbert Spaces of Analytic Functions was held at the Centre de recherches mathématiques (CRM), Montréal, from 8 to 12 December 2008. Even though this event was not a part of the CRM thematic year, 62 mathematicians attended the workshop. They formed a blend of researchers with a common interest in spaces of analytic functions, but seen from many different angles.

Hilbert spaces of analytic functions are currently a very active field of complex analysis. The Hardy space H^2 is the most senior member of this family. Its relatives, such as the Bergman space A^p, the Dirichlet space $\mathcal{D}$, the de Branges–Rovnyak spaces $\mathcal{H}(b)$, and various spaces of entire functions, have been extensively studied by prominent mathematicians since the beginning of the last century. These spaces have been exploited in different fields of mathematics and also in physics and engineering. For example, de Branges used them to solve the Bieberbach conjecture, and Zames, a late professor of McGill University, applied them to construct his theory of H^∞ control. But there are still many open problems, old and new, which attract a wide spectrum of mathematicians.

In this conference, 38 speakers talked about Hilbert spaces of analytic functions. In five days, a wide variety of applications were discussed. It was a lively atmosphere in which many mutual research projects were designed.

Javad Mashreghi
Thomas Ransford
Kristian Seip

Centre de Recherches Mathématiques
CRM Proceedings and Lecture Notes
Volume **51**, 2010

Canonical de Branges – Rovnyak Model Transfer-Function Realization for Multivariable Schur-Class Functions

Joseph A. Ball and Vladimir Bolotnikov

ABSTRACT. Associated with any Schur-class function $S(z)$ (i.e., a contractive holomorphic function on the the unit disk) is the de Branges – Rovnyak kernel $K_S(z,\zeta) = [I - S(z)S(\zeta)^*]/(1 - z\bar{\zeta})$ and the de Branges – Rovnyak reproducing kernel Hilbert space $\mathcal{H}(K_S)$. This space plays a prominent role in system theory as a canonical-model state space for a transfer-function realization of a given Schur-class function. There has been recent work extending the notion of Schur-class function to several multivariable settings. We here make explicit to what extent the role of de Branges – Rovnyak spaces as the canonical-model state space for transfer-function realizations of Schur-class functions extends to these multivariable settings.

1. Introduction

Let $\mathcal{U}$ and $\mathcal{Y}$ be two Hilbert spaces and let $\mathcal{L}(\mathcal{U},\mathcal{Y})$ be the space of all bounded linear operators between $\mathcal{U}$ and $\mathcal{Y}$. The operator-valued version of the classical Schur class $\mathcal{S}(\mathcal{U},\mathcal{Y})$ is defined to be the set of all holomorphic, contractive $\mathcal{L}(\mathcal{U},\mathcal{Y})$-valued functions on $\mathbb{D}$. The following equivalent characterizations of the Schur class are well known. Here we use the notation H^2 for the Hardy space over the unit disk and $H^2_{\mathcal{X}} = H^2 \otimes \mathcal{X}$ for the Hardy space with values in the auxiliary Hilbert space $\mathcal{X}$.

Theorem 1.1. *Let $S\colon \mathbb{D} \to \mathcal{L}(\mathcal{U},\mathcal{Y})$ be given. Then the following are equivalent:*

(1) (a) *$S \in \mathcal{S}(\mathcal{U},\mathcal{Y})$, i.e., S is holomorphic on $\mathbb{D}$ with $\|S(z)\| \le 1$ for all $z \in \mathbb{D}$.*
(b) *The operator $M_S\colon f(z) \mapsto S(z)f(z)$ of multiplication by S defines a contraction operator from $H^2_{\mathcal{U}}$ to $H^2_{\mathcal{Y}}$.*
(c) *S satisfies the von Neumann inequality: $\|S(T)\| \le 1$ for any strictly contractive operator T on a Hilbert space $\mathcal{H}$, where $S(T)$ is defined*

2000 *Mathematics Subject Classification.* 47A57.
Key words and phrases. Operator-valued functions, Schur multiplier, canonical functional model, reproducing kernel Hilbert space.
This is the final form of the paper.

by

$$S(T) = \sum_{n=0}^{\infty} S_n \otimes T^n \in \mathcal{L}(\mathcal{U} \otimes \mathcal{H}, \mathcal{Y} \otimes \mathcal{H}) \qquad \textit{if } S(z) = \sum_{n=0}^{\infty} S_n z^n.$$

(2) *The associated kernel*

$$K_S(z,\zeta) = \frac{I_{\mathcal{Y}} - S(z)S(\zeta)^*}{1 - z\bar{\zeta}} \tag{1.1}$$

is positive on $\mathbb{D} \times \mathbb{D}$, *i.e., there exists an operator-valued function* $H\colon \mathbb{D} \to \mathcal{L}(\mathcal{X}, \mathcal{Y})$ *for some auxiliary Hilbert space* $\mathcal{X}$ *so that*

$$K_S(z,\zeta) = H(z)H(\zeta)^*. \tag{1.2}$$

(3) *There is an auxiliary Hilbert space* $\mathcal{X}$ *and a unitary connecting operator*

$$\mathbf{U} = \begin{bmatrix} A & B \\ C & D \end{bmatrix} : \begin{bmatrix} \mathcal{X} \\ \mathcal{U} \end{bmatrix} \to \begin{bmatrix} \mathcal{X} \\ \mathcal{Y} \end{bmatrix}$$

so that $S(z)$ *can be expressed as*

$$S(z) = D + zC(I - zA)^{-1}B. \tag{1.3}$$

(4) $S(z)$ *has a realization as in* (1.3) *where the connecting operator* $\mathbf{U}$ *is any one of* (i) *isometric,* (ii) *coisometric, or* (iii) *contractive.*

We note that the equivalence of any of (1a), (1b), (1c) with (2) and (3) can be gleaned, e.g., from Lemma V.3.2, Proposition I.8.3, Proposition V.8.1 and Theorem V.3.1 in [26]. As for condition (4), it is trivial to see that (3) implies (4) and then it is easy to verify directly that (4) implies (1a). Alternatively, one can use Lemma 5.1 from Andô's notes [6] to see directly that (4iii) implies (4ii) (see Remark 2.2 below).

The reproducing kernel Hilbert space $\mathcal{H}(K_S)$ with the *de Branges – Rovnyak kernel* $K_S(z,\zeta)$ is the classical *de Branges – Rovnyak reproducing kernel Hilbert space* associated with the Schur-class function S which has been much studied over the years, both as an object in itself and as a tool for other types of applications (see [6, 11–13, 16–18, 20, 21, 24, 27, 28, 31]). The special role of the de Branges – Rovnyak space in connection with the transfer-function realization for Schur-class functions is illustrated in the following theorem; this form of the results appears at least implicitly in the work of de Branges – Rovnyak [20, 21].

Theorem 1.2. *Suppose that the function* S *is in the Schur class* $\mathcal{S}(\mathcal{U}, \mathcal{Y})$ *and let* $\mathcal{H}(K_S)$ *be the associated de Branges – Rovnyak space. Define operators* A, B, C, D *by*

$$A\colon f(z) \mapsto \frac{f(z) - f(0)}{z}, \qquad B\colon u \mapsto \frac{S(z) - S(0)}{z}u,$$
$$C\colon f(z) \mapsto f(0), \qquad D\colon u \mapsto S(0)u.$$

Then the operator-block matrix $\mathbf{U} = \left[\begin{smallmatrix} A & B \\ C & D \end{smallmatrix}\right]$ *has the following properties:*

(1) $\mathbf{U}$ *defines a coisometry from* $\mathcal{H}(K_S) \oplus \mathcal{U}$ *to* $\mathcal{H}(K_S) \oplus \mathcal{Y}$.

(2) (C, A) *is an* observable pair, *i.e.,*

$$CA^n f = 0 \textit{ for all } n = 0, 1, 2, \dots \implies f = 0 \textit{ as an element of } \mathcal{H}(K_S).$$

(3) *We recover* $S(z)$ *as* $S(z) = D + zC(I - zA)^{-1}B$.

(4) *If* $\left[\begin{smallmatrix} A' & B' \\ C' & D' \end{smallmatrix}\right] : \mathcal{X} \oplus \mathcal{U} \to \mathcal{X} \oplus \mathcal{Y}$ *is another colligation matrix with properties* (1), (2), (3) *above* (*with* $\mathcal{X}$ *in place of* $\mathcal{H}(K_S)$), *then there is a unitary operator* $U \colon \mathcal{H}(K_S) \to \mathcal{X}$ *so that*

$$\begin{bmatrix} U & 0 \\ 0 & I_{\mathcal{Y}} \end{bmatrix} \begin{bmatrix} A & B \\ C & D \end{bmatrix} = \begin{bmatrix} A' & B' \\ C' & D' \end{bmatrix} \begin{bmatrix} U & 0 \\ 0 & I_{\mathcal{U}} \end{bmatrix}.$$

It is easily seen from characterization (1a) in Theorem 1.1 that

$$S \in \mathcal{S}(\mathcal{U}, \mathcal{Y}) \iff \overline{S} \in \mathcal{S}(\mathcal{Y}, \mathcal{U}) \qquad \text{where } \overline{S}(z) := S(\bar{z})^*. \tag{1.4}$$

Hence for a given Schur-class function S there is also associated a dual de Branges–Rovnyak space $\mathcal{H}(K_{\overline{S}})$ with reproducing kernel $K_{\overline{S}}(z,\zeta) = [I - S(\bar{z})^* S(\bar{\zeta})]/(1 - z\bar{\zeta})$. The space $\mathcal{H}(K_{\overline{S}})$ plays the same role for isometric realizations of S as $\mathcal{H}(K_S)$ plays for coisometric realizations, as illustrated in the next theorem; this theorem is just the dual version of Theorem 1.2 upon application of the transformation (1.4).

Theorem 1.3. *Suppose that the function* S *is in the Schur class* $\mathcal{S}(\mathcal{U}, \mathcal{Y})$ *and let* $\mathcal{H}(K_{\overline{S}})$ *be the associated dual de Branges – Rovnyak space. Define operators* $A_{\rm d}, B_{\rm d}, C_{\rm d}, D_{\rm d}$ *by*

$$A_{\rm d} \colon g(z) \mapsto z g(z) - S(\bar{z})^* \tilde{g}(0), \qquad B_{\rm d} \colon u \mapsto \big(I - S(\bar{z})^* S(0)\big) u,$$
$$C_{\rm d} \colon g(z) \mapsto \tilde{g}(0), \qquad D_{\rm d} \colon u \mapsto S(0) u,$$

where $\tilde{g}(0)$ *is the unique vector in* $\mathcal{Y}$ *such that*

$$\langle \tilde{g}(0), y \rangle_{\mathcal{Y}} = \left\langle g(z), \frac{S(\bar{z})^* - S(0)^*}{z} y \right\rangle_{\mathcal{H}(K_{\overline{S}})} \qquad \text{for all } y \in \mathcal{Y}.$$

Then the operator-block matrix $\mathbf{U}_{\rm d} = \left[\begin{smallmatrix} A_{\rm d} & B_{\rm d} \\ C_{\rm d} & D_{\rm d} \end{smallmatrix}\right]$ *has the following properties*:

(1) $\mathbf{U}_{\rm d}$ *defines an isometry from* $\mathcal{H}(K_{\overline{S}}) \oplus \mathcal{U}$ *to* $\mathcal{H}(K_{\overline{S}}) \oplus \mathcal{Y}$,
(2) $(A_{\rm d}, B_{\rm d})$ *is a* controllable pair, *i.e.*, $\bigvee_{n \ge 0} \operatorname{Ran} A_{\rm d}^n B_{\rm d} = \mathcal{H}(K_{\overline{S}})$, *where* $\bigvee$ *stands for the closed linear span.*
(3) *We recover* $S(z)$ *as* $S(z) = D_{\rm d} + z C_{\rm d}(I - z A_{\rm d})^{-1} B_{\rm d}$.
(4) *If* $\left[\begin{smallmatrix} A' & B' \\ C' & D' \end{smallmatrix}\right] : \mathcal{X} \oplus \mathcal{U} \to \mathcal{X} \oplus \mathcal{Y}$ *is another colligation matrix with properties* (1), (2), (3) *above* (*with* $\mathcal{X}$ *in place of* $\mathcal{H}(K_{\overline{S}})$), *then there is a unitary operator* $U \colon \mathcal{H}(K_{\overline{S}}) \to \mathcal{X}$ *so that*

$$\begin{bmatrix} U & 0 \\ 0 & I_{\mathcal{Y}} \end{bmatrix} \begin{bmatrix} A_{\rm d} & B_{\rm d} \\ C_{\rm d} & D_{\rm d} \end{bmatrix} = \begin{bmatrix} A' & B' \\ C' & D' \end{bmatrix} \begin{bmatrix} U & 0 \\ 0 & I_{\mathcal{U}} \end{bmatrix}.$$

In addition to the kernels K_S and $K_{\overline{S}}$, there is a positive kernel $\widehat{K}_S$ which combines these two and is defined as follows:

$$\widehat{K}(z,\zeta) = \begin{bmatrix} K_S(z,\zeta) & \frac{S(z) - S(\bar{\zeta})}{z - \bar{\zeta}} \\ \frac{\overline{S}(z) - \overline{S}(\zeta)}{z - \bar{\zeta}} & K_{\overline{S}}(z,\zeta) \end{bmatrix} = \begin{bmatrix} \frac{I - S(z) S(\zeta)^*}{1 - z\bar{\zeta}} & \frac{S(z) - S(\bar{\zeta})}{z - \bar{\zeta}} \\ \frac{S(\bar{z})^* - S(\zeta)^*}{z - \bar{\zeta}} & \frac{I - S(\bar{z})^* S(\bar{\zeta})}{1 - z\bar{\zeta}} \end{bmatrix}. \tag{1.5}$$

It turns out that $\widehat{K}$ is also a positive kernel on $\mathbb{D} \times \mathbb{D}$ and the associated reproducing kernel Hilbert space $\mathcal{H}(\widehat{K}_S)$ is the canonical functional-model state space for unitary realizations of S, as summarized in the following theorem. This result also appears at least implicitly in the work of de Branges and Rovnyak [20,21] and more explicitly the paper of de Branges and Shulman [22], where the two-component space $\mathcal{H}(\widehat{K}_S)$ associated with the Schur-class function S is denoted as $\mathcal{D}(S)$; see also [11] for an explanation of the connections with the Sz.-Nagy – Foias model space.

Theorem 1.4. *Suppose that the function S is in the Schur class $\mathcal{S}(\mathcal{U},\mathcal{Y})$ and let $\widehat{K}(z,\zeta)$ be the positive kernel on $\mathbb{D}$ given by* (1.5). *Define operators $\hat{A},\widehat{B},\widehat{C},\widehat{D}$ by*

$$\hat{A}\colon \begin{bmatrix} f(z) \\ g(z) \end{bmatrix} \mapsto \begin{bmatrix} [f(z)-f(0)]/z \\ zg(z)-S(z)f(0) \end{bmatrix}, \qquad \widehat{B}\colon u \mapsto \begin{bmatrix} ([S(z)-S(0)]/z)u \\ (I-S(\bar{z})^*S(0))u \end{bmatrix}$$

$$\widehat{C}\colon \begin{bmatrix} f(z) \\ g(z) \end{bmatrix} \mapsto f(0), \qquad \widehat{D}\colon u \mapsto S(0)u.$$

Then the operator-block matrix $\widehat{\mathbf{U}} = \left[\begin{smallmatrix} \hat{A} & \widehat{B} \\ \widehat{C} & \widehat{D} \end{smallmatrix}\right]$ satisfies the following:

(1) *$\widehat{\mathbf{U}}$ defines a unitary operator from $\mathcal{H}(\widehat{K}_S)\oplus\mathcal{U}$ onto $\mathcal{H}(\widehat{K}_S)\oplus\mathcal{Y}$.*
(2) *$\widehat{\mathbf{U}}$ is a* closely connected *operator colligation, i.e.,*

$$\bigvee_{n\geq 0}\{\operatorname{Ran}\hat{A}^n\widehat{B}, \operatorname{Ran}\hat{A}^{*n}\widehat{C}^*\} = \mathcal{H}(\widehat{K}_S).$$

(3) *We recover $S(z)$ as $S(z) = \widehat{D} + z\widehat{C}(I-z\hat{A})^{-1}\widehat{B}$.*
(4) *If $\left[\begin{smallmatrix} A' & B' \\ C' & D' \end{smallmatrix}\right]\colon \mathcal{X}\oplus\mathcal{U}\to\mathcal{X}\oplus\mathcal{Y}$ is any other operator colligation satisfying conditions* (1), (2), (3) *above (with $\mathcal{X}$ in place of $\mathcal{H}(\widehat{K}_S)$), then there is a unitary operator $U\colon \mathcal{H}(\widehat{K}_S)\to\mathcal{X}$ so that*

$$\begin{bmatrix} U & 0 \\ 0 & I_{\mathcal{Y}} \end{bmatrix}\begin{bmatrix} \hat{A} & \widehat{B} \\ \widehat{C} & \widehat{D} \end{bmatrix} = \begin{bmatrix} A' & B' \\ C' & D' \end{bmatrix}\begin{bmatrix} U & 0 \\ 0 & I_{\mathcal{U}} \end{bmatrix}.$$

Our goal in this article is to present multivariable analogues of Theorem 1.2. The multivariable settings which we shall discuss are (1) the unit ball $\mathbb{B}^d$ in $\mathbb{C}^d$ and the associated Schur class of contractive multipliers between vector-valued Drury–Arveson spaces $\mathcal{H}_{\mathcal{U}}(k_{\rm d})$ and $\mathcal{H}_{\mathcal{Y}}(k_{\rm d})$, (2) the polydisk with the associated Schur class taken to be the class of contractive operator-valued functions on $\mathbb{D}^d$ which satisfy a von Neumann inequality, and (3) a more general setting where the underlying domain is characterized via a polynomial-matrix defining function and the Schur class is defined by the appropriate analogue of the von Neumann inequality. In these multivariable settings, the analogues of Theorem 1.1 have already been set down at length elsewhere (see [3, 15, 23] for the ball case, [1, 2, 14] for the polydisk case, and [4, 5, 9] for the case of domains with polynomial-matrix defining function — see [8] for a survey). Our emphasis here is to make explicit how Theorem 1.2 can be extended to these multivariable settings. While the reproducing kernel spaces themselves appear in a straightforward fashion, the canonical model operators on these spaces are more muddled: in the coisometric case, while the analogues of the output operator C and the feedthrough operator D are tied down, there is no canonical choice of the analogue of the state operator A and the input operator B: A and B are required to solve certain types of Gleason problems; we refer to [25] and [30, Section 6.6] for some perspective on the Gleason problems in general. The Gleason property can be formulated also in terms of the adjoint operators A^* and B^*: the actions of the adjoint operators are prescribed on a certain canonically prescribed proper subspace of the whole state space. From this latter formulation, one can see that the Gleason problem, although at first sight appearing to be rather complicated, always has solutions. Also, the adjoint of the colligation matrix, rather than being isometric, is required only to be isometric on a certain subspace of the whole space $\mathcal{X}\oplus\mathcal{Y}$. With these adjustments, Theorem 1.2 goes through in the three settings. Most of these results appear in [10] for the ball case and in more implicit form in [14] for the polydisk case, although not in the precise formulation presented here. The parallel results for the third setting are presented here for the first time. We plan to discuss multivariable analogs of Theorems 1.3 and 1.4 in a future publication.

The paper is organized as follows. After the present Introduction, Section 2 lays out the results for the ball case, Section 3 for the polydisk case, and Section 4 for the case of domains with polynomial-matrix defining function. At the end of Section 4 we indicate how the results of Sections 2 and 3 can be recovered as special cases of the general formalism in Section 4.

2. de Branges – Rovnyak kernel associated with a Schur multiplier on the Drury – Arveson space

A natural extension of the Szegő kernel is the Drury – Arveson kernel

$$k_d(z,\zeta) = \frac{1}{1 - z_1\overline{\zeta_1} - \cdots - z_d\overline{\zeta_d}} =: \frac{1}{1 - \langle z, \zeta\rangle_{\mathbb{C}^d}}.$$

The kernel $k_d(z,\zeta)$ is positive on $\mathbb{B}^d \times \mathbb{B}^d$ where

$$\mathbb{B}^d = \{z = (z_1, \dots, z_d) \in \mathbb{C}^d \colon \langle z, z\rangle = |z_1|^2 + \cdots + |z_d|^2 < 1\}$$

is the unit ball in $\mathbb{C}^d$, and the associated reproducing kernel Hilbert space $\mathcal{H}(k_{\rm d})$ is called the *Drury – Arveson space.* For $\mathcal{X}$ any auxiliary Hilbert space, we use the shorthand notation $\mathcal{H}_{\mathcal{X}}(k_d)$ for the space $\mathcal{H}(k_d) \otimes \mathcal{X}$ of vector-valued Drury – Arveson-space functions. A holomorphic operator-valued function $S \colon \mathbb{B}^d \to \mathcal{L}(\mathcal{U}, \mathcal{Y})$ is said to be a *Drury – Arveson space multiplier* if the multiplication operator $M_S \colon f(z) \mapsto S(z)f(z)$ defines a bounded operator from $\mathcal{H}_{\mathcal{U}}(k_d)$ to $\mathcal{H}_{\mathcal{Y}}(k_d)$. In case in addition M_S defines a contraction operator ($\|M_S\|_{\rm op} \le 1$), we say that S is in the Schur-multiplier class $\mathcal{S}_d(\mathcal{U}, \mathcal{Y})$. Then the following theorem is the analogue of Theorem 1.1 for this setting; this result appears in [10, 15, 23]. The alert reader will notice that there is no analogue of condition (1a) in Theorem 1.1 in the following theorem.

Theorem 2.1. *Let $S \colon \mathbb{B}^d \to \mathcal{L}(\mathcal{U}, \mathcal{Y})$ be given. Then the following are equivalent:*

(1) (b) *$S \in \mathcal{S}_d(\mathcal{U}, \mathcal{Y})$, i.e., the operator M_S of multiplication by S defines a contraction operator from $\mathcal{H}_{\mathcal{U}}(k_d)$ into $\mathcal{H}_{\mathcal{Y}}(k_d)$.*

(c) *S satisfies the von Neumann inequality: $\|S(T)\| \le 1$ for any commutative operator d-tuple $T = (T_1, \dots, T_d)$ of operators on a Hilbert space $\mathcal{K}$ such that the operator-block row matrix $\begin{bmatrix} T_1 & \dots & T_d \end{bmatrix}$ defines a strict contraction operator from $\mathcal{K}^d$ into $\mathcal{K}$, where*

$$S(T) = \sum_{n \in \mathbb{Z}^d_+} S_n \otimes T^n \in \mathcal{L}(\mathcal{U} \otimes \mathcal{H}, \mathcal{Y} \otimes \mathcal{H}) \qquad \text{if } S(z) = \sum_{n \in \mathbb{Z}^d_+} S_n z^n. \tag{2.1}$$

Here we use the standard multivariable notation:

$$z^n = z_1^{n_1} \dots z_d^{n_d} \quad \text{and} \quad T^n = T_1^{n_1} \dots T_d^{n_d} \qquad \text{if } n = (n_1, \dots, n_d) \in \mathbb{Z}^d_+.$$

(2) *The associated kernel*

$$K_S(z,\zeta) = \frac{I_{\mathcal{Y}} - S(z)S(\zeta)^*}{1 - \langle z, \zeta\rangle} \tag{2.2}$$

is positive on $\mathbb{B}\times\mathbb{B}$, *i.e., there exists an operator-valued function* $H\colon \mathbb{B}^d \to \mathcal{L}(\mathcal{X},\mathcal{Y})$ *for some auxiliary Hilbert space* $\mathcal{X}$ *so that* $K_S(z,\zeta) = H(z)H(\zeta)^*$.

(3) *There is an auxiliary Hilbert space* $\mathcal{X}$ *and a unitary connecting operator*

$$\mathbf{U} = \begin{bmatrix} A & B \\ C & D \end{bmatrix} = \begin{bmatrix} A_1 & B_1 \\ \vdots & \vdots \\ A_d & B_d \\ C & D \end{bmatrix} : \begin{bmatrix} \mathcal{X} \\ \mathcal{U} \end{bmatrix} \to \begin{bmatrix} \mathcal{X}^d \\ \mathcal{Y} \end{bmatrix} \tag{2.3}$$

so that $S(z)$ *can be expressed as*

$$S(z) = D + C(I - Z_{\text{row}}(z)A)^{-1}Z_{\text{row}}(z)B, \tag{2.4}$$

where we have set

$$Z_{\text{row}}(z) = \begin{bmatrix} z_1 I_{\mathcal{X}} & \dots & z_d I_{\mathcal{X}} \end{bmatrix}.$$

(4) $S(z)$ *has a realization as in* (2.4) *where the connecting operator* $\mathbf{U}$ *is any one of* (i) *isometric,* (ii) *coisometric, or* (iii) *contractive.*

Remark 2.2. Statement (4iii) concerning contractive realizations is not mentioned in [15] but is discussed in [10, 23]. The approach in [23] is to show that for S of the form (2.4) with $\mathbf{U} = \left[\begin{smallmatrix} A & B \\ C & D \end{smallmatrix}\right]$ contractive, the inequality $\|S(T)\| \le 1$ holds for any commutative operator d-tuple $T = (T_1, \dots, T_d)$ with $\|\begin{bmatrix} T_1 & \dots & T_d \end{bmatrix}\| < 1$, i.e, one verifies (4iii) $\implies$ (1c).

The idea of the second approach in [10] is to embed the contraction $\mathbf{U} = \left[\begin{smallmatrix} A & B \\ C & D \end{smallmatrix}\right]$ into a coisometry $\left[\begin{smallmatrix} A & \widetilde{B} \\ C & \widetilde{D} \end{smallmatrix}\right] = \left[\begin{smallmatrix} A & B & B_1 \\ C & D & D_1 \end{smallmatrix}\right]$ with associated transfer function of the form $\widetilde{S}(z) = \begin{bmatrix} S(z) & S_1(z) \end{bmatrix}$ equal to an extension of $S(z)$ with a larger input space. From the coisometry property of $\left[\begin{smallmatrix} A & \widetilde{B} \\ C & \widetilde{D} \end{smallmatrix}\right]$ one sees that $K_{\widetilde{S}}(z,w) = C(I - Z_{\text{row}}(z)A)^{-1}(I - A^*Z_{\text{row}}(\zeta)^{-1}C^*$, i.e., $\widetilde{S}$ meets condition (2) for the Schur-class $\mathcal{S}_d(\mathcal{U}\oplus\mathcal{U}_1,\mathcal{Y})$ with $H(z) = C(I - Z_{\text{row}}(z)A)^{-1}$. From the equivalence (1b) $\iff$ (2), it is easy now to read off that $S \in \mathcal{S}_d(\mathcal{U},\mathcal{Y})$.

A third approach worked out for the classical case but extendable to multivariable settings appears in Andô's notes [6, Lemma 5.1]. Given a contractive colligation $\mathbf{U} = \left[\begin{smallmatrix} A & B \\ C & D \end{smallmatrix}\right]$, one can keep the input and output spaces the same but enlarge the state space to construct a coisometric colligation $\widehat{\mathbf{U}} = \left[\begin{smallmatrix} \hat{A} & \widehat{B} \\ \widehat{C} & \widehat{D} \end{smallmatrix}\right]$ having the same transfer function, namely:

$$\hat{A} = \begin{bmatrix} A & Q_{11} & Q_{12} & 0 & 0 & \dots \\ 0 & 0 & 0 & I & 0 & \dots \\ 0 & 0 & 0 & 0 & I & \\ & & & & \ddots & \ddots \end{bmatrix}, \qquad \widehat{B} = \begin{bmatrix} B \\ 0 \\ 0 \\ \vdots \end{bmatrix},$$

$$\widehat{C} = \begin{bmatrix} C & Q_{21} & Q_{22} & 0 & 0 & \dots \end{bmatrix}, \qquad \widehat{D} = D$$

where

$$Q = \begin{bmatrix} Q_{11} & Q_{12} \\ Q_{12}^* & Q_{22} \end{bmatrix} = (I - \mathbf{U}\mathbf{U}^*)^{1/2}.$$

In this way one gets a direct proof of (4iii) $\implies$ (4ii).

For colligations $\mathbf{U}$ of the form (2.3), it turns out that a somewhat weaker notion of coisometry is more useful than simply requiring that $\mathbf{U}$ be coisometric.

Definition 2.3. The operator-block matrix $\mathbf{U}$ of the form (2.3) is *weakly coisometric* if the restriction of $\mathbf{U}^*$ to the subspace

$$\mathcal{D}_{\mathbf{U}^*} := \bigvee_{\substack{\zeta\in\mathbb{B}^d \\ y\in\mathcal{Y}}} \begin{bmatrix} Z_{\text{row}}(\zeta)^*(I - A^* Z_{\text{row}}(\zeta)^*)^{-1}C^* y \\ y \end{bmatrix} \subset \begin{bmatrix} \mathcal{X}^d \\ \mathcal{Y} \end{bmatrix} \tag{2.5}$$

is isometric.

It turns out that the weak-coisometry property of the colligation (2.3) is exactly what is needed to guarantee the decomposition

$$K_S(z,\zeta) = C(I - Z_{\text{row}}(z)A)(I - A^* Z_{\text{row}}(\zeta)^*)^{-1}C^* \tag{2.6}$$

of the de Branges – Rovnyak kernel K_S associated with S of the form (2.4) (see Proposition 1.5 in [10]).

2.1. Weakly coisometric canonical functional-model colligations. As the kernel K_S given by (2.2) is positive on $\mathbb{B} \times \mathbb{B}$, we can associate a reproducing kernel Hilbert space $\mathcal{H}(K_S)$ just as in the classical case, where now the elements of $\mathcal{H}(K_S)$ are holomorphic $\mathcal{Y}$-valued functions on $\mathbb{B}^d$. In the classical case, as we see from Theorem 1.2, there are canonically defined operators A, B, C, D so that the operator-block matrix $\mathbf{U} = \left[\begin{smallmatrix} A & B \\ C & D \end{smallmatrix}\right]$ is coisometric from $\mathcal{H}(K_S) \oplus \mathcal{U}$ to $\mathcal{H}(K_S) \oplus \mathcal{Y}$ and yields the essentially unique observable, coisometric realization for $S \in \mathcal{S}(\mathcal{U},\mathcal{Y})$. For the present Drury – Arveson space setting, a similar result holds, but the operators A, B in the colligation matrix $\mathbf{U}$ are not completely uniquely determined. To explain the result, we say that the operator $A\colon \mathcal{H}(K_S) \to \mathcal{H}(K_S)^d$ *solves the Gleason problem for* $\mathcal{H}(K_S)$ if the identity

$$f(z) - f(0) = \sum_{k=1}^{d} z_k (Af)_k(z) \qquad \text{holds for all } f \in \mathcal{H}(K_S), \tag{2.7}$$

where we write $(Af)(z) = \begin{bmatrix} (Af)_1(z) \\ \vdots \\ (Af)_d(z) \end{bmatrix} \in \mathcal{H}(K_S)^d$. We say that the operator $B\colon \mathcal{U} \to \mathcal{H}(K_S)^d$ *solves the* $\mathcal{H}(K_S)$*-Gleason problem for* S if the identity

$$S(z)u - S(0)u = \sum_{k=1}^{d} z_k (Bu)_k(z) \qquad \text{holds for all } u \in \mathcal{U}. \tag{2.8}$$

Solutions of such Gleason problems are easily characterized in terms of adjoint operators.

Proposition 2.4. *The operator* $A\colon \mathcal{H}(K_S) \to \mathcal{H}(K_S)^d$ *solves the Gleason problem for* $\mathcal{H}(K_S)$ *(2.7) if and only if* $A^*\colon \mathcal{H}(K_S)^d \to \mathcal{H}(K_S)$ *has the following action on special kernel functions:*

$$A^*\colon Z_{\text{row}}(\zeta)^* K_S(\cdot,\zeta)y \mapsto K_S(\cdot,\zeta)y - K_S(\cdot,0)y \qquad \textit{for all } \zeta \in \mathbb{B}^d,\ y \in \mathcal{Y}. \tag{2.9}$$

The operator $B\colon \mathcal{U} \to \mathcal{H}(K_S)^d$ *solves the* $\mathcal{H}(K_S)$*-Gleason problem for* S *(2.8) if and only if* $B^*\colon \mathcal{H}(K_S)^d \to \mathcal{U}$ *has the following action on special kernel functions:*

$$B^*\colon Z_{\text{row}}(\zeta)^* K_S(\cdot,\zeta)y \mapsto S(\zeta)^*y - S(0)^*y \qquad \textit{for all } \zeta \in \mathbb{B}^d,\ y \in \mathcal{Y}. \tag{2.10}$$

PROOF. By the reproducing kernel property, we have for $f \in \mathcal{H}(K_S)$,

$$\langle f(z) - f(0), y\rangle_{\mathcal{Y}} = \langle f, K_S(\cdot, z)y - K_S(\cdot, 0)y\rangle_{\mathcal{H}(K_S)}.$$

On the other hand,

$$\begin{aligned}\left\langle \sum_{k=1}^{d} z_k (Af)_k(z), y \right\rangle_{\mathcal{Y}} &= \sum_{k=1}^{d} \langle (Af)_k, \bar{z}_k K_S(\cdot, z)y\rangle_{\mathcal{H}(K_S)} \\ &= \langle Af, Z_{\text{row}}(z)^* K_S(\cdot, z)y\rangle_{\mathcal{H}(K_S)^d} \\ &= \langle f, A^* Z_{\text{row}}(z)^* K_S(\cdot, z)y\rangle_{\mathcal{H}(K_S)}\end{aligned}$$

and since the two latter equalities hold for all $f \in \mathcal{H}(K_S)$, $z \in \mathbb{B}^d$ and $y \in \mathcal{Y}$, the equivalence of (2.7) and (2.9) follows. Equivalence of (2.8) and (2.10) follows similarly from the computation:

$$\begin{aligned}\left\langle \sum_{k=1}^{d} z_k (Bu)_k(z), y \right\rangle_{\mathcal{Y}} &= \sum_{k=1}^{d} \langle (Bu)_k, \bar{z}_k K_S(\cdot, z)y\rangle_{\mathcal{H}(K_S)} \\ &= \langle Bu, Z_{\text{row}}(z)^* K_S(\cdot, z)y\rangle_{\mathcal{H}(K_S)^d} \\ &= \langle u, B^* Z_{\text{row}}(z)^* K_S(\cdot, z)y\rangle_{\mathcal{U}}.\end{aligned}$$

□

Let us introduce the notation

$$\mathcal{D} = \bigvee_{\substack{\zeta \in \mathbb{B}^d \\ y \in \mathcal{Y}}} Z_{\text{row}}(\zeta)^* K_S(\cdot, \zeta)y. \tag{2.11}$$

Definition 2.5. Given $S \in \mathcal{S}_d(\mathcal{U}, \mathcal{Y})$, we shall say that the block-operator matrix $\mathbf{U} = \left[\begin{smallmatrix} A & B \\ C & D \end{smallmatrix}\right]$ is a *canonical functional-model colligation* for S if

(1) $\mathbf{U}$ is contractive and the state space equals $\mathcal{H}(K_S)$.
(2) $A\colon \mathcal{H}(K_S) \to \mathcal{H}(K_S)^d$ solves the Gleason problem for $\mathcal{H}(K_S)$ (2.7).
(3) $B\colon \mathcal{U} \to \mathcal{H}(K_S)^d$ solves the $\mathcal{H}(K_S)$-Gleason problem for S (2.8).
(4) The operators $C\colon \mathcal{H}(K_S) \to \mathcal{Y}$ and $D\colon \mathcal{U} \to \mathcal{Y}$ are given by

$$C\colon f(z) \mapsto f(0), \quad D\colon u \mapsto S(0)u. \tag{2.12}$$

Remark 2.6. It is useful to have the formulas for the adjoints $C^*\colon \mathcal{Y} \to \mathcal{H}(K_S)$ and $D\colon \mathcal{Y} \to \mathcal{U}$:

$$C^*\colon y \mapsto K_S(\cdot, 0)y \quad D^*\colon y \mapsto S(0)^*y \tag{2.13}$$

which are equivalent to (2.12). The formula for D^* is obvious while the formula for C^* follows from equalities

$$\langle f, C^*y\rangle_{\mathcal{H}(K_S)} = \langle Cf, y\rangle_{\mathcal{Y}} = \langle f(0), y\rangle_{\mathcal{Y}} = \langle f, K_S(\cdot, 0)y\rangle_{\mathcal{H}(K_S)}$$

holding for every $f \in \mathcal{H}(K_S)$ and $y \in \mathcal{Y}$.

Theorem 2.7. *There exists a canonical functional-model realization for every $S \in \mathcal{S}_d(\mathcal{U}, \mathcal{Y})$.*

PROOF. Let S be in $\mathcal{S}_d(\mathcal{U}, \mathcal{Y})$ and let $\mathcal{H}(K_S)$ be the associated de Branges–Rovnyak space. Equality (2.2) can be rearranged as

$$\sum_{j=1}^{d} z_j \bar{\zeta}_j K_S(z, \zeta) + I_{\mathcal{Y}} = K_S(z, \zeta) + S(z)S(\zeta)^*$$

which in turn, can be written in the inner-product form as the identity

$$\left\langle \begin{bmatrix} Z_{\rm row}(\zeta)^*K_S(\cdot,\zeta)y \\ y \end{bmatrix}, \begin{bmatrix} Z_{\rm row}(z)^*K_S(\cdot,z)y' \\ y' \end{bmatrix} \right\rangle_{\mathcal{H}(K_S)^d\oplus\mathcal{Y}} = \left\langle \begin{bmatrix} K_S(\cdot,\zeta)y \\ S(\zeta)^*y \end{bmatrix}, \begin{bmatrix} K_S(\cdot,z)y' \\ S(z)^*y' \end{bmatrix} \right\rangle_{\mathcal{H}(K_S)\oplus\mathcal{U}} \tag{2.14}$$

holding for every $y, y' \in \mathcal{Y}$ and $\zeta, z \in \mathbb{B}^d$. The latter identity tells us that the linear map

$$V\colon \begin{bmatrix} Z_{\rm row}(\zeta)^*K_S(\cdot,\zeta)y \\ y \end{bmatrix} \mapsto \begin{bmatrix} K_S(\cdot,\zeta)y \\ S(\zeta)^*y \end{bmatrix} \tag{2.15}$$

extends to the isometry from $\mathcal{D}_V = \mathcal{D}\oplus\mathcal{Y} \subset \mathcal{H}(K_S)^d\oplus\mathcal{Y}$ (where $\mathcal{D}$ is given in (2.11)) onto

$$\mathcal{R}_V = \bigvee_{\substack{\zeta\in\mathbb{B}^d \\ y\in\mathcal{Y}}} \begin{bmatrix} K_S(\cdot,\zeta)y \\ S(\zeta)^*y \end{bmatrix} \subset \mathcal{H}(K_S)\oplus\mathcal{U}.$$

Extend V to a contraction $\mathbf{U}^*\colon \mathcal{H}(K_S)^d\oplus\mathcal{Y} \to \mathcal{H}(K_S)\oplus\mathcal{U}$. Thus,

$$\mathbf{U}^* = \begin{bmatrix} A^* & C^* \\ B^* & D^* \end{bmatrix} \colon \begin{bmatrix} Z_{\rm row}(\zeta)^*K_S(\cdot,\zeta)y \\ y \end{bmatrix} \to \begin{bmatrix} K_S(\cdot,\zeta)y \\ S(\zeta)^*y \end{bmatrix}. \tag{2.16}$$

Comparing the top and the bottom components in (2.16) gives

$$A^*Z_{\rm row}(\zeta)^*K_S(\cdot,\zeta)y + C^*y = K_S(\cdot,\zeta)y, \tag{2.17}$$

$$B^*Z_{\rm row}(\zeta)^*K_S(\cdot,\zeta)y + D^*y = S(\zeta)^*y. \tag{2.18}$$

Solving (2.17) for $K_S(\cdot,\zeta)y$ gives

$$K_S(\cdot,\zeta)y = (I - A^*Z_{\rm row}(\zeta)^*)^{-1}C^*y. \tag{2.19}$$

Substituting this into (2.18) then gives

$$B^*Z_{\rm row}(\zeta)^*(I - A^*Z_{\rm row}(\zeta)^*)^{-1}C^*y + D^*y = S(\zeta)^*y. \tag{2.20}$$

By taking adjoints and using the fact that $\zeta \in \mathbb{B}^d$ and $y \in \mathcal{Y}$ are arbitrary, we may then conclude that $\mathbf{U}$ is a contractive realization for S. It remains to show that $\mathbf{U}$ meets the requirements (2)–(4) in Definition 2.5. To this end, we let $\zeta = 0$ in (2.17) and (2.18) to get

$$C^*y = K_S(\cdot,0)y \quad\text{and}\quad D^*y = S(0)^*y. \tag{2.21}$$

Substituting (2.21) back into (2.17) and (2.18), we get equalities (2.9) and (2.10) which are equivalent (by Proposition 2.4) to A and B solving the Gleason problems (2.7) and (2.8), respectively. By Remark 2.6, equalities (2.21) are equivalent to (2.12). □

Remark 2.8. A consequence of the isometry property of V in (2.15) is that formulas (2.9) and (2.10) extend by linearity and continuity to give rise to uniquely determined well-defined linear operators $A^*_{\mathcal{D}}$ and $B^*_{\mathcal{D}}$ from $\mathcal{D}$ to $\mathcal{H}(K_S)$ and $\mathcal{U}$, respectively. In this way we see that the existence problem for operators A solving the Gleason problem is settled: $A\colon \mathcal{H}(K_S) \to \mathcal{H}(K_S)^d$ *solves the Gleason problem for* $\mathcal{H}(K_S)$ (2.7) *if and only if* A^* *is an extension to all of* $\mathcal{H}(K_S)^d$ *of the operator* $A^*_{\mathcal{D}}\colon \mathcal{D} \to \mathcal{H}(K_S)$ *uniquely determined by the formula* (2.9). Similarly, *the operator* $B\colon \mathcal{U} \to \mathcal{H}(K_S)^d$ *is a solution of the* $\mathcal{H}(K_S)$*-Gleason problem for* S (2.8) *if and only* *if the operator* $B^*\colon \mathcal{H}(K_S)^d \to \mathcal{U}$ *is an extension to all of* $\mathcal{H}(K_S)^d$ *of the operator* $B^*_{\mathcal{D}}\colon \mathcal{D} \to \mathcal{U}$ *uniquely determined from the formula* (2.10).

The following result is essentially contained in [10]. For the ball setting, we use the following definition of observability: given an operator pair (C, A) with output operator $C\colon \mathcal{X} \to \mathcal{Y}$ and with $A\colon \mathcal{X} \to \mathcal{X}^d$, we say that (C, A) is *observable* if $C(I - Z_{\text{row}}(z)A)^{-1}x = 0$ for all z in a neighborhood of 0 in $\mathbb{C}^d$ implies that $x = 0$ in $\mathcal{X}$. Equivalently, this means that

$$\bigvee_{\substack{z\in\Delta\\ y\in\mathcal{Y}}} (I - A^* Z_{\text{row}}(z)^*)^{-1} C^* y = \mathcal{X}$$

for some neighborhood Δ of 0 in $\mathbb{C}^d$.

Theorem 2.9. *Let S be a Schur-class multiplier in $\mathcal{S}_d(\mathcal{U}, \mathcal{Y})$ and suppose that $\mathbf{U} = \left[\begin{smallmatrix} A & B \\ C & D \end{smallmatrix}\right]$ is any canonical functional-model colligation for S. Then*:

1. $\mathbf{U}$ *is weakly coisometric.*
2. *The pair (C, A) is* observable.
3. *We recover $S(z)$ as $S(z) = D + C(I - Z_{\text{row}}(z)A)^{-1} Z_{\text{row}}(z)B$.*
4. *If $\mathbf{U}' = \left[\begin{smallmatrix} A' & B' \\ C' & D' \end{smallmatrix}\right] : \mathcal{X} \oplus \mathcal{U} \to \mathcal{X}^d \oplus \mathcal{Y}$ is any other colligation matrix enjoying properties* (1), (2), (3), *then there is a canonical functional-model colligation $\mathbf{U} = \left[\begin{smallmatrix} A & B \\ C & D \end{smallmatrix}\right] : \mathcal{H}(K_S) \oplus \mathcal{U} \to \mathcal{H}(K_S)^d \oplus \mathcal{Y}$ so that $\mathbf{U}$ is unitarily equivalent to $\mathbf{U}'$, i.e., there is a unitary operator $U\colon \mathcal{X} \to \mathcal{H}(K_S)$ so that*
$$\begin{bmatrix} A & B \\ C & D \end{bmatrix} \begin{bmatrix} U & 0 \\ 0 & I_{\mathcal{Y}} \end{bmatrix} = \begin{bmatrix} \bigoplus_{j=1}^d U & 0 \\ 0 & I_{\mathcal{U}} \end{bmatrix} \begin{bmatrix} A' & B' \\ C' & D' \end{bmatrix}.$$

PROOF. Since $\mathbf{U}$ is a canonical functional-model colligation for S, the operators A and B solve the Gleason problems (2.7) and (2.8), respectively. By Proposition 2.4, this is equivalent to identities

$$A^* Z_{\text{row}}(\zeta)^* K_S(\cdot, \zeta) y = K_S(\cdot, \zeta) y - K_S(\cdot, 0) y$$
$$B^* Z_{\text{row}}(\zeta)^* K_S(\cdot, \zeta) y = S(\zeta)^* y - S(0)^* y.$$

Besides, C and D are defined by formulas (2.12). Substituting their adjoints from (2.13) into the two latter equalities we arrive at (2.17) and (2.18). As we have seen, equalities (2.17) and (2.18) imply (2.19) and (2.20). Equality (2.20) proves statement (3). Equality (2.19) gives

$$\bigvee_{\substack{\zeta\in\mathbb{B}^d\\ y\in\mathcal{Y}}} (I - A^* Z_{\text{row}}(\zeta)^* C^* y = \bigvee_{\substack{\zeta\in\mathbb{B}^d\\ y\in\mathcal{Y}}} K_S(\cdot, \zeta) y = \mathcal{H}(K_S). \tag{2.22}$$

Thus the identity $C(I - Z_{\text{row}}(z)A)^{-1} f \equiv 0$ leads to $\langle f, (I - A^* Z_{\text{row}}(z)^*)^{-1} C^* y\rangle = 0$ for every $z \in \mathbb{B}^d$ and $y \in \mathcal{Y}$; this together with equality (2.22) implies $f \equiv 0$, and it follows that the pair (C, A) is observable.

On the other hand, equalities (2.17) and (2.18) are equivalent to (2.16). Substituting (2.19) into (2.16) and in (2.14) (for $z = \zeta$ and $y = y'$) gives

$$\mathbf{U}^* \begin{bmatrix} Z_{\text{row}}(\zeta)^*(I - A^* Z_{\text{row}}(\zeta)^*)^{-1} C^* y \\ y \end{bmatrix} = \begin{bmatrix} (I - A^* Z_{\text{row}}(\zeta)^*)^{-1} C^* y \\ S(\zeta)^* y \end{bmatrix}$$

and

$$\left\| \begin{bmatrix} Z_{\text{row}}(\zeta)^*(I - A^* Z_{\text{row}}(\zeta)^*)^{-1} C^* y \\ y \end{bmatrix} \right\| = \left\| \begin{bmatrix} (I - A^* Z_{\text{row}}(\zeta)^*)^{-1} C^* y \\ S(\zeta)^* y \end{bmatrix} \right\|,$$

respectively. The two latter equalities tell us that $\mathbf{U}^*$ is isometric on the space $\mathcal{D}_{\mathbf{U}^*}$ (see (2.5)) and therefore $\mathbf{U}$ is weakly coisometric. For the proof of part (4) we refer to [10, Theorem 3.4]. □

Definition 2.5 does not require $\mathbf{U}$ to be a realization for S: representation (2.4) is automatic once the operators A, B, C and D are of the required form. We can look at this from a different point of view as follows. Let us say that $A\colon \mathcal{H}(K_S) \to \mathcal{H}(K_S)^d$ is a *contractive solution* of the Gleason problem (2.7) if in addition to (2.7), inequality

$$\sum_{k=1}^{d} \|(Af)_k\|^2_{\mathcal{H}(K_S)} \le \|f\|^2_{\mathcal{H}(K_S)} - \|f(0)\|^2_{\mathcal{Y}} \tag{2.23}$$

holds for every $f \in \mathcal{H}(K_S)$. It is readily seen that inequality (2.23) can be equivalently written in operator form as

$$A^*A + C^*C \le I$$

where the operator $C\colon \mathcal{H}(K_S) \to \mathcal{Y}$ is given in (2.12). It therefore follows from Definition 2.5 that for every canonical functional-model colligation $\mathbf{U} = \left[\begin{smallmatrix} A & B \\ C & D \end{smallmatrix}\right]$ for S, the operator A is a contractive solution of the Gleason problem (2.7). The following theorem provides a converse to this statement.

Theorem 2.10. *Let $S \in \mathcal{S}_d(\mathcal{U}, \mathcal{Y})$ be given and let us assume that C, D are given by formulas* (2.12). *Then*

(1) *For every contractive solution A of the Gleason problem* (2.7) *for $\mathcal{H}(K_S)$, there exists an operator $B : \mathcal{U} \to \mathcal{H}(K_S)$ such that $\mathbf{U} = \left[\begin{smallmatrix} A & B \\ C & D \end{smallmatrix}\right]$ is contractive and S is realized as in* (2.4).
(2) *Every such B solves the $\mathcal{H}(K_S)$-Gleason problem* (2.8) *so that $\mathbf{U}$ is a canonical functional-model colligation.*

PROOF. Since A solves the Gleason problem (2.7) and since C is defined as in (2.12), we conclude as in the proof of Theorem 2.9 that identity (2.17) holds which is equivalent to (2.19). On account of (2.19), it is readily seen that (2.18) and (2.20) are equivalent. But (2.20) is just the adjoint form of (2.4) whereas (2.18) coincides with (2.10) (since $D = S(0)$) which in turn, is equivalent to (2.8) by Proposition 2.4. Thus, it remains to show that there exists an operator $B^*\colon \mathcal{H}(K_S) \to \mathcal{U}$ completely determined on the subspace $\mathcal{D} \subset \mathcal{H}(K_S)$ by formula (2.10) and such that $\mathbf{U}^* = \left[\begin{smallmatrix} A^* & C^* \\ B^* & D^* \end{smallmatrix}\right]$ is contractive. This demonstration can be found in [10, Theorem 2.4]. □

3. de Branges – Rovnyak kernels associated with a Schur – Agler-class function on the polydisk

Here we introduce a generalized Schur class, called Schur – Agler class, associated with the unit polydisk

$$\mathbb{D}^d = \{z = (z_1, \dots, z_d) \in \mathbb{C}^d : |z_k| < 1 \text{ for } k = 1, \dots, d\}.$$

We define the *Schur – Agler class* $\mathcal{SA}_d(\mathcal{U}, \mathcal{Y})$ to consist of holomorphic functions $S\colon \mathbb{D}^d \to \mathcal{L}(\mathcal{U}, \mathcal{Y})$ such that $\|S(T)\| \le 1$ for any collection of d commuting operators $T = (T_1, \dots, T_d)$ on a Hilbert space $\mathcal{K}$ with $\|T_k\| < 1$ for each $k = 1, \dots, d$ where the operator $S(T)$ is defined as in (2.1).

The following result appears in [1, 2, 14] and is another multivariable analogue of Theorem 1.1. The reader will notice that analogues of both (1a) and (1b) from Theorem 1.1 are missing in this theorem.

Theorem 3.1. *Let S be a $\mathcal{L}(\mathcal{U},\mathcal{Y})$-valued function defined on $\mathbb{D}^d$. The following statements are equivalent:*

(1) (c) *S belongs to the class $\mathcal{SA}_d(\mathcal{U},\mathcal{Y})$, i.e., S satisfies the von Neumann inequality $\|S(T_1,\dots,T_d)\| \le 1$ for any commutative d-tuple $T = (T_1,\dots,T_d)$ of strict contraction operators on an auxiliary Hilbert space $\mathcal{K}$.*

(2) *There exist positive kernels $K_1,\dots,K_d\colon \mathbb{D}^d \times \mathbb{D}^d \to \mathcal{L}(\mathcal{Y})$ such that for every $z = (z_1,\dots,z_d)$ and $\zeta = (\zeta_1,\dots,\zeta_d)$ in $\mathbb{D}^d$,*

$$I_{\mathcal{Y}} - S(z)S(\zeta)^* = \sum_{k=1}^{d}(1 - z_i\bar{\zeta}_i)K_i(z,\zeta). \tag{3.1}$$

(3) *There exist Hilbert spaces $\mathcal{X}_1,\dots,\mathcal{X}_d$ and a unitary connecting operator $\mathbf{U}$ of the structured form*

$$\mathbf{U} = \begin{bmatrix} A & B \\ C & D \end{bmatrix} = \begin{bmatrix} A_{11} & \dots & A_{1d} & B_1 \\ \vdots & & \vdots & \vdots \\ A_{d1} & \dots & A_{dd} & B_d \\ C_1 & \dots & C_d & D \end{bmatrix} : \begin{bmatrix} \mathcal{X}_1 \\ \vdots \\ \mathcal{X}_d \\ \mathcal{U} \end{bmatrix} \to \begin{bmatrix} \mathcal{X}_1 \\ \vdots \\ \mathcal{X}_d \\ \mathcal{Y} \end{bmatrix} \tag{3.2}$$

so that $S(z)$ can be realized in the form

$$S(z) = D + C\big(I - Z_{\text{diag}}(z)A\big)^{-1}Z_{\text{diag}}(z)B \qquad \text{for all } z \in \mathbb{D}^d \tag{3.3}$$

where we have set

$$Z_{\text{diag}}(z) = \begin{bmatrix} z_1 I_{\mathcal{X}_1} & & 0 \\ & \ddots & \\ 0 & & z_d I_{\mathcal{X}_d} \end{bmatrix}. \tag{3.4}$$

(4) *There exist Hilbert spaces $\mathcal{X}_1,\dots,\mathcal{X}_d$ and a contractive connecting operator $\mathbf{U}$ of the form (3.2) so that $S(z)$ can be realized in the form (3.3)*

Remark 3.2. Although statement (4) in Theorem 3.1 concerning contractive realizations does not appear in [1, 2, 14], its equivalence to statements (1)–(3) can be seen by any one of the three approaches mentioned in Remark 2.2.

Similar to the notion introduced above for the unit-ball case, there is a notion of weak coisometry for the polydisk setting as follows.

Definition 3.3. The operator-block matrix $\mathbf{U}$ of the form (3.2) is *weakly coisometric* if the restriction of $\mathbf{U}^*$ to the subspace

$$\mathcal{D}_{\mathbf{U}^*} := \bigvee_{\substack{\zeta\in\mathbb{D}^d \\ y\in\mathcal{Y}}} \begin{bmatrix} Z_{\text{diag}}(\zeta)^*(I - A^*Z_{\text{diag}}(\zeta)^*)^{-1}C^*y \\ y \end{bmatrix} \subset \begin{bmatrix} \mathcal{X}^d \\ \mathcal{Y} \end{bmatrix} \tag{3.5}$$

is isometric.

When $\mathbf{U}$ is given by (3.2) and $S(z)$ is given by (3.3), it is immediate that we have the equality

$$\begin{bmatrix} C(I - Z_{\text{diag}}(z)A)^{-1}Z_{\text{diag}}(z) & I \end{bmatrix} \mathbf{U} = \begin{bmatrix} C(I - Z_{\text{diag}}(z)A)^{-1} & S(z) \end{bmatrix}.$$

From this it is easy to verify the following general identity:

$$\begin{aligned} (3.6) \quad I - S(z)S(\zeta)^* &= C(I - Z(z)A)^{-1}(I - Z(z)Z(\zeta)^*)(I - A^*Z(\zeta)^*)^{-1}C^* \\ &+ \begin{bmatrix} C(I - Z(z)A)^{-1}Z(z) & I \end{bmatrix} (I - \mathbf{U}\mathbf{U}^*) \begin{bmatrix} Z(\zeta)^*(I - A^*Z(\zeta)^*)^{-1}C^* \\ I \end{bmatrix} \end{aligned}$$

where here we set $Z(z) = Z_{\text{diag}}(z)$ for short. It is readily seen from (3.6) that the weak-coisometry property of the colligation (3.2) is exactly what is needed to guarantee the representation

$$\begin{aligned} (3.7) \quad I &- S(z)S(\zeta)^* \\ &= C(I - Z_{\text{diag}}(z)A)^{-1}(I - Z_{\text{diag}}(z)Z_{\text{diag}}(\zeta)^*)(I - A^*Z_{\text{diag}}(\zeta)^*)^{-1}C^*. \end{aligned}$$

Note that the representation (3.7) has the form (3.1) if we take

$$(3.8) \qquad K_k(z, \zeta) = C(I - Z_{\text{diag}}(z)A)^{-1}P_{\mathcal{X}_k}(I - A^*Z_{\text{diag}}(\zeta)^*)^{-1}C^*$$

for $k = 1, \ldots, d$, where $P_{\mathcal{X}_k}$ is the orthogonal projection of $\mathcal{X} := \bigoplus_{i=1}^d \mathcal{X}_i$ onto $\mathcal{X}_k$.

3.1. Weakly coisometric canonical functional-model colligations. Let us say that a collection of positive kernels $\{K_1(z, \zeta), \ldots, K_d(z, \zeta)\}$ for which the decomposition (3.1) holds is an *Agler decomposition* for S. In view of (3.7), we see that a realization (3.3) for S arising from a weakly coisometric colligation matrix $\mathbf{U}$ (3.2) determines a particular Agler decomposition, namely that given by (3.8).

Our next goal is to find a canonical weakly coisometric realization for S compatible with the given Agler decomposition. Toward this goal we make the following definitions.

Suppose that we are given a Schur–Agler class function $S \in \mathcal{SA}_d(\mathcal{U}, \mathcal{Y})$ together with an Agler decomposition $\{K_1(z, \zeta), \ldots, K_d(z, \zeta)\}$ for S. We set

$$\mathbb{K}(z, \zeta) = K_1(z, \zeta) + \cdots + K_d(z, \zeta).$$

Then $\mathbb{K}$ is also a positive kernel on $\mathbb{D}^d$ and the associated reproducing kernel Hilbert space $\mathcal{H}(\mathbb{K})$ can be characterized as

$$\mathcal{H}(\mathbb{K}) = \left\{ \sum_{i=1}^d f_i : f_i \in \mathcal{H}(K_i) \text{ for } i = 1, \ldots, d \right\}$$

with norm given by

$$\left\| \sum_{i=1}^d f_i \right\|_{\mathcal{H}(\mathbb{K})} = \|P_{(\ker \mathbf{s})^\perp} f\|_{\bigoplus_{i=1}^d \mathcal{H}(K_i)},$$

where $\mathbf{s}\colon \bigoplus_{i=1}^d \mathcal{H}(K_i) \to \mathcal{H}(\mathbb{K})$ is the linear map defined by

$$(3.9) \qquad \mathbf{s}f = f_1 + \cdots + f_d, \qquad \text{where } f = \bigoplus_{i=1}^d f_i := \begin{bmatrix} f_1 \\ \vdots \\ f_d \end{bmatrix}.$$

It is clear that $\ker \mathbf{s} = \{f \in \bigoplus_{i=1}^d \mathcal{H}(K_i) : f_1(z) + \cdots + f_d(z) \equiv 0\}$. If we let

$$\mathbb{T}(z,\zeta) := \begin{bmatrix} K_1(z,\zeta) \\ \vdots \\ K_d(z,\zeta) \end{bmatrix}, \tag{3.10}$$

we observe that by the reproducing kernel property,

$$\begin{aligned} \langle f, \mathbb{T}(\cdot,\zeta)y\rangle_{\bigoplus_{i=1}^d \mathcal{H}(K_i)} &= \sum_{i=1}^d \langle f_i, K_i(\cdot,\zeta)y\rangle_{\mathcal{H}(K_i)} \\ &= \left\langle \sum_{i=1}^d f_i(\zeta), y \right\rangle_{\mathcal{Y}} = \langle \mathbf{s}f, \mathbb{K}(\cdot,\zeta)y\rangle_{\mathcal{H}(\mathbb{K})} \end{aligned} \tag{3.11}$$

so that

$$\mathbf{s}^* : \mathbb{K}(\cdot,\zeta)y \to \mathbb{T}(\cdot,\zeta)y. \tag{3.12}$$

Furthermore,

$$(\ker \mathbf{s})^\perp = \bigvee_{\substack{\zeta\in\mathbb{D}^d \\ y\in\mathcal{Y}}} \mathbb{T}(\cdot,\zeta)y \subset \bigoplus_{k=1}^d \mathcal{H}(K_k).$$

We next introduce the subspace

$$\mathcal{D} = \bigvee_{\substack{\zeta\in\mathbb{D}^d \\ y\in\mathcal{Y}}} Z_{\rm diag}(\zeta)^*\mathbb{T}(\cdot,\zeta)y \tag{3.13}$$

of $\bigoplus_{k=1}^d \mathcal{H}(K_k)$ and observe that its orthogonal complement can be described as

$$\mathcal{D}^\perp = \left\{ f = \bigoplus_{i=1}^d f_i \in \bigoplus_{i=1}^d \mathcal{H}(K_i) : \sum_{i=1}^d z_i f_i(z) \equiv 0 \right\}.$$

In addition, the straightforward computation

$$\|\mathbb{T}(\cdot,\zeta)y\|^2_{\bigoplus_{k=1}^d \mathcal{H}(K_k)} = \sum_{k=1}^d \langle K_k(\zeta,\zeta)y,y\rangle_{\mathcal{Y}} = \langle \mathbb{K}(\zeta,\zeta)y,y\rangle_{\mathcal{Y}} = \|\mathbb{K}(\cdot,\zeta)y\|^2_{\mathcal{H}(\mathbb{K})}$$

combined with (3.12) shows that $\mathbf{s}^*$ is an isometry, i.e., that $\mathbf{s}$ is a coisometry. We remark that all the items introduced so far are uniquely determined from decomposition (3.1).

Given an operator-block matrix $A = [A_{ij}]_{i,j=1}^d$ acting on $\bigoplus_{i=1}^d \mathcal{H}(K_i)$, we say that A *solves the structured Gleason problem for the kernel collection* $\{K_1,\dots,K_d\}$ if the identity

$$f_1(z) + \cdots + f_d(z) - [f_1(0) + \cdots + f_d(0)] = \sum_{i=1}^d z_i (Af)_i(z) \tag{3.14}$$

holds for all $f = \bigoplus_{i=1}^d f_i \in \bigoplus_{i=1}^d \mathcal{H}(K_i)$, where we write $(Af)(z) = \bigoplus_{i=1}^d (Af)_i(z) \in \bigoplus_{i=1}^d \mathcal{H}(K_i)$. Note that (3.14) can be written more compactly as

$$(\mathbf{s}f)(z) - (\mathbf{s}f)(0) = \sum_{i=1}^d z_i (A_{i\bullet}f)(z) \qquad \text{for all } f \in \bigoplus_{k=1}^d \mathcal{H}(K_k), \tag{3.15}$$

where $\mathbf{s}$ is given in (3.9) and where

$$A_{i\bullet} = \begin{bmatrix} A_{i1} & \dots & A_{id} \end{bmatrix} : \bigoplus_{k=1}^{d} \mathcal{H}(K_k) \to \mathcal{H}(K_i) \qquad (i = 1, \dots, d) \tag{3.16}$$

so that

$$A = \bigoplus_{i=1}^{d} A_{i\bullet} = [A_{ij}]_{i,j=1}^{d} \colon \bigoplus_{k=1}^{d} \mathcal{H}(K_k) \to \bigoplus_{i=1}^{d} \mathcal{H}(K_i). \tag{3.17}$$

We say that the operator $B \colon \mathcal{U} \to \bigoplus_{k=1}^{d} \mathcal{H}(K_k)$ *solves the structured* $\mathcal{H}(K_k)$-*Gleason problem for* S if the identity

$$S(z)u - S(0)u = z_1(Bu)_1(z) + \cdots + z_d(Bu)_d(z) \qquad \text{holds for all } u \in \mathcal{U}. \tag{3.18}$$

The following is the parallel to Proposition 2.4 for the polydisk setting.

Proposition 3.4. *The operator* $A\colon \bigoplus_{i=1}^{d} \mathcal{H}(K_i) \to \bigoplus_{i=1}^{d} \mathcal{H}(K_i)$ *solves the structured Gleason problem* (3.15) *if and only if the adjoint operator* A^* *has the following action on special kernel functions*:

$$A^* \colon \begin{bmatrix} \bar{\zeta}_1 K_1(\cdot, \zeta)y \\ \vdots \\ \bar{\zeta}_d K_d(\cdot, \zeta)y \end{bmatrix} \mapsto \begin{bmatrix} K_1(\cdot, \zeta)y \\ \vdots \\ K_d(\cdot, \zeta)y \end{bmatrix} - \begin{bmatrix} K_1(\cdot, 0)y \\ \vdots \\ K_d(\cdot, 0)y \end{bmatrix} \qquad \textit{for all } \zeta \in \mathbb{D}^d \textit{ and } y \in \mathcal{Y}.$$

The operator $B\colon \mathcal{U} \to \bigoplus_{i=1}^{d} \mathcal{H}(K_i)$ *solves the structured Gleason problem* (3.18) *for* S *if and only if the adjoint operator* $B^*\colon \bigoplus_{i=1}^{d} \mathcal{H}(K_i) \to \mathcal{U}$ *has the following action on special kernel functions*:

$$B^* \colon \begin{bmatrix} \bar{\zeta}_1 K_1(\cdot, \zeta)y \\ \vdots \\ \bar{\zeta}_d K_d(\cdot, \zeta)y \end{bmatrix} \mapsto S(\zeta)^*y - S(0)^*y \qquad \textit{for all } \zeta \in \mathbb{D}^d \textit{ and } y \in \mathcal{Y}.$$

PROOF. Making use of notation (3.10) and (3.4) we can write the definitions of A^* and B^* more compactly as

$$A^* Z_{\rm diag}(\zeta)^* \mathbb{T}(\cdot, \zeta)y = \mathbb{T}(\cdot, \zeta)y - \mathbb{T}(\cdot, 0)y, \tag{3.19}$$

$$B^* Z_{\rm diag}(\zeta)^* \mathbb{T}(\cdot, \zeta)y = S(\zeta)^*y - S(0)^*y. \tag{3.20}$$

By calculation (3.11),

$$\langle f, \mathbb{T}(\cdot, \zeta)y - \mathbb{T}(\cdot, 0)y \rangle_{\bigoplus_{i=1}^{d} \mathcal{H}(K_i)} = \langle (\mathbf{s}f)(\zeta) - (\mathbf{s}f)(0), y \rangle_{\mathcal{Y}}.$$

On the other hand, it follows by the reproducing kernel property that

$$\begin{aligned} \langle f, A^* Z_{\rm diag}(z)^* \mathbb{T}(\cdot, z)y \rangle_{\bigoplus_{i=1}^{d} \mathcal{H}(K_i)} &= \langle Z_{\rm diag}(z) A f, \mathbb{T}(\cdot, z)y \rangle_{\bigoplus_{i=1}^{d} \mathcal{H}(K_i)} \\ &= \left\langle \sum_{i=1}^{d} [Z_{\rm diag}(z) A f]_i(z), y \right\rangle_{\mathcal{Y}} \\ &= \left\langle \sum_{i=1}^{d} z_i [A f]_i(z), y \right\rangle_{\mathcal{Y}} \end{aligned}$$

and the two latter equalities show that (3.15) holds if and only if (3.19) is in force for every $y \in \mathcal{Y}$. Equivalence of (3.18) and (3.20) is verified quite similarly. □

The following definition of a canonical functional-model colligation is the analogue of Definition 2.5 for the polydisk setting.

Definition 3.5. Given $S \in \mathcal{SA}_d(\mathcal{U}, \mathcal{Y})$, we shall say that the block-operator matrix $\mathbf{U} = \left[\begin{smallmatrix} A & B \\ C & D \end{smallmatrix}\right]$ of the form (3.2) is a *canonical functional-model colligation* associated with the Agler decomposition (3.1) for S if

(1) $\mathbf{U}$ is contractive and the state space equals $\bigoplus_{i=1}^d \mathcal{H}(K_i)$.
(2) $A \colon \bigoplus_{i=1}^d \mathcal{H}(K_i) \to \bigoplus_{i=1}^d \mathcal{H}(K_i)$ solves the structured Gleason problem (3.15).
(3) $B \colon \mathcal{U} \to \bigoplus_{i=1}^d \mathcal{H}(K_i)$ solves the structured Gleason problem (3.18) for S.
(4) The operators $C \colon \bigoplus_{i=1}^d \mathcal{H}(K_i) \to \mathcal{Y}$ and $D \colon \mathcal{U} \to \mathcal{Y}$ are given by

$$C \colon f(z) \mapsto (\mathbf{s}f)(0), \qquad D \colon u \mapsto S(0)u. \tag{3.21}$$

Remark 3.6. For C and D defined in (3.21), the adjoint operators are given by

$$C^* : y \mapsto \mathbb{T}(\cdot, 0)y \qquad D^* : y \mapsto S(0)^* y. \tag{3.22}$$

The formula for D^* is obvious while the formula for C^* follows from equalities

$$\langle f, C^* y\rangle_{\bigoplus_{i=1}^d \mathcal{H}(K_i)} = \langle Cf, y\rangle_{\mathcal{Y}} = \langle (\mathbf{s}f)(0), y\rangle_{\mathcal{Y}} = \langle f, \mathbb{T}(\cdot, 0)y\rangle_{\bigoplus_{i=1}^d \mathcal{H}(K_i)}$$

holding for every $f \in \bigoplus_{i=1}^d \mathcal{H}(K_i)$.

Theorem 3.7. *Let S be a given function in the Schur – Agler class $\mathcal{S}A_d(\mathcal{U}, \mathcal{Y})$ and suppose that we are given an Agler decomposition* (3.1) *for S. Then there exists a canonical functional-model colligation associated with $\{K_1, \dots, K_d\}$.*

Proof. Let us represent a given Agler decomposition (3.1) in the inner product form as

$$\begin{aligned}\sum_{i=1}^d \langle \bar{\zeta}_i K_i(\cdot, \zeta)y, \bar{z}_i K_i(\cdot, z)y'\rangle_{\mathcal{H}(K_i)} + \langle y, y'\rangle_{\mathcal{Y}} \\ = \sum_{i=1}^d \langle K_i(\cdot, \zeta)y, K_i(\cdot, z)y'\rangle_{\mathcal{H}(K_i)} + \langle S(\zeta)^* y, S(z)^* y'\rangle_{\mathcal{U}},\end{aligned}$$

or equivalently, as

$$\begin{aligned}\left\langle \begin{bmatrix} Z_{\text{diag}}(\zeta)^* \mathbb{T}(\cdot, \zeta)y \\ y \end{bmatrix}, \begin{bmatrix} Z_{\text{diag}}(z)^* \mathbb{T}(\cdot, z)y' \\ y' \end{bmatrix} \right\rangle_{(\bigoplus_{i=1}^d \mathcal{H}(K_i)) \oplus \mathcal{Y}} \\ = \left\langle \begin{bmatrix} \mathbb{T}(\cdot, \zeta)y \\ S(\zeta)^* y \end{bmatrix}, \begin{bmatrix} \mathbb{T}(\cdot, z)y' \\ S(z)^* y' \end{bmatrix} \right\rangle_{(\bigoplus_{i=1}^d \mathcal{H}(K_i)) \oplus \mathcal{U}}\end{aligned} \tag{3.23}$$

where $\mathbb{T}$ is given in (3.10). The latter identity implies that the map

$$\widetilde{V} \colon \begin{bmatrix} Z_{\text{diag}}(\zeta)^* \mathbb{T}(\cdot, \zeta)y \\ y \end{bmatrix} \mapsto \begin{bmatrix} \mathbb{T}(\cdot, \zeta)y \\ S(\zeta)^* y \end{bmatrix} \tag{3.24}$$

extends by linearity and continuity to an isometry from $\mathcal{D}_{\widetilde{V}} = \mathcal{D} \oplus \mathcal{Y}$ (a subspace of $(\bigoplus_{i=1}^d \mathcal{H}(K_i)) \oplus \mathcal{Y}$ — (3.13) for definition of $\mathcal{D}$) onto

$$\mathcal{R}_{\widetilde{V}} = \bigvee_{\zeta \in \mathbb{D}^d, y \in \mathcal{Y}} \begin{bmatrix} \mathbb{T}(\cdot, \zeta)y \\ S(\zeta)^* y \end{bmatrix} \subset \begin{bmatrix} \bigoplus_{i=1}^d \mathcal{H}(K_i) \\ \mathcal{U} \end{bmatrix}.$$

Let us extend $\widetilde{V}$ to a contraction $\mathbf{U}^*\colon \begin{bmatrix} \bigoplus_{i=1}^d \mathcal{H}(K_i) \\ \mathcal{Y} \end{bmatrix} \to \begin{bmatrix} \bigoplus_{i=1}^d \mathcal{H}(K_i) \\ \mathcal{U} \end{bmatrix}$. Thus,

$$\mathbf{U}^* = \begin{bmatrix} A^* & C^* \\ B^* & D^* \end{bmatrix} : \begin{bmatrix} Z_{\text{diag}}(\zeta)^*\mathbb{T}(\cdot,\zeta)y \\ y \end{bmatrix} \mapsto \begin{bmatrix} \mathbb{T}(\cdot,\zeta)y \\ S(\zeta)^*y \end{bmatrix}. \tag{3.25}$$

Computing the top and bottom components in (3.25) gives

$$A^*Z_{\text{diag}}(\zeta)^*\mathbb{T}(\cdot,\zeta)y + C^*y = \mathbb{T}(\cdot,\zeta)y, \tag{3.26}$$

$$B^*Z_{\text{diag}}(\zeta)^*\mathbb{T}(\cdot,\zeta)y + D^*y = S(\zeta)^*y. \tag{3.27}$$

Letting $\zeta = 0$ in the latter equalities yields (3.22) which means that C and D are of the requisite form (3.21). By substituting (3.22) into (3.26) and (3.27), we arrive at (3.19) and (3.20) which in turn are equivalent to (3.15) and (3.18), respectively. Thus, $\mathbf{U}$ meets all the requirements of Definition 3.5. □

We have the following parallel of Remark 2.8 for the polydisk setting.

Remark 3.8. As a consequence of the isometric property of the operator $\widetilde{V}$ (3.24) introduced in the proof of Theorem 3.7, formulas (3.19) and (3.20) can be extended by linearity and continuity to define uniquely determined operators $A_{\mathcal{D}}^*\colon \mathcal{D} \to \bigoplus_{i=1}^d \mathcal{H}(K_i)$ and $B_{\mathcal{D}}^*\colon \mathcal{D} \to \mathcal{U}$ where the subspace $\mathcal{D}$ of $\bigoplus_{i=1}^d \mathcal{H}(K_i)$ is defined in (3.13). In view of Proposition 3.4, we see that the existence question is then settled: *any operator* $A\colon \bigoplus_{i=1}^d \mathcal{H}(K_i) \to \bigoplus_{i=1}^d \mathcal{H}(K_i)$ *such that* A^* *is an extension of* $A_{\mathcal{D}}^*$ *from* $\mathcal{D}$ *to all of* $\bigoplus_{i=1}^d \mathcal{H}(K_i)$ *is a solution of the structured Gleason problem* (3.15) and *any operator* $B\colon \mathcal{U} \to \bigoplus_{i=1}^d \mathcal{H}(K_i)$ *so that* B^* *is an extension of the operator* $B_{\mathcal{D}}^*\colon \mathcal{D} \to \mathcal{U}$ *is a solution of the structured Gleason problem* (3.18) *for* S.

In the polydisk setting we use the following definition of observability: given an operator A on $\bigoplus_{i=1}^d \mathcal{X}_i$ and an operator $C\colon \bigoplus_{i=1}^d \mathcal{X}_i \to \mathcal{Y}$, the pair (C, A) will be called *observable* if equalities $C(I - Z_{\text{diag}}(z)A)^{-1}P_{\mathcal{X}_i}x = 0$ for all z in a neighborhood of the origin and for all $i = 1, \ldots, d$ forces $x = 0$ in $\bigoplus_{i=1}^d \mathcal{X}_i$. The latter is equivalent to the equality

$$\bigvee_{z\in\Delta, y\in\mathcal{Y}} P_{\mathcal{X}_i}(I - A^*Z_{\text{diag}}(z)^*)^{-1}C^*y = \mathcal{X}_i \qquad \text{for } i = 1, \ldots, d \tag{3.28}$$

for some neighborhood Δ of the origin in $\mathbb{C}^d$. The following theorem is the analogue of Theorem 1.2 for the polydisk setting; portions of this theorem appear already in [14, Section 3.3.1].

Theorem 3.9. *Let* S *be a function in the Schur–Agler class* $\mathcal{SA}_d(\mathcal{U}, \mathcal{Y})$ *with a given Agler decomposition* $\{K_1, \ldots, K_d\}$ *for* S *and let us suppose that*

$$\mathbf{U} = \begin{bmatrix} A & B \\ C & D \end{bmatrix} : \begin{bmatrix} \bigoplus_{i=1}^d \mathcal{H}(K_i) \\ \mathcal{U} \end{bmatrix} \to \begin{bmatrix} \bigoplus_{i=1}^d \mathcal{H}(K_i) \\ \mathcal{Y} \end{bmatrix} \tag{3.29}$$

is a canonical functional-model colligation associated with this decomposition. Then:

(1) $\mathbf{U}$ *is weakly coisometric.*
(2) *The pair* (C, A) *is observable in the sense of* (3.28).
(3) *We recover* $S(z)$ *as* $S(z) = D + C(I - Z_{\text{diag}}(z)A)^{-1}Z_{\text{diag}}(z)B$.

(4) *If* $\widetilde{\mathbf{U}} = \left[\begin{smallmatrix} \tilde{A} & \widetilde{B} \\ \widetilde{C} & \widetilde{D} \end{smallmatrix}\right] : (\bigoplus_{i=1}^d \mathcal{X}_i) \oplus \mathcal{U} \to (\bigoplus_{i=1}^d \mathcal{X}_i) \oplus \mathcal{Y}$ *is any other colligation matrix enjoying properties* (1), (2), (3) *above, then there is a canonical functional-model colligation* $\mathbf{U} = \left[\begin{smallmatrix} A & B \\ C & D \end{smallmatrix}\right]$ *as in* (3.29) *which is unitarily equivalent to* $\widetilde{\mathbf{U}}$ *in the sense that there are unitary operators* $U_i \colon \mathcal{X}_i \to \mathcal{H}(K_i)$ *so that*

$$\begin{bmatrix} A & B \\ C & D \end{bmatrix} \begin{bmatrix} \bigoplus_{i=1}^d U_i & 0 \\ 0 & I_{\mathcal{Y}} \end{bmatrix} = \begin{bmatrix} \bigoplus_{i=1}^d U_i & 0 \\ 0 & I_{\mathcal{U}} \end{bmatrix} \begin{bmatrix} \tilde{A} & \widetilde{B} \\ \widetilde{C} & \widetilde{D} \end{bmatrix}. \tag{3.30}$$

PROOF. Let $\mathbf{U} = \left[\begin{smallmatrix} A & B \\ C & D \end{smallmatrix}\right]$ be a canonical functional-model realization of S associated with a fixed Agler decomposition (3.1). Then combining equalities (3.19), (3.20) (equivalent to the given (3.15) and (3.18) by Proposition 3.4) and also formulas (3.22) (equivalent to the given (3.21)) leads us to

$$\mathbb{T}(\cdot, \zeta)y = (I - A^* Z_{\rm diag}(\zeta)^*)^{-1}\mathbb{T}(\cdot, 0)y = (I - A^* Z_{\rm diag}(\zeta)^*)^{-1}C^* y \tag{3.31}$$

and

$$S(\zeta)^* y = S(0)^* y + B^* Z_{\rm diag}(\zeta)^* \mathbb{T}(\cdot, \zeta)y = D^* y + B^* Z_{\rm diag}(\zeta)^* \mathbb{T}(\cdot, \zeta)y. \tag{3.32}$$

Substituting (3.31) into (3.32) and taking into account that $y \in \mathcal{Y}$ is arbitrary, we get

$$S(\zeta)^* = S(0)^* + B^* Z_{\rm diag}(\zeta)^* (I - A^* Z_{\rm diag}(\zeta)^*)^{-1} C^* \tag{3.33}$$

which proves part (3) of the theorem. Also we have from (3.31) and (3.10),

$$\begin{aligned} \bigvee_{\substack{\zeta\in\mathbb{D}^d \\ y\in\mathcal{Y}}} P_{\mathcal{H}(K_i)}(I - A^* Z_{\rm diag}(\zeta)^*)^{-1} C^* y &= \bigvee_{\substack{\zeta\in\mathbb{D}^d \\ y\in\mathcal{Y}}} P_{\mathcal{H}(K_i)} \mathbb{T}(\cdot, \zeta) y \\ &= \bigvee_{\substack{\zeta\in\mathbb{D}^d \\ y\in\mathcal{Y}}} K_i(\cdot, \zeta) y = \mathcal{H}(K_i) \end{aligned}$$

and the pair (C, A) is observable in the sense of (3.28). On the other hand, equalities (3.19), (3.20) are equivalent to (3.25). Substituting (3.31) into (3.25) and into identity (3.23) (for $z = \zeta$ and $y = y'$) gives

$$\mathbf{U}^* \begin{bmatrix} Z(\zeta)^*(I - A^* Z(\zeta)^*)^{-1} C^* y \\ y \end{bmatrix} = \begin{bmatrix} (I - A^* Z(\zeta)^*)^{-1} C^* y \\ S(\zeta)^* y \end{bmatrix}$$

and

$$\left\| \begin{bmatrix} Z(\zeta)^*(I - A^* Z(\zeta)^*)^{-1} C^* y \\ y \end{bmatrix} \right\| = \left\| \begin{bmatrix} (I - A^* Z(\zeta)^*)^{-1} C^* y \\ S(\zeta)^* y \end{bmatrix} \right\|,$$

respectively. The two latter equalities show that $\mathbf{U}^*$ is isometric on the space $\mathcal{D}_{\mathbf{U}^*}$ (see (3.5)) and therefore $\mathbf{U}$ is weakly coisometric.

To prove part (4), let us assume that

$$S(z) = S(0) + \widetilde{C}(I - Z_{\rm diag}(z)\tilde{A})^{-1} Z_{\rm diag}(z) \widetilde{B} \tag{3.34}$$

is a weakly coisometric realization of S with the state space $\bigoplus_{i=1}^d X_i$ and such that the pair $(\widetilde{C}, \tilde{A})$ is observable in the sense of (3.28). Then S admits an Agler decomposition (3.1) with kernels K_i defined as in (3.8):

$$K_i(z, \zeta) = \widetilde{C}(I - Z_{\rm diag}(z)\tilde{A})^{-1} P_{\widetilde{X}_i} (I - \tilde{A}^* Z_{\rm diag}(\zeta)^*)^{-1} \widetilde{C}^*$$

for $i = 1, \dots, d$. Let $\mathcal{H}(K_i)$ be the associated reproducing kernel Hilbert spaces and let $\mathcal{I}_i : \mathcal{X}_i \to \mathcal{X} = \bigoplus_{i=1}^d \mathcal{X}_j$ be the inclusion maps

$$\mathcal{I}_i \colon x_i \to 0 \oplus \cdots \oplus 0 \oplus x_i \oplus 0 \oplus \cdots \oplus 0.$$

Since the pair $(\widetilde{C}, \tilde{A})$ is observable, the operators $U_i \colon \mathcal{X}_i \to \mathcal{H}(K_i)$ given by

$$U_i \colon x_i \to \widetilde{C}(I - Z_{\text{diag}}(z)\tilde{A})^{-1}\mathcal{I}_i x_i \tag{3.35}$$

are unitary. Let us define $A \in \mathcal{L}\big(\bigoplus_{i=1}^d \mathcal{H}(K_i)\big)$ and $B \in \mathcal{L}\big(\mathcal{U}, \bigoplus_{i=1}^d \mathcal{H}(K_i)\big)$ by

$$A\Big(\bigoplus_{i=1}^d U_i\Big) = \Big(\bigoplus_{i=1}^d U_i\Big)\tilde{A} \quad \text{and} \quad B = \Big(\bigoplus_{i=1}^d U_i\Big)\widetilde{B}. \tag{3.36}$$

In more detail: $A = [A_{ij}]_{i,j=1}^d$ where

$$A_{ij} \colon \widetilde{C}(I - Z_{\text{diag}}(z)\tilde{A})^{-1}\mathcal{I}_j x_j \to \widetilde{C}(I - Z_{\text{diag}}(z)\tilde{A})^{-1}\mathcal{I}_i \tilde{A}_{ij} x_j. \tag{3.37}$$

Define the operators $A_{i\bullet}$ as in (3.16) and similarly the operators $\tilde{A}_{i\bullet}$ for $i = 1, \dots, d$. Take the generic element f of $\bigoplus_{i=1}^d \mathcal{H}(K_i)$ and $x \in \mathcal{X}$ in the form

$$f(z) = \bigoplus_{j=1}^d \widetilde{C}(I_{\mathcal{X}} - Z_{\text{diag}}(z)\tilde{A})^{-1}\mathcal{I}_j x_j, \qquad x = \bigoplus_{j=1}^d x_j \in \mathcal{X}. \tag{3.38}$$

By (3.37), we have

$$\begin{aligned}(A_{i\bullet} f)(z) &= \begin{bmatrix} A_{i1} & \dots & A_{id}\end{bmatrix} \begin{bmatrix} \widetilde{C}(I - Z_{\text{diag}}(z)\tilde{A})^{-1}\mathcal{I}_1 x_1 \\ \vdots \\ \widetilde{C}(I - Z_{\text{diag}}(z)\tilde{A})^{-1}\mathcal{I}_d x_d \end{bmatrix} \\ &= \sum_{j=1}^d A_{ij}(\widetilde{C}(I - Z_{\text{diag}}(z)\tilde{A})^{-1}\mathcal{I}_j x_j) \\ &= \sum_{j=1}^d \widetilde{C}(I - Z_{\text{diag}}(z)\tilde{A})^{-1}\mathcal{I}_i \tilde{A}_{ij} x_j \\ &= \widetilde{C}(I - Z_{\text{diag}}(z)\tilde{A})^{-1}\mathcal{I}_i \tilde{A}_{i\bullet} x.\end{aligned} \tag{3.39}$$

For f and x as in (3.38), we have

$$\begin{aligned}(\mathbf{s}f)(z) = \sum_{j=1}^d \widetilde{C}(I_{\mathcal{X}} - Z_{\text{diag}}(z)\tilde{A})^{-1}\mathcal{I}_j x_j &= \widetilde{C}(I_{\mathcal{X}} - Z_{\text{diag}}(z)\tilde{A})^{-1} \sum_{j=1}^d \mathcal{I}_j x_j \\ &= \widetilde{C}(I_{\mathcal{X}} - Z_{\text{diag}}(z)\tilde{A})^{-1} x\end{aligned}$$

which together with (3.39) gives

$$\begin{aligned}(\mathbf{s}f)(z) - (\mathbf{s}f)(0) &= \widetilde{C}(I - Z_{\text{diag}}(z)\tilde{A})^{-1}x - \widetilde{C}x \\ &= \widetilde{C}(I - Z_{\text{diag}}(z)\tilde{A})^{-1} Z_{\text{diag}}(z)\tilde{A}x \\ &= \sum_{j=1}^d z_j \cdot \widetilde{C}(I - Z_{\text{diag}}(z)\tilde{A})^{-1}\mathcal{I}_j \tilde{A}_{j\bullet} x = \sum_{j=1}^d z_j \cdot (A_{j\bullet} f)(z),\end{aligned}$$

which means (since f is the generic element of $\bigoplus_{i=1}^d \mathcal{H}(K_i)$) that the operators $A_{1\bullet}, \dots, A_{d\bullet}$ satisfy identity (3.15). Furthermore, on account of (3.38), (3.35) and (3.34),

$$\begin{aligned}\sum_{i=1}^d z_i(Bu)_i(z) &= \sum_{i=1}^d z_i\widetilde{C}(I - Z_{\rm diag}(z)\tilde{A})^{-1}\mathcal{I}_i\widetilde{B}_i u \\ &= \widetilde{C}(I - Z_{\rm diag}(z)\tilde{A})^{-1}\sum_{i=1}^d z_i\mathcal{I}_i\widetilde{B}_i u \\ &= \widetilde{C}(I - Z_{\rm diag}(z)\tilde{A})^{-1}Z_{\rm diag}(z)Bu = S(z)u - S(0)u\end{aligned}$$

and thus, B solves the Gleason problem (3.18) for S. On the other hand, for an x of the form (3.38), for operators U_i defined in (3.35), and for the operator C defined on $\bigoplus_{i=1}^d \mathcal{H}(K_i)$ by formula (3.21), we have

$$C\left(\bigoplus_{i=1}^d U_i\right)x = \sum_{i=1}^d (U_i x_i)(0) = \sum_{i=1}^d \widetilde{C}(I - Z_{\rm diag}(0)\tilde{A})^{-1}\mathcal{I}_i x_i = \widetilde{C}\sum_{i=1}^d \mathcal{I}_i x_i = \widetilde{C}x$$

and thus $C(\bigoplus_{i=1}^d U_i) = \widetilde{C}$. The latter equality together with definitions (3.36) implies (3.30). Thus the realization $\mathbf{U} = \left[\begin{smallmatrix} A & B \\ C & D\end{smallmatrix}\right]$ is unitarily equivalent to the original realization $\widetilde{\mathbf{U}} = \left[\begin{smallmatrix} \tilde{A} & \widetilde{B} \\ \widetilde{C} & D\end{smallmatrix}\right]$ via the unitary operator $\bigoplus_{i=1}^d U_i$. This realization is a canonical functional-model realization associated with the Agler decomposition $\{K_1, \dots, K_d\}$ of S since all the requirements in Definition 3.5 are met. □

We conclude this section with a theorem parallel to Theorem 2.9. In analogy with the ball setting, we say that the operator A on $\bigoplus_{i=1}^d \mathcal{H}(K_i)$ is a *contractive solution* of the structured Gleason problem for the kernel collection $\{K_1, \dots, K_d\}$ if in addition to identity (3.15) the inequality

$$\|Af\|^2_{\bigoplus_{i=1}^d \mathcal{H}(K_i)} := \sum_{i=1}^d \|A_{i\bullet}f\|^2_{\mathcal{H}(K_i)} \le \|f\|^2_{\bigoplus_{i=1}^d \mathcal{H}(K_i)} - \|(\mathbf{s}f)(0)\|^2_{\mathcal{Y}}$$

holds for every function $f \in \bigoplus_{i=1}^d \mathcal{H}(K_i)$ or equivalently, the pair (C, A) is contractive:

$$A^*A + C^*C \le I,$$

where $C\colon \bigoplus_{i=1}^d \mathcal{H}(K_i) \to \mathcal{Y}$ is the operator given in (3.21). By Definition 3.5, for every canonical functional-model colligation $\mathbf{U} = \left[\begin{smallmatrix} A & B \\ C & D\end{smallmatrix}\right]$ associated with a given Agler decomposition of S, the operator A is a contractive solution of the structured Gleason problem (3.15).

Theorem 3.10. *Let* (3.1) *be a fixed Agler decomposition of a given function* $S \in \mathcal{SA}_d(\mathcal{U}, \mathcal{Y})$ *and let* C *and* D *be defined as in* (3.21). *Then*

(1) *For every contractive solution* A *of the structured Gleason problem* (3.15), *there is an operator* $B = \bigoplus B_i \colon \mathcal{U} \to \bigoplus_{i=1}^d \mathcal{H}(K_i)$ *such that* $\mathbf{U} = \left[\begin{smallmatrix} A & B \\ C & D\end{smallmatrix}\right]$ *is a canonical functional-model colligation for* S.
(2) *Every such* B *solves the Gleason problem* (3.18) *for* S.

PROOF. We start the proof with two preliminary steps.

Step 1. *Let A of the form* (3.17) *solve the Gleason problem* (3.15). *Then*

$$C(I - Z_{\text{diag}}(z)A)^{-1}f = (\mathbf{s}f)(z) \qquad \left(z \in \mathbb{D}^d;\, f \in \bigoplus_{i=1}^{d} \mathcal{H}(K_i)\right) \tag{3.40}$$

where $\mathbf{s}$ *and* C *are defined in* (3.9) *and* (3.21), *respectively.*

Proof of Step 1. To show that identity (3.15) is equivalent to (3.40) we take A in the form (3.17) and define the operators

$$\hat{A}_{1\bullet} = \begin{bmatrix} A_{1\bullet} \\ 0 \\ \vdots \\ 0 \end{bmatrix}, \qquad \hat{A}_{2\bullet} = \begin{bmatrix} 0 \\ A_{2\bullet} \\ \vdots \\ 0 \end{bmatrix}, \qquad \ldots, \qquad \hat{A}_{d\bullet} = \begin{bmatrix} 0 \\ \vdots \\ 0 \\ A_{d\bullet} \end{bmatrix}, \tag{3.41}$$

so that $\hat{A}_{i\bullet} \colon \bigoplus_{i=1}^d \mathcal{H}(K_i) \to \bigoplus_{i=1}^d \mathcal{H}(K_i)$ and $A = \sum_{i=1}^{d} \hat{A}_{i\bullet}$. On account of (3.41) and due to the block structure (3.4) of $Z_{\text{diag}}(z)$ we have

$$\begin{aligned}(I - Z_{\text{diag}}(z)A)^{-1} &= (I - z_1\hat{A}_{1\bullet} - \cdots - z_d\hat{A}_{d\bullet})^{-1} \\ &= \sum_{k=0}^{\infty} (z_1\hat{A}_{1\bullet} + \cdots + z_d\hat{A}_{d\bullet})^k.\end{aligned}$$

Applying the operator $C(I - Z_{\text{diag}}(z)A)^{-1}$ to an arbitrary $f \in \bigoplus_{i=1}^d \mathcal{H}(K_i)$ and making use of formula (3.21) for C, we get

$$\begin{aligned} C(I &- Z_{\text{diag}}(z)A)^{-1}f \\ &= C\sum_{k=0}^{\infty}(z_1\hat{A}_{1\bullet} + \cdots + z_d\hat{A}_{d\bullet})^k f \\ &= (\mathbf{s}f)(0) + \sum_{i=1}^{d} z_i(\mathbf{s}\hat{A}_{i\bullet}f)(0) + \sum_{i,j=1}^{d} z_iz_j(\mathbf{s}\hat{A}_{i\bullet}\hat{A}_{j\bullet}f)(0) + \cdots . \end{aligned} \tag{3.42}$$

On the other hand, by writing (3.15) in the form

$$(\mathbf{s}f)(z) = (\mathbf{s}f)(0) + \sum_{i=1}^{d} z_i(\mathbf{s}\hat{A}_{i\bullet}f)(z)$$

and iterating the latter formula for each $f \in \bigoplus_{i=1}^d \mathcal{H}(K_i)$, we get

$$\begin{aligned} (\mathbf{s}f)(z) &\\ = (\mathbf{s}f)(0) &+ \sum_{j_1=1}^{d} z_{j_1}\Bigg[(\mathbf{s}\hat{A}_{j_1\bullet}f)(0) + \sum_{j_2=1}^{d} z_{j_2}\Bigg[(\mathbf{s}A_{j_2\bullet}A_{j_1\bullet}f)(0) + \cdots \\ &+ \sum_{j_k=1}^{d} z_{j_k}[(\mathbf{s}A_{j_k\bullet}\ldots A_{j_2\bullet}A_{j_1\bullet}f)(0) + \cdots]\cdots\Bigg]\Bigg]. \end{aligned} \tag{3.43}$$

Since the right-hand side expressions in (3.42) and (3.43) are identical, (3.40) follows. Now we have from (3.40)

$$\begin{aligned}\langle f, (I - A^* Z_{\text{diag}}(z)^*)^{-1} C^* y\rangle = \langle C(I - Z_{\text{diag}}(z)A)^{-1} f, y\rangle &= \langle (\mathbf{s}f)(z), y\rangle \\ &= \langle\langle f, \mathbb{T}(\cdot, z) y\rangle\end{aligned}$$

for every $z \in \mathbb{D}^d$ and $f \in \bigoplus_{i=1}^d \mathcal{H}(K_i)$, and thus, equality (3.31) holds. □

Step 2. *Given operators A, C and D with A of the form (3.17) equal to a contractive solution of the structured Gleason problem for the kernel collection $\{K_1, \dots, K_d\}$ and with C and D given by (3.21), if $\mathbf{U} = \left[\begin{smallmatrix} A & B \\ C & D\end{smallmatrix}\right]$ is a contractive realization of S for some operator $B = \bigoplus B_i \colon \mathcal{U} \to \bigoplus_{i=1}^d \mathcal{H}(K_i)$, then B solves the $\bigoplus_{k=1}^d \mathcal{H}(K_k)$-Gleason problem for S, i.e., B satisfies identity* (3.18).

Proof of Step 2. Since $\mathbf{U} = \left[\begin{smallmatrix} A & B \\ C & D\end{smallmatrix}\right]$ is a realization for S, equality (3.33) holds. Making use of equality (3.31) (which holds by Step 1) one can write (3.33) as

$$B^* Z_{\text{diag}}(\zeta)^* \mathbb{T}(\cdot, \zeta) y + D^* y = S(\zeta)^* y$$

or, in view of formula (3.21) for D, as

$$B^* Z_{\text{diag}}(\zeta)^* \mathbb{T}(\cdot, \zeta) y = S(\zeta)^* y - S(0)^* y. \tag{3.44}$$

Taking the inner product of both parts in (3.44) with an arbitrary function $f \in \bigoplus_{i=1}^d \mathcal{H}(K_i)$ leads us to $Z_{\text{diag}}(z)Bu = S(z)u - S(0)u$ which is the same as (3.18). □

To complete the proof of the theorem, it suffices to show that there exists an operator $B : \mathcal{U} \to \bigoplus_{i=1}^d \mathcal{H}(K_i)$ such that equality (3.33) holds for every $u \in \mathcal{U}$ and the operator matrix

$$\mathbf{U}^* = \begin{bmatrix} A^* & C^* \\ B^* & D^* \end{bmatrix} : \begin{bmatrix} \bigoplus_{i=1}^d \mathcal{H}(K_i) \\ \mathcal{Y} \end{bmatrix} \to \begin{bmatrix} \bigoplus_{i=1}^d \mathcal{H}(K_i) \\ \mathcal{U} \end{bmatrix} \tag{3.45}$$

is a contraction. As we have seen, equality (3.33) is equivalent to (3.44), which in turn, defines B^* on the space $\mathcal{D}$ introduced in (3.13). Let us define $\widetilde{B} \colon \mathcal{D} \to \bigoplus_{i=1}^d \mathcal{H}(K_i)$ by the formula

$$\widetilde{B} \colon Z(\zeta)^* \mathbb{T}(\cdot, \zeta) y = S(\zeta)^* y - S(0)^* y$$

and subsequent extension by linearity and continuity; it is a consequence of the isometric property of the operator V in (3.24) that the extension is well-defined and bounded. We arrive at the following contractive matrix-completion problem: *find $B \colon \mathcal{U} \to \bigoplus_{i=1}^d \mathcal{H}(K_i)$ such that $B^*|_{\mathcal{D}} = \widetilde{B}$ and such that U^* of the form* (3.45) *is a contraction.* Following [10] we convert this problem to a standard matrix-completion problem as follows. Define operators

$$T_{11} \colon \mathcal{D}^\perp \to \bigoplus_{i=1}^d \mathcal{H}(K_i), \qquad T_{12} \colon \mathcal{D} \oplus \mathcal{Y} \to \bigoplus_{i=1}^d \mathcal{H}(K_i), \qquad T_{22} \colon \mathcal{D} \oplus \mathcal{Y} \to \mathcal{U}$$

by

$$T_{11} = A^*|_{\mathcal{D}^\perp}, \qquad T_{12} = \begin{bmatrix} A^*|_{\mathcal{D}} & C^* \end{bmatrix}, \quad T_{22} = \begin{bmatrix} \widetilde{B} & D^* \end{bmatrix}. \tag{3.46}$$

Identifying $\left[\begin{smallmatrix} \mathcal{D}^{\perp} \\ \mathcal{D}\oplus\mathcal{Y} \end{smallmatrix}\right]$ with $\left[\begin{smallmatrix} \bigoplus_{i=1}^{d}\mathcal{H}(K_i) \\ \mathcal{Y} \end{smallmatrix}\right]$ we then can represent $\mathbf{U}^*$ from (3.45) as

$$\mathbf{U}^* = \begin{bmatrix} T_{11} & T_{12} \\ X & T_{22} \end{bmatrix} : \begin{bmatrix} \mathcal{D}^{\perp} \\ \mathcal{D}\oplus\mathcal{Y} \end{bmatrix} \to \begin{bmatrix} \bigoplus_{i=1}^{d}\mathcal{H}(K_i) \\ \mathcal{U} \end{bmatrix} \tag{3.47}$$

where $X = B^*|_{\mathcal{D}^{\perp}}$ is unknown. Thus, an operator B gives rise to a canonical functional-model realization $\mathbf{U} = \left[\begin{smallmatrix} A & B \\ C & D \end{smallmatrix}\right]$ of S if and only if it is of the form

$$B = \begin{bmatrix} X^* \\ \widetilde{B}^* \end{bmatrix} : \mathcal{U} \to \begin{bmatrix} \mathcal{D}^{\perp} \\ \mathcal{D} \end{bmatrix} \cong \bigoplus_{i=1}^{d} \mathcal{H}(K_i)$$

where X is any solution of the contractive matrix-completion problem (3.47). But this is a standard matrix-completion problem which can be handled by the well-known Parrott's result [29]: it has a solution X if and only if the obvious necessary conditions hold:

$$\left\| \begin{bmatrix} T_{11} & T_{12} \end{bmatrix} \right\| \le 1, \qquad \left\| \begin{bmatrix} T_{12} \\ T_{22} \end{bmatrix} \right\| \le 1. \tag{3.48}$$

Making use of the definitions of T_{11}, T_{12}, T_{22} from (3.46), we get more explicitly

$$\begin{bmatrix} T_{11} & T_{12} \end{bmatrix} = \begin{bmatrix} A^* & C^* \end{bmatrix}, \qquad \begin{bmatrix} T_{12} \\ T_{22} \end{bmatrix} = \begin{bmatrix} A^*|_{\mathcal{D}} & C^* \\ \widetilde{B} & D^* \end{bmatrix}.$$

Thus the first expression in (3.48) is contractive since A is a contractive solution of the structured Gleason problem (3.15), while the second expression collapses to $\widetilde{V}$ (see formula (3.24)) which is isometric by (3.1). We conclude that the necessary conditions (3.48) are satisfied and hence, by the result of [29], there exists a solution X to problem (3.47). This completes the proof of the theorem. □

4. de Branges – Rovnyak kernels associated with a Schur – Agler-class function on a domain with matrix-polynomial defining function

A generalized Schur class containing all those discussed in the previous sections as special cases was introduced and studied in [4, 9] (see also [5] for the scalar-valued case) and can be defined as follows. Let $\mathbf{Q}$ be a $p \times q$ matrix-valued polynomial

$$\mathbf{Q}(z) = \begin{bmatrix} \mathbf{q}_{11}(z) & \dots & \mathbf{q}_{1q}(z) \\ \vdots & & \vdots \\ \mathbf{q}_{p1}(z) & \dots & \mathbf{q}_{pq}(z) \end{bmatrix} : \mathbb{C}^n \to \mathbb{C}^{p\times q} \tag{4.1}$$

such that

$$\mathbf{Q}(0) = 0 \tag{4.2}$$

and let $\mathcal{D}_{\mathbf{Q}} \in \mathbb{C}^n$ be the domain defined by

$$\mathcal{D}_{\mathbf{Q}} = \{ z \in \mathbb{C}^n \colon \|\mathbf{Q}(z)\| < 1 \}.$$

Now we recall the *Schur – Agler class* $\mathcal{SA}_{\mathbf{Q}}(\mathcal{U}, \mathcal{Y})$ that consists, by definition, of $\mathcal{L}(\mathcal{U}, \mathcal{Y})$-valued functions $S(z) = S(z_1, \dots, z_n)$ analytic on $\mathcal{D}_{\mathbf{Q}}$ and such that $\|S(T)\| \le 1$ for any collection of n commuting operators $T = (T_1, \dots, T_n)$ on a Hilbert space $\mathcal{K}$, subject to $\|\mathbf{Q}(T)\| < 1$. By [5, Lemma 1], the Taylor joint spectrum of the commuting n-tuple $T = (T_1, \dots, T_n)$ is contained in $\mathcal{D}_{\mathbf{Q}}$ whenever $\|\mathbf{Q}(T)\| < 1$, and hence $S(T)$ is well defined by the Taylor functional calculus (see [19]) for any $\mathcal{L}(\mathcal{U}, \mathcal{Y})$-valued function S which is analytic on $\mathcal{D}_{\mathbf{Q}}$. Upon using $\mathcal{K} = \mathbb{C}$ and $T_j = z_j$

for $j = 1, \dots, n$ where $(z_1, \dots, z_n)$ is a point in $\mathcal{D}_{\mathbf{Q}}$ we conclude that any function in $\mathcal{SA}_{\mathbf{Q}}(\mathcal{U}, \mathcal{Y})$ is contractive-valued, and thus, the class $\mathcal{SA}_{\mathbf{Q}}(\mathcal{U}, \mathcal{Y})$ is the subclass of the *Schur class* $\mathcal{S}_{\mathcal{D}_{\mathbf{Q}}}(\mathcal{U}, \mathcal{Y})$ of contractive valued functions analytic on $\mathcal{D}_{\mathbf{Q}}$. By the von Neumann result, in the case when $\mathbf{Q}(z) = z$, these classes coincide; in general, $\mathcal{SA}_{\mathbf{Q}}(\mathcal{U}, \mathcal{Y})$ is a proper subclass of $\mathcal{S}_{\mathcal{D}_{\mathbf{Q}}}(\mathcal{U}, \mathcal{Y})$. The following result appears in [4,9] (see also [5] for the scalar-valued case $\mathcal{U} = \mathcal{Y} = \mathbb{C}$) and is yet another multivariable analogue of Theorem 1.1. We will often abuse notation and will write $\mathbf{Q}(z)$ instead of $\mathbf{Q}(z) \otimes I$ where I is the identity operator on an appropriate Hilbert space clear from the context. When the following theorem is viewed as a parallel of Theorem 1.1 we see that, just as in the polydisk setting, there is no parallel to conditions (1a) and (1b).

Theorem 4.1. *Let S be a $\mathcal{L}(\mathcal{U}, \mathcal{Y})$-valued function defined on $\mathcal{D}_{\mathbf{Q}}$. The following statements are equivalent:*

(1) (c) *S belongs to $\mathcal{SA}_{\mathbf{Q}}(\mathcal{U}, \mathcal{Y})$.*
(2) *There exists a positive kernel*

$$\mathbb{K} = \begin{bmatrix} \mathbb{K}_{11} & \dots & \mathbb{K}_{1p} \\ \vdots & & \vdots \\ \mathbb{K}_{p1} & \dots & \mathbb{K}_{pp} \end{bmatrix} : \mathcal{D}_{\mathbf{Q}} \times \mathcal{D}_{\mathbf{Q}} \to \mathcal{L}(\mathcal{Y}^p) \tag{4.3}$$

which provides a **Q**-Agler decomposition *for S, i.e., such that for every $z, \zeta \in \mathcal{D}_{\mathbf{Q}}$,*

$$I_{\mathcal{Y}} - S(z)S(\zeta)^* = \sum_{k=1}^{p} \mathbb{K}_{kk}(z,\zeta) - \sum_{k=1}^{q} \sum_{i,l=1}^{p} \mathbf{q}_{ik}(z)\overline{\mathbf{q}_{lk}(\zeta)}\mathbb{K}_{il}(z,\zeta). \tag{4.4}$$

(2′) *There exist an auxiliary Hilbert space $\mathcal{X}$ and a function*

$$H(z) = \begin{bmatrix} H_1(z) & \dots & H_p(z) \end{bmatrix} \tag{4.5}$$

analytic on $\mathcal{D}_{\mathbf{Q}}$ with values in $\mathcal{L}(\mathcal{X}^p, \mathcal{Y})$ so that for every $z, \zeta \in \mathcal{D}_{\mathbf{Q}}$,

$$I_{\mathcal{Y}} - S(z)S(\zeta)^* = H(z)(I_{\mathcal{X}^p} - \mathbf{Q}(z)\mathbf{Q}(\zeta)^*)H(\zeta)^*. \tag{4.6}$$

(3) *There exist an auxiliary Hilbert space $\mathcal{X}$ and a unitary connecting operator* **U** *of the form*

$$\mathbf{U} = \begin{bmatrix} A & B \\ C & D \end{bmatrix} : \begin{bmatrix} \mathcal{X}^p \\ \mathcal{U} \end{bmatrix} \to \begin{bmatrix} \mathcal{X}^q \\ \mathcal{Y} \end{bmatrix} \tag{4.7}$$

so that $S(z)$ can be realized in the form

$$S(z) = D + C(I_{\mathcal{X}^p} - \mathbf{Q}(z)A)^{-1}\mathbf{Q}(z)B \quad \textit{for all } z \in \mathcal{D}_{\mathbf{Q}}. \tag{4.8}$$

(4) *There exist an auxiliary Hilbert space $\mathcal{X}$ and a contractive connecting operator* **U** *of the form* (4.7) *so that $S(z)$ can be realized in the form* (4.8).

Remark 4.2. If $S = \left[\begin{smallmatrix} S_{11} & S_{12} \\ S_{21} & S_{22} \end{smallmatrix}\right] \in \mathcal{SA}_{\mathbf{Q}}(\mathcal{U}_1 \oplus \mathcal{U}_2, \mathcal{Y}_1 \oplus \mathcal{Y}_2)$, then the block entry S_{ij} belongs to the Schur–Agler class $\mathcal{SA}_{\mathbf{Q}}(\mathcal{U}_i, \mathcal{Y}_j)$ for $i, j = 1, 2$. For the proof, it suffices to note that $\|S_{ij}(T)\| \le \|S(T)\|$.

Remark 4.3. The equivalence (2) $\iff$ (2′) can be seen by using the Kolmogorov decomposition for the positive kernel $\mathbb{K}$:

$$\mathbb{K}(z,\zeta) = \begin{bmatrix} H_1(z) \\ \vdots \\ H_p(z) \end{bmatrix} \begin{bmatrix} H_1(\zeta)^* & \dots & H_p(\zeta)^* \end{bmatrix}. \tag{4.9}$$

The implication (4) $\Longrightarrow$ (1) can be handled by any of the three approaches sketched in Remark 2.2. Following the approach from [10], we first handle the case where $\mathbf{U}$ is coisometric, using the identity

$$\begin{aligned} I - S(z)S(\zeta)^* &= C(I - \mathbf{Q}(z)A)^{-1}(I - \mathbf{Q}(z)\mathbf{Q}(\zeta)^*)(I - A^*\mathbf{Q}(\zeta)^*)^{-1}C^* \\ &+ \begin{bmatrix} C(I - \mathbf{Q}(z)A)^{-1}\mathbf{Q}(z) & I \end{bmatrix}(I - \mathbf{U}\mathbf{U}^*)\begin{bmatrix} \mathbf{Q}(\zeta)^*(I - A^*\mathbf{Q}(\zeta)^*)^{-1}C^* \\ I \end{bmatrix}. \end{aligned} \tag{4.10}$$

holding for S of the form (4.8) and $\mathbf{U}$ given by (4.7), the straightforward verification of which is based on the identity

$$\begin{bmatrix} C(I - \mathbf{Q}(z)A)^{-1}\mathbf{Q}(z) & I \end{bmatrix}\mathbf{U} = \begin{bmatrix} C(I - \mathbf{Q}(z)A)^{-1} & S(z) \end{bmatrix}.$$

Then the general (contractive) case follows by extension arguments and Remark 4.2.

Remark 4.4. With no assumptions on the polynomial matrix $\mathbf{Q}(z)$ some degeneracies occur which can be eliminated with proper normalizations. We note first of all that it is natural to assume that no row of $\mathbf{Q}(z)$ vanishes identically; otherwise one can cross out any vanishing column to get a new matrix polynomial $\widetilde{\mathbf{Q}}(z)$ of smaller size which defines the same domain $\mathcal{D}_{\mathbf{Q}}$ in $\mathbb{C}^n$. Secondly, in the second term of the $\mathbf{Q}$-Agler decomposition (4.4), the (i,l)-entry $\mathbb{K}_{i,l}$ of $\mathbb{K}$ is irrelevant for any pair of indices i,l such that at least one of $\mathbf{q}_{ik}(z)$ and $\mathbf{q}_{lk}(z)$ vanish identically for each $k = 1,\dots,q$. Note that if the first reduction has been carried out, then all diagonal entries $\mathbb{K}_{ii}$ are relevant in the second term of (4.4) in this sense. It follows that, without loss of generality, we may assume that $\mathbb{K}_{il}(z,\zeta) \equiv 0$ for each such pair of indices (i,l). To organize the bookkeeping, we may multiply $\mathbf{Q}(z)$ on the left and right by a permutation matrices Π and Π' (of respective sizes $p \times p$ and $q \times q$) so that $\widetilde{\mathbf{Q}}(z) = \Pi Q(z)\Pi'$ has a block diagonal form

$$\widetilde{\mathbf{Q}}(z) = \begin{bmatrix} \mathbf{Q}^{(1)}(z) & & 0 \\ & \ddots & \\ 0 & & \mathbf{Q}^{(d)}(z) \end{bmatrix} \tag{4.11}$$

with the αth block $\mathbf{Q}^{(\alpha)}$ $(\alpha = 1,\dots,d)$ of say size $p_\alpha \times q_\alpha$ and of the form

$$\mathbf{Q}^{(\alpha)}(z) = [\mathbf{q}_{i,j}^{(\alpha)}(z)]_{i=1}^{p_\alpha}{}_{j=1}^{q_\alpha}$$

and *irreducible* in the sense that $\widetilde{\mathbf{Q}}$ has no finer block-diagonal decomposition after permutation equivalence, i.e., for each α for which $\mathbf{Q}^{(\alpha)}$ is nonzero and for any pair of indices i,l $(1 \le i,l \le p_\alpha)$, there is some k $(1 \le k \le q_j)$ so that either $\mathbf{q}_{ik}^{(\alpha)}(z)$ or $\mathbf{q}_{lk}^{(\alpha)}(z)$ does not vanish identically. Without loss of generality we may assume that the original matrix polynomial $\mathbf{Q}$ is normalized so that $\widetilde{\mathbf{Q}} = \mathbf{Q}$. We may then assume that the positive kernel in (4.3) and (4.4) has the block diagonal decomposition

$$\mathbb{K}(z,\zeta) = \begin{bmatrix} \mathbb{K}^{(1)}(z,\zeta) & & 0 \\ & \ddots & \\ 0 & & \mathbb{K}^{(d)}(z,\zeta) \end{bmatrix} \tag{4.12}$$

where $\mathbb{K}^{(\alpha)}$ in turn has the form

$$\mathbb{K}^{(\alpha)} = \begin{bmatrix} \mathbb{K}^{(\alpha)}_{11} & \dots & \mathbb{K}^{(\alpha)}_{1p_\alpha} \\ \vdots & & \vdots \\ \mathbb{K}^{(\alpha)}_{p_\alpha 1} & \dots & \mathbb{K}^{(\alpha)}_{p_\alpha p_\alpha} \end{bmatrix} : \mathcal{D}_{\mathbf{Q}} \times \mathcal{D}_{\mathbf{Q}} \to \mathcal{L}(\mathcal{Y}^{p_\alpha}).$$

Under the normalizing assumption that $\mathbb{K}$ has this block diagonal form (4.12) once $\mathbf{Q}(z)$ is written as a direct sum of irreducible pieces (4.11), the constructions to follow can be done with more efficient labeling but at the cost of an additional layer of notation. We therefore shall assume in the sequel that this diagonal structure has not been taken into account (or that the matrix polynomial $\mathbf{Q}$ is already irreducible) until the very end of the paper where we explain how the polydisk setting can be seen as an instance of the general setting.

As in the previous particular settings of the ball and of the polydisk, we introduce the weak-coisometry property as the property equivalent to (4.10) collapsing to

$$I - S(z)S(\zeta)^* = C(I - \mathbf{Q}(z)A)^{-1}(I - \mathbf{Q}(z)\mathbf{Q}(\zeta)^*)(I - A^*\mathbf{Q}(\zeta)^*)^{-1}C^*.$$

Definition 4.5. The operator-block matrix $\mathbf{U}$ of the form (4.7) is *weakly coisometric* if the restriction of $\mathbf{U}^*$ to the subspace

$$\mathcal{D}_{\mathbf{U}^*} := \bigvee_{\substack{\zeta \in \mathcal{D}_{\mathbf{Q}} \\ y \in \mathcal{Y}}} \begin{bmatrix} \mathbf{Q}(\zeta)^*(I - A^*\mathbf{Q}(\zeta)^*)^{-1}C^*y \\ y \end{bmatrix} \subset \begin{bmatrix} \mathcal{X}^q \\ \mathcal{Y} \end{bmatrix} \tag{4.13}$$

is isometric.

Due to assumption (4.2), the space $\mathcal{D}_{\mathbf{U}^*}$ splits in the form $\mathcal{D}_{\mathbf{U}^*} = \mathcal{D} \oplus \mathcal{Y}$ where

$$\mathcal{D} = \bigvee_{\zeta \in \mathcal{D}_{\mathbf{Q}}, y \in \mathcal{Y}} \mathbf{Q}(\zeta)^*(I - A^*\mathbf{Q}(\zeta)^*)^{-1}C^*y \subset \mathcal{X}^q. \tag{4.14}$$

4.1. Weakly coisometric canonical functional-model Q-realizations. Let us suppose that we are given a function S in the Schur – Agler class $\mathcal{SA}_{\mathbf{Q}}(\mathcal{U},\mathcal{Y})$ together with an Agler decomposition $\mathbb{K}$ as in (4.3) (so (4.4) is satisfied). We will use the notation $\mathbf{Q}_{\bullet k}(\zeta)$ for the k-th column of the polynomial matrix $\mathbf{Q}$. What actually comes up often is the transpose:

$$\mathbf{Q}_{\bullet k}(\zeta)^\top = \begin{bmatrix} \mathbf{q}_{1k}(\zeta) & \mathbf{q}_{2k}(\zeta) & \dots & \mathbf{q}_{pk}(\zeta) \end{bmatrix}. \tag{4.15}$$

Note that with this notation the $\mathbf{Q}$-*Agler decomposition for* S (4.4) can be written more compactly as

$$I_{\mathcal{Y}} - S(z)S(\zeta)^* = \sum_{k=1}^{p} K_{k,k}(z,\zeta) - \sum_{j=1}^{q} \mathbf{Q}^\top_{\bullet j}(z)\mathbb{K}(z,\zeta)\mathbf{Q}^\top_{\bullet j}(\zeta)^*, \tag{4.16}$$

an expression more suggestive of the Agler decomposition (3.1) for the polydisk case.

We say that the operator $A\colon \mathcal{H}(\mathbb{K})^p \to \mathcal{H}(\mathbb{K})^q$ *solves the* $\mathbf{Q}$*-coupled Gleason problem for* $\mathcal{H}(\mathbb{K})$ if

$$\sum_{k=1}^{p}\big(f_{k,k}(z) - f_{k,k}(0)\big) = \sum_{k=1}^{q} \mathbf{Q}_{\bullet k}(z)^{\top}[Af]_k(z) \qquad \text{for all } f \in \mathcal{H}(\mathbb{K})^p \tag{4.17}$$

so each $f \in \mathcal{H}(\mathbb{K})^p$ has the form

$$f = \begin{bmatrix} f_1 \\ \vdots \\ f_p \end{bmatrix} \qquad \text{where } f_k = \begin{bmatrix} f_{k,1} \\ \vdots \\ f_{k,p} \end{bmatrix} \in \mathcal{H}(\mathbb{K}).$$

Similarly, we say that the operator $B\colon \mathcal{U} \to \mathcal{H}(\mathbb{K})^q$ *solves the* $\mathbf{Q}$*-coupled* $\mathcal{H}(\mathbb{K})$*-Gleason problem for* S if the identity

$$S(z)u - S(0)u = \sum_{k=1}^{q} \mathbf{Q}_{\bullet k}(z)^{\top}[Bu]_k(z) \qquad \text{holds for all } u \in \mathcal{U}. \tag{4.18}$$

The following proposition gives the reformulation of Gleason-problem solutions in terms of the adjoint operators. In what follows, we let $\{\mathbf{e}_1, \dots, \mathbf{e}_p\}$ to be the standard basis for $\mathbb{C}^p$.

Proposition 4.6. *The operator* $A\colon \mathcal{H}(\mathbb{K})^p \to \mathcal{H}(\mathbb{K})^q$ *solves the* $\mathbf{Q}$*-coupled Gleason problem* (4.17) *if and only if the adjoint* A^* *of* A *has the following action on special kernel functions:*

$$A^*\colon \begin{bmatrix} \mathbb{K}(\cdot,\zeta)\mathbf{Q}_{\bullet 1}(\zeta)^{\top *}y \\ \vdots \\ \mathbb{K}(\cdot,\zeta)\mathbf{Q}_{\bullet q}(\zeta)^{\top *}y \end{bmatrix} \mapsto \begin{bmatrix} \mathbb{K}(\cdot,\zeta)E_1 y \\ \vdots \\ \mathbb{K}(\cdot,\zeta)E_p y \end{bmatrix} - \begin{bmatrix} \mathbb{K}(\cdot,0)E_1 y \\ \vdots \\ \mathbb{K}(\cdot,0)E_p y \end{bmatrix} \tag{4.19}$$

for all $\zeta \in \mathcal{D}_{\mathbf{Q}}$ *and* $y \in \mathcal{Y}$*, where* $E_i = I_{\mathcal{Y}} \otimes \mathbf{e}_i$ *for* $i = 1, \dots, p$*:*

$$E_1 = \begin{bmatrix} I_{\mathcal{Y}} \\ 0 \\ \vdots \\ 0 \end{bmatrix}, \qquad E_2 = \begin{bmatrix} 0 \\ I_{\mathcal{Y}} \\ \vdots \\ 0 \end{bmatrix}, \qquad \dots, \qquad E_p = \begin{bmatrix} 0 \\ \vdots \\ 0 \\ I_{\mathcal{Y}} \end{bmatrix}. \tag{4.20}$$

The operator $B\colon \mathcal{U} \to \mathcal{H}(\mathbb{K})^q$ *solves the* $\mathbf{Q}$*-coupled* $\mathcal{H}(\mathbb{K})$*-Gleason problem* (4.18) *for* S *if and only if* $B^*\colon \mathcal{H}(\mathbb{K})^q \to \mathcal{U}$ *has the following action on special kernel functions:*

$$B^*\colon \begin{bmatrix} \mathbb{K}(\cdot,\zeta)\mathbf{Q}_{\bullet 1}(\zeta)^{\top *}y \\ \vdots \\ \mathbb{K}(\cdot,\zeta)\mathbf{Q}_{\bullet q}(\zeta)^{\top *}y \end{bmatrix} \mapsto S(\zeta)^*y - S(0)^*y \quad \textit{for all } \zeta \in \mathcal{D}_{\mathbf{Q}} \textit{ and } y \in \mathcal{Y}. \tag{4.21}$$

PROOF. We start with the identity

$$\mathbf{Q}(\zeta)^* \begin{bmatrix} \mathbb{K}(z,\zeta)E_1 y \\ \vdots \\ \mathbb{K}(z,\zeta)E_p y \end{bmatrix} = \begin{bmatrix} \mathbb{K}(z,\zeta)\mathbf{Q}_{\bullet 1}(\zeta)^{\top *}y \\ \vdots \\ \mathbb{K}(z,\zeta)\mathbf{Q}_{\bullet q}(\zeta)^{\top *}y \end{bmatrix} \tag{4.22}$$

which holds for all $z, \zeta \in \mathcal{D}_{\mathbf{Q}}$ and $y \in \mathcal{Y}$; once $\mathbf{Q}(\zeta)^*$ is interpreted as $\mathbf{Q}(\zeta)^* \otimes I_{\mathcal{Y}}$ and similarly for $\mathbf{Q}_{\bullet k}(\zeta)^{\top *}$, this can be seen as a direct consequence of the definitions

(4.1), (4.3), (4.15) and (4.20). Letting

$$\mathbb{T}(z,\zeta) := \begin{bmatrix} \mathbb{K}(z,\zeta)E_1 \\ \vdots \\ \mathbb{K}(z,\zeta)E_p \end{bmatrix} \tag{4.23}$$

for short, we then can write formulas (4.19), (4.21) more compactly as

$$A^*\mathbf{Q}(\zeta)^*\mathbb{T}(\cdot,\zeta)y = \mathbb{T}(\cdot,\zeta)y - \mathbb{T}(\cdot,0)y, \tag{4.24}$$

$$B^*\mathbf{Q}(\zeta)^*\mathbb{T}(\cdot,\zeta)y = S(\zeta)^*y - S(0)^*y \tag{4.25}$$

where now $\mathbf{Q}(\zeta)^*$ is to be interpreted as $\mathbf{Q}(\zeta)^* \otimes I_{\mathcal{H}(\mathbb{K})}$. In the following computations, $\mathbf{Q}(\zeta)^*$ is either $\mathbf{Q}(\zeta)^* \otimes I_{\mathcal{Y}}$ or $\mathbf{Q}(\zeta)^* \otimes \mathbf{I}_{\mathcal{H}(\mathbb{K})}$ according to the context. By the reproducing kernel property, we have for every $f = \bigoplus_{k=1}^p f_k \in \mathcal{H}(\mathbb{K})^p$,

$$\begin{aligned} \langle f, \mathbb{T}(\cdot,\zeta)y\rangle_{\mathcal{H}(\mathbb{K})^p} &= \sum_{k=1}^p \langle f_k, \mathbb{K}(\cdot,\zeta)E_k y\rangle_{\mathcal{H}(\mathbb{K})} = \sum_{k=1}^p \langle E_k^* f_k(\zeta), y\rangle_{\mathcal{Y}} \\ &= \left\langle \sum_{k=1}^p f_{k,k}(\zeta), y\right\rangle_{\mathcal{Y}}. \end{aligned} \tag{4.26}$$

Therefore,

$$\langle f, \mathbb{T}(\cdot,\zeta)y - \mathbb{T}(\cdot,0)y\rangle_{\mathcal{H}(\mathbb{K})^p} = \left\langle \sum_{k=1}^p (f_{k,k}(\zeta) - f_{k,k}(0)), y\right\rangle_{\mathcal{Y}}. \tag{4.27}$$

On the other hand, it follows again from (4.26) that

$$\langle f, A^*\mathbf{Q}(z)^*\mathbb{T}(\cdot,z)y\rangle_{\mathcal{H}(\mathbb{K})^p} = \langle \mathbf{Q}(z)Af, \mathbb{T}(\cdot,z)y\rangle_{\mathcal{H}(\mathbb{K})^p} = \left\langle \sum_{j=1}^p [\mathbf{Q}(z)Af]_{j,j}(z), y\right\rangle_{\mathcal{Y}}$$

and since

$$\begin{aligned} \sum_{j=1}^p [\mathbf{Q}(z)Af]_{j,j}(z) &= \sum_{j=1}^p \sum_{k=1}^q \mathbf{q}_{jk}(z)[Af]_{k,j}(z) \\ &= \sum_{k=1}^q \left(\sum_{j=1}^p \mathbf{q}_{jk}(z)\, [Af]_{k,j}(z) \right) = \sum_{k=1}^q \mathbf{Q}_{\bullet k}(z)^\top [Af]_k(z), \end{aligned} \tag{4.28}$$

we get

$$\langle f, A^*\mathbf{Q}(z)^*\mathbb{T}(\cdot,z)y\rangle_{\mathcal{H}(\mathbb{K})^p} = \left\langle \sum_{k=1}^q \mathbf{Q}_{\bullet k}(z)^\top [Af]_k(z), y\right\rangle_{\mathcal{Y}}.$$

Since the last equality and (4.27) hold for every $f \in \mathcal{H}(\mathbb{K})^p$, $\zeta \in \mathcal{D}_{\mathbf{Q}}$ and $y \in \mathcal{Y}$, the equivalence of (4.17) and (4.24) (which is the same as (4.19)) follows. Equivalence of (4.18) and (4.25) follows by the same argument from equalities

$$\langle u, S(\zeta)^*y - S(0)^*y\rangle_{\mathcal{U}} = \langle S(\zeta)u - S(0)u, y\rangle_{\mathcal{Y}}$$

and

$$\begin{aligned} \langle u, B^*\mathbf{Q}(z)^*\mathbb{T}(\cdot,z)y\rangle_{\mathcal{U}} &= \langle \mathbf{Q}(z)Bu, \mathbb{T}(\cdot,z)y\rangle_{\mathcal{H}(\mathbb{K})^p} \\ &= \left\langle \sum_{j=1}^p [\mathbf{Q}(z)Bu]_{j,j}(z), y\right\rangle_{\mathcal{Y}} = \left\langle \sum_{k=1}^q \mathbf{Q}_{\bullet k}(z)^\top [Bu]_k(z), y\right\rangle_{\mathcal{Y}} \end{aligned}$$

holding for all $u \in \mathcal{U}$ and $y \in \mathcal{Y}$. □

Just as in the particular cases discussed in the previous sections, it turns out that the formulas (4.19) and (4.21) can be extended by linearity and continuity to define uniquely determined bounded well-defined operators

$$A_{\mathcal{D}}^*: \mathcal{D} \to \mathcal{H}(\mathbb{K})^p, \qquad B_{\mathcal{D}}^*: \mathcal{D} \to \mathcal{U}$$

as a consequence of the isometric property of the operator V defined below in (4.33).

Definition 4.7. We say that the operator-block matrix $\mathbf{U} = \left[\begin{smallmatrix} A & B \\ C & D \end{smallmatrix}\right] : \mathcal{H}(\mathbb{K})^p \oplus \mathcal{U} \to \mathcal{H}(\mathbb{K})^q \oplus \mathcal{Y}$ is a *canonical functional-model colligation matrix* for the given function S and Agler decomposition $\mathbb{K}$ if

(1) $\mathbf{U}$ is contractive.
(2) The operator A solves the $\mathbf{Q}$-coupled Gleason problem (4.17) for $\mathcal{H}(\mathbb{K})$.
(3) The operator B solves the $\mathbf{Q}$-coupled $\mathcal{H}(\mathbb{K})$-Gleason problem (4.18) for S.
(4) The operators $C: \mathcal{H}(\mathbb{K})^p \to \mathcal{Y}$ and $D: \mathcal{U} \to \mathcal{Y}$ are given by

$$C: \begin{bmatrix} f_1(z) \\ \vdots \\ f_p(z) \end{bmatrix} \mapsto f_{1,1}(0) + \cdots + f_{p,p}(0), \quad D: u \mapsto S(0)u. \tag{4.29}$$

Formulas (4.25) can be written equivalently in terms of adjoint operators as follows:

$$C^*: y \mapsto \mathbb{T}(\cdot, 0)y \qquad D^*: y \mapsto S(0)^*y \tag{4.30}$$

where $\mathbb{T}$ is defined in (4.23). The next theorem is the analogue of Theorem 3.7.

Theorem 4.8. *Let S be a given function in the Schur–Agler class $\mathcal{SA}_{\mathbf{Q}}(\mathcal{U}, \mathcal{Y})$ and suppose that we are given an Agler decomposition (4.4) for S. Then there exists a canonical functional-model colligation associated with the kernel $\mathbb{K}$.*

Proof. Let us rearrange the given Agler decomposition (4.4) or (4.16) as

$$I_{\mathcal{Y}} + \sum_{k=1}^{q} \mathbf{Q}_{\bullet k}(z)^{\top} \mathbb{K}(z, \zeta) \mathbf{Q}_{\bullet k}(\zeta)^{\top *} = S(z)S(\zeta)^* + \sum_{j=1}^{p} E_j^* \mathbb{K}(z, \zeta) E_j,$$

and then invoke the reproducing kernel property to rewrite the latter identity in the inner product form as

$$\begin{aligned} \sum_{k=1}^{q} \langle \mathbb{K}(\cdot, \zeta) \mathbf{Q}_{\bullet k}(\zeta)^{\top *} y, \mathbb{K}(\cdot, z) \mathbf{Q}_{\bullet k}(z)^{\top *} y' \rangle_{\mathcal{H}(\mathbb{K})} + \langle y, y' \rangle_{\mathcal{Y}} \\ = \sum_{j=1}^{p} \langle \mathbb{K}(\cdot, \zeta) E_j y, \mathbb{K}(\cdot, z) E_j y' \rangle_{\mathcal{H}(\mathbb{K})} + \langle S(\zeta)^* y, S(z)^* y' \rangle_{\mathcal{U}}. \end{aligned} \tag{4.31}$$

The latter can be written in the matrix form as

$$\left\langle \begin{bmatrix} \mathbb{K}(\cdot,\zeta)\mathbf{Q}_{\bullet 1}(\zeta)^{\top *}y \\ \vdots \\ \mathbb{K}(\cdot,\zeta)\mathbf{Q}_{\bullet q}(\zeta)^{\top *}y \\ y \end{bmatrix}, \begin{bmatrix} \mathbb{K}(\cdot,z)\mathbf{Q}_{\bullet 1}(z)^{\top *}y' \\ \vdots \\ \mathbb{K}(\cdot,z)\mathbf{Q}_{\bullet q}(z)^{\top *}y' \\ y' \end{bmatrix} \right\rangle_{\mathcal{H}(\mathbb{K})^q\oplus\mathcal{Y}}$$

$$= \left\langle \begin{bmatrix} \mathbb{K}(\cdot,\zeta)E_1 y \\ \vdots \\ \mathbb{K}(\cdot,\zeta)E_p y \\ S(\zeta)^* y \end{bmatrix}, \begin{bmatrix} \mathbb{K}(\cdot,z)E_1 y' \\ \vdots \\ \mathbb{K}(\cdot,z)E_p y' \\ S(z)^* y' \end{bmatrix} \right\rangle_{\mathcal{H}(\mathbb{K})^p\oplus\mathcal{U}}$$

or, upon making use of notation (4.23) and of identity (4.22), as

$$\left\langle \begin{bmatrix} \mathbf{Q}(\zeta)^*\mathbb{T}(\cdot,\zeta)y \\ y \end{bmatrix}, \begin{bmatrix} \mathbf{Q}(z)^*\mathbb{T}(\cdot,z)y' \\ y' \end{bmatrix} \right\rangle_{\mathcal{H}(\mathbb{K})^q\oplus\mathcal{Y}} = \left\langle \begin{bmatrix} \mathbb{T}(\cdot,\zeta)y \\ S(\zeta)^*y \end{bmatrix}, \begin{bmatrix} \mathbb{T}(\cdot,z)y' \\ S(z)^*y' \end{bmatrix} \right\rangle_{\mathcal{H}(\mathbb{K})^p\oplus\mathcal{U}}. \tag{4.32}$$

The latter identity implies that the formula

$$V\colon \begin{bmatrix} \mathbf{Q}(\zeta)^*\mathbb{T}(\cdot,\zeta)y \\ y \end{bmatrix} \mapsto \begin{bmatrix} \mathbb{T}(\cdot,\zeta)y \\ S(\zeta)^*y \end{bmatrix} \tag{4.33}$$

extends by continuity to define the isometry from $\mathcal{D}_V = \mathcal{D}\oplus\mathcal{Y} \subset \mathcal{H}(\mathbb{K})^q\oplus\mathcal{Y}$ (see (4.14) for definition of $\mathcal{D}$) onto

$$\mathcal{R}_V = \bigvee_{\substack{\zeta\in\mathcal{D}_{\mathbf{Q}} \\ y\in\mathcal{Y}}} \begin{bmatrix} \mathbb{T}(\cdot,\zeta)y \\ S(\zeta)^*y \end{bmatrix} \subset \begin{bmatrix} \mathcal{H}(\mathbb{K})^p \\ \mathcal{U} \end{bmatrix}.$$

Let us extend V to a contraction $\mathbf{U}^*\colon \begin{bmatrix} \mathcal{H}(\mathbb{K})^q \\ \mathcal{Y} \end{bmatrix} \to \begin{bmatrix} \mathcal{H}(\mathbb{K})^p \\ \mathcal{U} \end{bmatrix}$. Thus,

$$\mathbf{U}^* = \begin{bmatrix} A^* & C^* \\ B^* & D^* \end{bmatrix}\colon \begin{bmatrix} \mathbf{Q}(\zeta)^*\mathbb{T}(\cdot,\zeta)y \\ y \end{bmatrix} \mapsto \begin{bmatrix} \mathbb{T}(\cdot,\zeta)y \\ S(\zeta)^*y \end{bmatrix}. \tag{4.34}$$

Computation of the top and bottom components in (3.25) gives

$$A^*\mathbf{Q}(\zeta)^*\mathbb{T}(\cdot,\zeta)y + C^*y = \mathbb{T}(\cdot,\zeta)y, \tag{4.35}$$

$$B^*\mathbf{Q}(\zeta)^*\mathbb{T}(\cdot,\zeta)y + D^*y = S(\zeta)^*y. \tag{4.36}$$

Letting $\zeta = 0$ in the latter equalities and taking into account (4.2) leads us to (4.30) from which we see that C and D are of the requisite form (4.29). Substitution of (4.30) into (4.35) and (4.36) then leads us to (4.24) and (4.25) which are equivalent to (4.17) and (4.18), respectively. Thus we conclude that $\mathbf{U}$ is a canonical functional-model colligation as wanted. □

For this general setting we define observability as follows: given an operator $A\colon \mathcal{X}^p \to \mathcal{X}^q$ and an operator $C\colon \mathcal{X}^p \to \mathcal{Y}$, the pair (C, A) will be called **Q**-*observable* if the identities $C(I-\mathbf{Q}(z)A)^{-1}\mathcal{I}_i x = 0$ for all z in a neighborhood of the origin and for all $i = 1,\dots,p$ forces $x = 0$ in $\mathcal{X}$. By $\mathcal{I}_i : \mathcal{X} \to \mathcal{X}^p$ we denote the inclusion map which embeds $\mathcal{X}$ into the i-th component of $\mathcal{X}^p = \mathcal{X} \oplus \cdots \oplus \mathcal{X}$. Thus

$$\mathcal{I}_i \colon x_i \mapsto \begin{bmatrix} 0 \\ \vdots \\ x_i \\ \vdots \\ 0 \end{bmatrix} \quad \text{and} \quad \mathcal{I}_i^* \colon \begin{bmatrix} x_1 \\ \vdots \\ x_i \\ \vdots \\ x_p \end{bmatrix} \mapsto x_i. \tag{4.37}$$

The $\mathbf{Q}$-observability can be equivalently defined in terms of adjoint operators as

$$\bigvee \{\mathcal{I}_i^*(I - A^*\mathbf{Q}(z)^*)^{-1}C^*y : z \in \Delta, y \in \mathcal{Y}, i = 1, \dots, p\} = \mathcal{X} \tag{4.38}$$

where Δ is some neighborhood of the origin in $\mathbb{C}^n$. The following theorem is the analogue of Theorem 1.2 for the present general setting.

Theorem 4.9. *Let S be a function in the Schur – Agler class $\mathcal{SA}_{\mathbf{Q}}(\mathcal{U}, \mathcal{Y})$, let the positive kernel $\mathbb{K}$ of the form* (4.3) *provide an Agler decomposition* (4.4) *for S and suppose that $\mathbf{U} = \left[\begin{smallmatrix} A & B \\ C & D \end{smallmatrix}\right] : \mathcal{H}(\mathbb{K})^p \oplus \mathcal{U} \to \mathcal{H}(\mathbb{K})^q \oplus \mathcal{Y}$ is a canonical functional-model colligation associated with S and $\mathbb{K}$. Then the following hold*:

(1) *$\mathbf{U}$ is weakly coisometric.*
(2) *The pair (C, A) is $\mathbf{Q}$-observable in the sense of* (4.38).
(3) *We recover S as $S(z) = D + C(I - \mathbf{Q}(z)A)^{-1}\mathbf{Q}(z)B$.*
(4) *If $\widetilde{\mathbf{U}} = \left[\begin{smallmatrix} \widetilde{A} & \widetilde{B} \\ \widetilde{C} & \widetilde{D} \end{smallmatrix}\right] : \mathcal{X}^p \oplus \mathcal{U} \to \mathcal{X}^q \oplus \mathcal{Y}$ is another colligation matrix enjoying properties* (1), (2), (3) *above, then there is a canonical functional-model colligation $\mathbf{U}$ for $(S, \mathbb{K})$ such that $\mathbf{U}$ and $\widetilde{\mathbf{U}}$ are unitarily equivalent in the sense that there is a unitary operator $U \colon \mathcal{X} \to \mathcal{H}(\mathbb{K})$ so that*

$$\begin{bmatrix} A & B \\ C & D \end{bmatrix} \begin{bmatrix} \bigoplus_{i=1}^p U & 0 \\ 0 & I_{\mathcal{Y}} \end{bmatrix} = \begin{bmatrix} \bigoplus_{i=1}^q U & 0 \\ 0 & I_{\mathcal{U}} \end{bmatrix} \begin{bmatrix} \widetilde{A} & \widetilde{B} \\ \widetilde{C} & \widetilde{D} \end{bmatrix}. \tag{4.39}$$

Proof. Let $\mathbf{U} = \left[\begin{smallmatrix} A & B \\ C & D \end{smallmatrix}\right]$ be a canonical functional-model realization of S associated with a fixed Agler decomposition (4.4). Then combining equalities (4.24), (4.25) (equivalent to the given (4.17) and (4.18) by Proposition 4.6) and also formulas (4.30) (equivalent to the given (4.29)) gives

$$\mathbb{T}(\cdot, \zeta)y = (I - A^*\mathbf{Q}(\zeta)^*)^{-1}\mathbb{T}(\cdot, 0)y = (I - A^*\mathbf{Q}(\zeta)^*)^{-1}C^*y \tag{4.40}$$

and

$$S(\zeta)^*y = S(0)^*y + B^*\mathbf{Q}(\zeta)^*\mathbb{T}(\cdot, \zeta)y = D^*y + B^*\mathbf{Q}(\zeta)^*\mathbb{T}(\cdot, \zeta)y. \tag{4.41}$$

Substituting (4.40) into (4.41) and taking into account that $y \in \mathcal{Y}$ is arbitrary, we get

$$S(\zeta)^* = S(0)^* + B^*\mathbf{Q}(\zeta)^*(I - A^*\mathbf{Q}(\zeta)^*)^{-1}C^* \tag{4.42}$$

which proves part (3) of the theorem. Also we have from (4.40)

$$\bigvee_{\substack{\zeta \in \mathcal{D}_{\mathbf{Q}}, y \in \mathcal{Y}, \\ i=1,\dots,p}} \mathcal{I}_i^*(I - A^*\mathbf{Q}(\zeta)^*)^{-1}C^*y = \bigvee_{\substack{\zeta \in \mathcal{D}_{\mathbf{Q}}, y \in \mathcal{Y}, \\ i=1,\dots,p}} \mathcal{I}_i^*\mathbb{T}(\cdot, \zeta)y$$

and we can proceed due to (4.37) and (4.23) as follows:

$$\bigvee_{\substack{\zeta\in\mathcal{D}_{\mathbf{Q}},y\in\mathcal{Y},\\ i=1,\dots,p}} \mathcal{I}_i^*\mathbb{T}(\cdot,\zeta)y = \bigvee_{\substack{\zeta\in\mathcal{D}_{\mathbf{Q}},y\in\mathcal{Y},\\ i=1,\dots,p}} \mathbb{K}(\cdot,\zeta)E_iy = \bigvee_{\substack{\zeta\in\mathcal{D}_{\mathbf{Q}}\\ \mathbf{y}\in\mathcal{Y}^p}} \mathbb{K}(\cdot,\zeta)\mathbf{y} = \mathcal{H}(\mathbb{K}).$$

Thus the pair (C, A) is $\mathbf{Q}$-observable in the sense of (4.38). On the other hand, equalities (4.24), (4.25) are equivalent to (4.34). Substituting (4.40) into (4.34) and into identity (4.32) (for $z = \zeta$ and $y = y'$) gives

$$\mathbf{U}^*\begin{bmatrix}\mathbf{Q}(\zeta)^*(I - A^*\mathbf{Q}(\zeta)^*)^{-1}C^*y\\ y\end{bmatrix} = \begin{bmatrix}(I - A^*\mathbf{Q}(\zeta)^*)^{-1}C^*y\\ S(\zeta)^*y\end{bmatrix}$$

and

$$\left\|\begin{bmatrix}\mathbf{Q}(\zeta)^*(I - A^*\mathbf{Q}(\zeta)^*)^{-1}C^*y\\ y\end{bmatrix}\right\| = \left\|\begin{bmatrix}(I - A^*\mathbf{Q}(\zeta)^*)^{-1}C^*y\\ S(\zeta)^*y\end{bmatrix}\right\|,$$

respectively. The two latter equalities show that $\mathbf{U}^*$ is isometric on the space $\mathcal{D}_{\mathbf{U}^*}$ (see (4.13)) and therefore $\mathbf{U}$ is weakly coisometric.

To prove part (4), let us assume that

$$S(z) = S(0) + \widetilde{C}(I - \mathbf{Q}(z)\tilde{A})^{-1}\mathbf{Q}(z)\widetilde{B} \tag{4.43}$$

is a weakly coisometric realization of S with the state space $\mathcal{X}$ and such that the pair $(\widetilde{C}, \tilde{A})$ is $\mathbf{Q}$-observable in the sense of (4.38). Then

$$I - S(z)S(\zeta)^* = \widetilde{C}(I - \mathbf{Q}(z)\tilde{A})^{-1}(I - \mathbf{Q}(z)\mathbf{Q}(\zeta)^*)(I - \tilde{A}^*\mathbf{Q}(\zeta)^*)^{-1}\widetilde{C}^*$$

which means that S admits a representation (4.6) with $H(z) = \widetilde{C}(I - \mathbf{Q}(z)\tilde{A})^{-1}$. Let $\mathcal{I}_i$ be given as in (4.37). Representing H in the form (4.5) with

$$H_i(z) = \widetilde{C}(I - \mathbf{Q}(z)\tilde{A})^{-1}\mathcal{I}_i \tag{4.44}$$

we then conclude from Remark 4.3 that S admits the Agler decomposition (4.4) with

$$\mathbb{K}_{ij}(z,\zeta) = \widetilde{C}(I - \mathbf{Q}(z)\tilde{A})^{-1}\mathcal{I}_i\mathcal{I}_j^*(I - \tilde{A}^*\mathbf{Q}(\zeta)^*)^{-1}\widetilde{C}^* \qquad \text{for } i,j = 1,\dots,p.$$

Let $\mathcal{H}(\mathbb{K})$ be the reproducing kernel Hilbert space associated with the positive kernel $\mathbb{K} = [\mathbb{K}_{ij}]_{i,j=1}^p$. Let us arrange the functions (4.44) as follows

$$\mathbb{G}(z) := \begin{bmatrix}H_1(z)\\ \vdots\\ H_p(z)\end{bmatrix} = \begin{bmatrix}\widetilde{C}(I - \mathbf{Q}(z)\tilde{A})^{-1}\mathcal{I}_1\\ \vdots\\ \widetilde{C}(I - \mathbf{Q}(z)\tilde{A})^{-1}\mathcal{I}_p\end{bmatrix}. \tag{4.45}$$

Since by construction $\mathbb{K}(z,\zeta) = \mathbb{G}(z)\mathbb{G}(\zeta)^*$ and since $(\widetilde{C}, \tilde{A})$ is $\mathbf{Q}$-observable, the formula

$$U\colon x \mapsto \mathbb{G}(z)x \tag{4.46}$$

defines a unitary map from $\mathcal{X}$ onto $\mathcal{H}(\mathbb{K})$. Let us define the operators $A\colon \mathcal{H}(\mathbb{K})^p \to \mathcal{H}(\mathbb{K})^q$ and $B\colon \mathcal{U} \to \mathcal{H}(\mathbb{K})^q$ by

$$A\Big(\bigoplus_{i=1}^p U\Big) = \Big(\bigoplus_{i=1}^q U\Big)\tilde{A} \qquad \text{and} \qquad B = \Big(\bigoplus_{i=1}^q U\Big)\widetilde{B}. \tag{4.47}$$

In more detail, using representations

$$\tilde{A} = \begin{bmatrix} \tilde{A}_{11} & \dots & \tilde{A}_{1p} \\ \vdots & & \vdots \\ \tilde{A}_{q1} & \dots & \tilde{A}_{qp} \end{bmatrix} : \mathcal{X}^p \to \mathcal{X}^q \quad \text{and} \quad \widetilde{B} = \begin{bmatrix} \widetilde{B}_1 \\ \vdots \\ \widetilde{B}_q \end{bmatrix} : \mathcal{U} \to \mathcal{X}^q,$$

we define

$$A = \begin{bmatrix} A_{11} & \dots & A_{1p} \\ \vdots & & \vdots \\ A_{q1} & \dots & A_{qp} \end{bmatrix} : \mathcal{H}(\mathbb{K})^p \to \mathcal{H}(\mathbb{K})^q \quad \text{and} \quad B = \begin{bmatrix} B_1 \\ \vdots \\ B_q \end{bmatrix} : \mathcal{U} \to \mathcal{H}(\mathbb{K})^q$$

block-entrywise by

$$A_{ij} \colon \mathbb{G}(z)x \to \mathbb{G}(z)\tilde{A}_{ij}x \quad \text{and} \quad B_i u = \mathbb{G}(z)\widetilde{B}_i u \tag{4.48}$$

for $i = 1, \dots, q$ and $j = 1, \dots, p$. We next show that the operators A and B solve the Gleason problems (4.17) and (4.18), respectively. To this end, take the generic element f of $\mathcal{H}(\mathbb{K})^p$ in the form

$$f(z) = \begin{bmatrix} \mathbb{G}(z)x_1 \\ \vdots \\ \mathbb{G}(z)x_p \end{bmatrix} \quad \text{where } x := \begin{bmatrix} x_1 \\ \vdots \\ x_p \end{bmatrix} \in \mathcal{X}^p. \tag{4.49}$$

On account of (4.45), we have for f and x as in (4.49),

$$\begin{aligned} \sum_{k=1}^{p} f_{k,k}(z) &= \sum_{k=1}^{p} \widetilde{C}(I - \mathbf{Q}(z)\tilde{A})^{-1}\mathcal{I}_k x_k \\ &= \widetilde{C}(I - \mathbf{Q}(z)\tilde{A})^{-1}\sum_{k=1}^{p}\mathcal{I}_k x_k = \widetilde{C}(I - \mathbf{Q}(z)\tilde{A})^{-1}x. \end{aligned}$$

Therefore, and since $\mathbf{Q}(0) = 0$, we have

$$\begin{aligned} \sum_{k=1}^{p}\big(f_{k,k}(z) - f_{k,k}(0)\big) &= \widetilde{C}(I - \mathbf{Q}(z)\tilde{A})^{-1}x - \widetilde{C}x \\ &= \widetilde{C}(I - \mathbf{Q}(z)\tilde{A})^{-1}\mathbf{Q}(z)\tilde{A}x. \end{aligned} \tag{4.50}$$

On the other hand, we have by (4.45) and (4.48),

$$\begin{aligned} [Af]_{k,j}(z) &= \left[\sum_{i=1}^{p} A_{ki}\mathbb{G}(z)x_i\right]_j \\ &= \left[\mathbb{G}(z)\sum_{i=1}^{p}\tilde{A}_{ki}x_i\right]_j = \widetilde{C}(I - \mathbf{Q}(z)\tilde{A})^{-1}\mathcal{I}_j\sum_{i=1}^{p}\tilde{A}_{ki}x_i \end{aligned}$$

and it follows directly from (4.37) and (4.1) that

$$\sum_{j=1}^{p}\sum_{k=1}^{q}\mathbf{q}_{jk}(z)\mathcal{I}_j\sum_{i=1}^{p}\tilde{A}_{ki}x_i = \mathbf{Q}(z)\tilde{A}x.$$

Making use of the two last equalities and of (4.28) we get

$$\begin{aligned}\sum_{k=1}^{q}\mathbf{Q}_{\bullet k}(z)^{\top}[Af]_k(z) &= \sum_{j=1}^{p}\sum_{k=1}^{q}\mathbf{q}_{jk}(z)[Af]_{k,j}(z)\\ &= \sum_{j=1}^{p}\sum_{k=1}^{q}\mathbf{q}_{jk}(z)\widetilde{C}(I-\mathbf{Q}(z)\tilde{A})^{-1}\mathcal{I}_j\sum_{i=1}^{p}\tilde{A}_{ki}x_i\\ &= \widetilde{C}(I-\mathbf{Q}(z)\tilde{A})^{-1}\sum_{j=1}^{p}\sum_{k=1}^{q}\mathbf{q}_{jk}(z)\mathcal{I}_j\sum_{i=1}^{p}\tilde{A}_{ki}x_i\\ &= \widetilde{C}(I-\mathbf{Q}(z)\tilde{A})^{-1}\mathbf{Q}(z)\tilde{A}x\end{aligned}$$

which together with (4.50) implies (4.17). Similarly we conclude from (4.15), (4.45), (4.47) and (4.43) that

$$\begin{aligned}\sum_{k=1}^{q}\mathbf{Q}_{\bullet k}(z)^{\top}[Bu]_k(z) &= \sum_{j=1}^{p}\sum_{k=1}^{q}\mathbf{q}_{jk}(z)[B_k u]_j(z)\\ &= \sum_{j=1}^{p}\sum_{k=1}^{q}\mathbf{q}_{jk}(z)[\mathbb{G}(z)\widetilde{B}_k u]_j\\ &= \sum_{j=1}^{p}\sum_{k=1}^{q}\mathbf{q}_{jk}(z)\widetilde{C}(I-\mathbf{Q}(z)\tilde{A})^{-1}\mathcal{I}_j\widetilde{B}_k u\\ &= \widetilde{C}(I-\mathbf{Q}(z)\tilde{A})^{-1}\sum_{j=1}^{p}\sum_{k=1}^{q}\mathbf{q}_{jk}(z)\mathcal{I}_j\widetilde{B}_k u\\ &= \widetilde{C}(I-\mathbf{Q}(z)\tilde{A})^{-1}\mathbf{Q}(z)\widetilde{B}u = S(z)u - S(0)u\end{aligned}$$

and thus, B solves the Gleason problem (4.18) for S. Finally, for f and x of the form (4.49), for the operator U defined in (4.46) and for the operator C defined on $\mathcal{H}(\mathbb{K})^p$ by formula (4.29), we have

$$C(\bigoplus_{i=1}^{p}U)x = Cf = \sum_{k=1}^{p}f_{k,k}(0) = \sum_{k=1}^{p}\widetilde{C}(I-\mathbf{Q}(0)\tilde{A})^{-1}\mathcal{I}_k x_k = \sum_{k=1}^{p}\widetilde{C}\mathcal{I}_k x_k = \widetilde{C}x,$$

and thus,

$$C\left(\bigoplus_{i=1}^{p}U\right) = \widetilde{C}.$$

The latter equality together with (4.47) implies (4.39). According to Definition 4.7, the colligation $\mathbf{U} = \left[\begin{smallmatrix} A & B \\ C & D\end{smallmatrix}\right]$ is a canonical functional-model colligation associated with the Agler decomposition $\mathbb{K}$ of S. □

Let us say that the operator $A\colon \mathcal{H}(\mathbb{K})^p \to \mathcal{H}(\mathbb{K})^q$ is a *contractive solution* of the $\mathbf{Q}$-coupled Gleason problem for $\mathcal{H}(\mathbb{K})$ if in addition to identity (4.17) the inequality

$$\|Af\|^2_{\mathcal{H}(\mathbb{K})^q} \le \|f\|^2_{\mathcal{H}(\mathbb{K})^p} - \sum_{i=1}^{k}\|f_{k,k}(0)\|^2_{\mathcal{Y}}$$

holds for every function $f \in \mathcal{H}(\mathbb{K})^p$ or equivalently, if the pair (C, A) is contractive where $C\colon \mathcal{H}(\mathbb{K})^p \to \mathcal{Y}$ is the operator given in (4.29).

Theorem 4.10. *Let* (4.4) *be a fixed Agler decomposition of a given function* $S \in \mathcal{SA}_{\mathbf{Q}}(\mathcal{U}, \mathcal{Y})$ *and let* C *and* D *be defined as in* (4.29). *Then*

(1) *For every contractive solution* A *of the* $\mathbf{Q}$*-coupled Gleason problem* (4.17), *there is an operator* $B \colon \mathcal{U} \to \mathcal{H}(\mathbb{K})^q$ *such that* $\mathbf{U} = \left[\begin{smallmatrix} A & B \\ C & D \end{smallmatrix}\right]$ *is a canonical functional-model colligation for* S.
(2) *Every such* B *solves the Gleason problem* (4.18) *for* S.

The proof is very much similar to the proof of Theorem 3.10 and will be omitted.

In conclusion, we compare functional model $\mathbf{Q}$-realizations obtained in this section with particular cases considered in Sections 2 and 3.

The unit ball setting. In this case, $\mathbf{Q} = Z_{\text{row}}$ (in particular, $p = 1$) and definition (2.2) can be interpreted as the (uniquely determined) Agler decomposition of the form (4.4) with the kernel $\mathbb{K} = K_S$. Then (4.23) gives $\mathbb{T}(z, \zeta) = K_S(z, \zeta)$ and (4.33) coincides with (2.15). Since all canonical functional-model colligations are obtained via contractive extensions of isometries V (from (2.15) for the unit ball setting or from (4.33) for the general $\mathbf{Q}$-setting), it follows that realizations constructed in Section 2 can be obtained from those in Section 4 by letting $\mathbf{Q} = Z_{\text{row}}$. Moreover, if $\mathbf{Q} = Z_{\text{row}}$, then observability in the sense of (4.38) collapses to observability defined in part (2) of Theorem 2.9.

The unit polydisk setting. In this case, $\mathbf{Q} = Z_{\text{diag}}$, $p = q = d$, and the Agler representation (3.1) for an $S \in \mathcal{SA}_d(\mathcal{U}, \mathcal{Y})$ can be written in the form (4.4) with the kernel $\mathbb{K} = \begin{bmatrix} K_1 & & 0 \\ & \ddots & \\ 0 & & K_d \end{bmatrix}$. Then (4.23) takes the form

$$\mathbb{T}(z, \zeta) = \begin{bmatrix} K_1(z, \zeta) \otimes \mathbf{e}_1 \\ \vdots \\ K_d(z, \zeta) \otimes \mathbf{e}_d \end{bmatrix} \tag{4.51}$$

where $\{\mathbf{e}_1, \ldots, \mathbf{e}_d\}$ is the standard basis for $\mathbb{C}^d$. Observe that (4.51) is not the same as (3.10). Now (4.33) collapses to

$$V \colon \begin{bmatrix} \bar{\zeta}_1 K_1(z, \zeta) \otimes \mathbf{e}_1 \\ \vdots \\ \bar{\zeta}_d K_d(z, \zeta) \otimes \mathbf{e}_d \\ y \end{bmatrix} \mapsto \begin{bmatrix} K_1(z, \zeta) \otimes \mathbf{e}_1 \\ \vdots \\ K_d(z, \zeta) \otimes \mathbf{e}_d \\ S(\zeta)^* y \end{bmatrix} \tag{4.52}$$

whereas (3.24) can be written as

$$\widetilde{V} \colon \begin{bmatrix} \bar{\zeta}_1 K_1(z, \zeta) \\ \vdots \\ \bar{\zeta}_d K_d(z, \zeta) \\ y \end{bmatrix} \mapsto \begin{bmatrix} K_1(z, \zeta) \\ \vdots \\ K_d(z, \zeta) \\ S(\zeta)^* y \end{bmatrix} \tag{4.53}$$

To get canonical functional-model realizations as in Definition 3.5, we extend $\widetilde{V}$ to a contraction $\widetilde{\mathbf{U}}^* = \begin{bmatrix} \widetilde{A}^* & \widetilde{C}^* \\ \widetilde{B}^* & D^* \end{bmatrix} : \bigoplus_{i=1}^d \mathcal{H}(K_i) \oplus \mathcal{Y} \to \bigoplus_{i=1}^d \mathcal{H}(K_i) \oplus \mathcal{U}$. If for such a contraction we let $\mathbf{U}$ to be of the form (3.2) with

$$A_{ij} = \widetilde{A}_{ij} \otimes \mathbf{e}_i \mathbf{e}_j^*, \qquad B_i = \widetilde{B}_i \otimes \mathbf{e}_i, \qquad C_j = \widetilde{C}_j \otimes \mathbf{e}_j^*,$$

then $\mathbf{U}^*$ will be a contraction from $\bigoplus_{i=1}^d (\mathcal{H}(K_i))^p \oplus \mathcal{Y}$ to $\bigoplus_{i=1}^d (\mathcal{H}(K_i))^p \oplus \mathcal{U}$ extending the isometry V given in (4.53). It is not hard to see that $\mathbf{U}$ is a canonical functional-model $\mathbf{Q}$-realization for S in the sense of Definition 4.7. Thus, any "polydisk" canonical functional-model realization gives rise to a canonical functional-model $\mathbf{Q}$-realization for S. Of course, the converse is not true.

To see the polydisk setting as a particular instance of the general $\mathbf{Q}$-setting, we need to make use of the block-diagonal decomposition of $\mathbf{Q}$ into irreducible parts discussed in Remark 4.4. For the polydisk setting with $\mathbf{Q}(z) = Z_{\text{diag}}(z)$, this diagonal structure is nontrivial and already apparent. Thus we assume that $\mathbf{Q}(z)$ has the form (4.11) and the positive kernel $\mathbb{K}$ giving rise to the $\mathbf{Q}$-Agler decomposition (4.4) has the compatible block decomposition (4.12). The $\mathbf{Q}$-Agler decomposition (4.4) now has the form

$$I - S(z)S(\zeta)^* = \sum_{\alpha=1}^{d} \left[\sum_{k=1}^{p_\alpha} \mathbb{K}_{kk}^{(\alpha)}(z,\zeta) - \sum_{k=1}^{q_\alpha} \sum_{i,l=1}^{p_\alpha} q_{ik}^{(\alpha)}(z)\overline{q_{lk}^{(\alpha)}(\zeta)}\mathbb{K}_{il}^{(\alpha)}(z,\zeta) \right] \tag{4.54}$$

and can be rewritten in inner-product form as

$$\begin{aligned} &\sum_{\alpha=1}^{d}\sum_{k=1}^{q_\alpha} \langle \mathbb{K}^{(\alpha)}(\cdot,\zeta)\mathbf{Q}_{\bullet k}^{(\alpha)}(\zeta)^{\top *}y, \mathbb{K}^{(\alpha)}(\cdot,z)\mathbf{Q}_{\bullet k}(z)^{\top *}y'\rangle_{\mathcal{H}(\mathbb{K}^{(\alpha)})} \\ &\quad = \sum_{\alpha=1}^{d}\sum_{j=1}^{p_\alpha} \langle \mathbb{K}^{(\alpha)}(\cdot,\zeta)E_j^{(\alpha)}y, \mathbb{K}^{(\alpha)}(\cdot,z)E_j^{(\alpha)}y'\rangle_{\mathcal{H}(\mathbb{K}^{(\alpha)})} + \langle S(\zeta)^*y, S(z)^*y'\rangle_{\mathcal{U}} \end{aligned}$$

where $E_j^{(\alpha)} = I_{\mathcal{Y}} \otimes \mathbf{e}_j$ and where $\{\mathbf{e}_1, \dots, \mathbf{e}_{p_\alpha}\}$ is the standard basis for $\mathbb{C}^{p_\alpha}$. Then the isometry V in (4.33) has the form

$$V\colon \begin{bmatrix} \mathbf{Q}^{(1)}(\zeta)^*\mathbb{T}^{(1)}(\cdot,\zeta)y \\ \vdots \\ \mathbf{Q}^{(d)}(\zeta)^*\mathbb{T}^{(d)}(\cdot,\zeta)y \\ y \end{bmatrix} \mapsto \begin{bmatrix} \mathbb{T}^{(1)}(\cdot,\zeta)y \\ \vdots \\ \mathbb{T}^{(d)}(\cdot,\zeta)y \\ S(\zeta)^*y \end{bmatrix}$$

where $\mathbb{T}^{(\alpha)}(z,\zeta) = \begin{bmatrix} \mathbb{K}^{(\alpha)}(z,\zeta)E_1^{(\alpha)} \\ \vdots \\ \mathbb{K}^{(\alpha)}(z,\zeta)E_{p_\alpha}^{(\alpha)} \end{bmatrix}$ and where V has domain equal to $\mathcal{D}_V = \mathcal{D} \oplus \mathcal{Y}$ where

$$\mathcal{D} = \bigvee_{\zeta\in\mathcal{D}_{\mathbf{Q}}, y\in\mathcal{Y}} \begin{bmatrix} \mathbf{Q}^{(1)}(\zeta)^*\mathbb{T}^{(1)}(\cdot,\zeta)y \\ \vdots \\ \mathbf{Q}^{(\alpha)}(\zeta)^*\mathbb{T}^{(d)}(\cdot,\zeta)y \end{bmatrix} \subset \bigoplus_{\alpha=1}^{d} \mathcal{H}(\mathbb{K}^{(\alpha)})^{q_\alpha}$$

and where V has range $\mathcal{R}_V$ given by

$$\mathcal{R}_V = \bigvee_{\zeta\in\mathcal{D}_{\mathbf{Q}}, y\in\mathcal{Y}} \begin{bmatrix} \mathbb{T}^{(1)}(\cdot,\zeta)y \\ \vdots \\ \mathbb{T}^{(d)}(\cdot,\zeta)y \\ S(\zeta)^*y \end{bmatrix} \subset \bigoplus_{\alpha=1}^{d} \mathcal{H}(\mathbb{K}^{(\alpha)})^{p_\alpha} \oplus \mathcal{U};$$

all this specializes to (4.53) for the polydisk case. We say that the operator $A\colon \bigoplus_{\alpha=1}^d \mathcal{H}(\mathbb{K}^{(\alpha)})^{p_\alpha} \to \bigoplus_{\alpha=1}^d \mathcal{H}(\mathbb{K}^{(\alpha)})^{q_\alpha}$ solves the $\mathbf{Q}$-*structured Gleason problem for the kernel collection* $\{\mathbb{K}^{(1)}, \dots, \mathbb{K}^{(d)}\}$ if

$$\sum_{\alpha=1}^{d} \sum_{k=1}^{p_\alpha} (f_{k,k}^{(\alpha)}(z) - f_{k,k}^{(\alpha)}(0)) = \sum_{\alpha=1}^{d} \sum_{k=1}^{q_\alpha} \mathbf{Q}_{\cdot k}^{(\alpha)}(z)^\top [Af]_k^{(\alpha)}(z)$$

for all $f = \begin{bmatrix} f^{(1)} \\ \vdots \\ f^{(d)} \end{bmatrix} \in \bigoplus_{\alpha=1}^{d} \mathcal{H}(\mathbb{K}^{(\alpha)})$; so

$$f^{(\alpha)} = \begin{bmatrix} f_1^{(\alpha)} \\ \vdots \\ f_{p_\alpha}^{(\alpha)} \end{bmatrix} \qquad \text{with each } f_k^{(\alpha)} = \begin{bmatrix} f_{k,1}^{(\alpha)} \\ \vdots \\ f_{k,p_\alpha}^{(\alpha)} \end{bmatrix} \in \mathcal{H}(\mathbb{K}^{(\alpha)}).$$

We say that the operator $B \colon \mathcal{U} \to \bigoplus_{\alpha=1}^{d} \mathcal{H}(\mathbb{K}^{(\alpha)})^{q_\alpha}$ solves the $\mathbf{Q}$*-structured* $\{\mathbb{K}^{(1)}, \dots, \mathbb{K}^{(d)}\}$*-Gleason problem for* S if

$$S(z)u - S(0)u = \sum_{\alpha=1}^{d} \sum_{k=1}^{q_\alpha} \mathbf{Q}_{\bullet k}^{(\alpha)}(z)[Bu]_k^{(\alpha)}(z)$$

for all $u \in \mathcal{U}$, where we write $Bu = \begin{bmatrix} [Bu]^{(1)} \\ \vdots \\ [Bu]^{(d)} \end{bmatrix} \in \bigoplus_{\alpha=1}^{d} \mathcal{H}(\mathbb{K}^{(\alpha)})^{q_\alpha}$ with each $[Bu]^{(\alpha)} = \begin{bmatrix} [Bu]_1^{(\alpha)} \\ \vdots \\ [Bu]_{q_\alpha}^{(\alpha)} \end{bmatrix}$ where in turn $[Bu]_k^{(\alpha)} = \begin{bmatrix} [Bu]_{k,1}^{(\alpha)} \\ \vdots \\ [Bu]_{k,p_\alpha}^{(\alpha)} \end{bmatrix} \in \mathcal{H}(K^{(\alpha)})$. We define a *canonical functional-model colligation matrix* $\mathbf{U}$ for a given function $S \in \mathcal{SA}_{\mathbf{Q}}(\mathcal{U}, \mathcal{Y})$ and left $\mathbf{Q}$-Agler decomposition $\{\mathbb{K}_1, \dots, \mathbb{K}_d\}$ (so (4.54) holds) to be any operator-matrix $\mathbf{U} = \left[\begin{smallmatrix} A & B \\ C & D \end{smallmatrix}\right] : \bigoplus_{\alpha=1}^{d} \mathcal{H}(\mathbb{K}^{(\alpha)})^{p_\alpha} \oplus \mathcal{U} \to \bigoplus_{\alpha=1}^{d} \mathcal{H}(\mathbb{K}^{(\alpha)})^{q_\alpha} \oplus \mathcal{Y}$ so that

(1) $\mathbf{U}$ is a contraction,
(2) the operator A solves the $\mathbf{Q}$-structured Gleason problem for the kernel collection $\{\mathbb{K}^{(1)}, \dots, \mathbb{K}^{(d)}\}$,
(3) the operator B solves the $\mathbf{Q}$-structured $\{\mathbb{K}^{(1)}, \dots, \mathbb{K}^{(d)}\}$-Gleason problem for S, and
(4) the operators C and D are given by

$$C \colon \bigoplus_{\alpha=1}^{d} \begin{bmatrix} f_1^{(\alpha)} \\ \vdots \\ f_{p_\alpha}^{(\alpha)} \end{bmatrix} \mapsto \sum_{\alpha=1}^{d} \sum_{k=1}^{p_\alpha} f_{k,k}^{(\alpha)}(0), \qquad D \colon u \mapsto S(0)u.$$

Then we leave it to the industrious reader to check that Theorems 4.8, 4.9 and 4.10 all go through with this block-diagonal modification. Specializing this formalism to the polydisk case picks up exactly the results of Section 3.

References

1. J. Agler, *On the representation of certain holomorphic functions defined on a polydisc*, Topics in Operator Theory: Ernst D. Hellinger Memorial Volume, Oper. Theory Adv. Appl., vol. 48, Birkhäuser, Basel, 1990, pp. 47–66.
2. J. Agler and J. E. McCarthy, *Nevanlinna–Pick interpolation on the bidisk*, J. Reine Angew. Math. **506** (1999), 191–204.

3. ______, *Complete Nevanlinna – Pick kernels*, J. Funct. Anal. **175** (2000), no. 1, 111 – 124.
4. C. Ambrozie and J. Eschmeier, *A commutant lifting theorem on analytic polyhedra*, Topological Algebras, their Applications, and Related Topics, Banach Center Publ., vol. 67, Polish Acad. Sci., Warsaw, 2005, pp. 83 – 108.
5. C.-G. Ambrozie and D. Timotin, *A von Neumann type inequality for certain domains in* $\mathbf{C}^n$, Proc. Amer. Math. Soc. **131** (2003), no. 3, 859 – 869.
6. T. Andô, *de Branges spaces and analytic operator functions*, Hokkaido University, Sapporo, 1990, lecture notes.
7. N. Aronszajn, *Theory of reproducing kernels*, Trans. Amer. Math. Soc. **68** (1950), 337 – 404.
8. J. A. Ball, *Linear systems, operator model theory and scattering: multivariable generalizations*, Operator Theory and its Applications (Winnipeg, MB, 1998), Fields Inst. Commun., vol. 25, Amer. Math. Soc., Providence, RI, 2000, pp. 151 – 178.
9. J. A. Ball and V. Bolotnikov, *Realization and interpolation for Schur – Agler-class functions on domains with matrix polynomial defining function in* $\mathbb{C}^n$, J. Funct. Anal. **213** (2004), no. 1, 45 – 87.
10. J. A. Ball, V. Bolotnikov, and Q. Fang, *Transfer-function realization for multipliers of the Arveson space*, J. Math. Anal. Appl. **333** (2007), no. 1, 68 – 92.
11. J. A. Ball and T. L. Kriete III, *Operator-valued Nevanlinna – Pick kernels and the functional models for contraction operators*, Integral Equations Operator Theory **10** (1987), no. 1, 17 – 61.
12. J. A. Ball, D. S. Kaliuzhnyĭ-Verbovetskiĭ, C. Sadosky, and V. Vinnikov, *Scattering systems with several evolutions and formal reproducing kernel Hilbert spaces*, in preparation.
13. J. A. Ball, C. Sadosky, and V. Vinnikov, *Scattering systems with several evolutions and multidimensional input/state/output systems*, Integral Equations Operator Theory **52** (2005), no. 3, 323 – 393.
14. J. A. Ball and T. T. Trent, *Unitary colligations, reproducing kernel Hilbert spaces, and Nevanlinna – Pick interpolation in several variables*, J. Funct. Anal. **157** (1998), no. 1, 1 – 61.
15. J. A. Ball, T. T. Trent, and V. Vinnikov, *Interpolation and commutant lifting for multipliers on reproducing kernel Hilbert spaces*, Operator Theory and Analysis (Amsterdam, 1997), Oper. Theory Adv. Appl., vol. 122, Birkhäuser, Basel, 2001, pp. 89 – 138.
16. V. Bolotnikov and A. Kheifets, *A higher order analogue of the Carathéodory – Julia theorem*, J. Funct. Anal. **237** (2006), no. 1, 350 – 371.
17. ______, *The higher order Carathéodory – Julia theorem and related boundary interpolation problems*, Recent Advances in Matrix and Operator Theory, Oper. Theory Adv. Appl., vol. 179, Birkhäuser, Basel, 2008, pp. 63 – 102.
18. ______, *Carathéodory – Julia type theorems for operator valued Schur functions*, J. Anal. Math. **106** (2008), 237 – 270.
19. R. E. Curto, *Applications of several complex variables to multiparameter spectral theory*, Surveys of Some Recent Results in Operator Theory, Vol. II, Pitman Res. Notes Math. Ser., vol. 192, Longman Sci. Tech., Harlow, 1988, pp. 25 – 90.
20. L. de Branges and J. Rovnyak, *Canonical models in quantum scattering theory*, Perturbation Theory and its Applications in Quantum Mechanics (Madison, WI, 1965), Wiley, New York, 1966, pp. 295 – 392.
21. ______, *Square summable power series*, Holt, Rinehart and Winston, New York, 1966.
22. L. de Branges and L. Shulman, *Perturbations of unitary transformations*, J. Math. Anal. Appl. **23** (1968), 294 – 326.
23. J. Eschmeier and M. Putinar, *Spherical contractions and interpolation problems on the unit ball*, J. Reine Angew. Math. **542** (2002), 219 – 236.
24. E. Fricain and J. Mashreghi, *Boundary behavior of functions in the de Branges – Rovnyak spaces*, Complex Anal. Oper. Theory **2** (2008), no. 1, 87 – 97.
25. A. M. Gleason, *Finitely generated ideals in Banach algebras*, J. Math. Mech. **13** (1964), 125 – 132.
26. B. Sz.-Nagy and C. Foias, *Harmonic analysis of operators on Hilbert space*, North-Holland, Amsterdam, 1970.
27. N. K. Nikolskiĭ and V. I. Vasyunin, *A unified approach to function models, and the transcription problem*, The Gohberg Anniversary Collection, Vol. II (Calgary, AB, 1988), Oper. Theory Adv. Appl., vol. 41, Birkhäuser, Basel, 1989, pp. 405 – 434.

28. ———, *Elements of spectral theory in terms of the free function model.* I: *Basic constructions*, Holomorphic Spaces (Berkeley, CA, 1995), Math. Sci. Res. Inst. Publ., vol. 33, Cambridge Univ. Press, Cambridge, 1998, pp. 211–302.
29. S. Parrott, *On a quotient norm and the Sz.-Nagy–Foias lifting theorem*, J. Funct. Anal. **30** (1978), no. 3, 311–328.
30. W. Rudin, *Function theory in the unit ball of* $\mathbf{C}^n$, Grundlehren Math. Wiss., vol. 241, Springer, New York, 1980.
31. D. Sarason, *Sub-Hardy Hilbert spaces in the unit disk*, Univ. Arkansas Lecture Notes Math. Sci., vol. 10, Wiley, New York, 1994.

Department of Mathematics, Virginia Tech, Blacksburg, VA 24061-0123, USA
E-mail address: ball@math.vt.edu

Department of Mathematics, The College of William and Mary, Williamsburg VA 23187-8795, USA
E-mail address: vladi@math.wm.edu

Centre de Recherches Mathématiques
CRM Proceedings and Lecture Notes
Volume **51**, 2010

Two Variations on the Drury – Arveson Space

Nicola Arcozzi, Richard Rochberg, and Eric Sawyer

1. Introduction

The Drury – Arveson space DA is a Hilbert space of holomorphic functions on $\mathbb{B}_{n+1}$, the unit ball of $\mathbb{C}^{n+1}$. It was introduced by Drury [11] in 1978 in connection with the multivariable von Neumann inequality. Interest in the space grew after an influential article by Arveson [7], and expanded further when Agler and McCarthy [1] proved that DA is universal among the reproducing kernel Hilbert spaces having the complete Nevanlinna – Pick property. The multiplier algebra of DA plays an important role in these studies. Recently the authors obtained explicit and rather sharp estimates for the norms of function acting as multipliers of DA [3], an alternative proof is given in [17].

In our work we made use of a discretized version of the reproducing kernel for DA, or, rather, of its real part. In this note we consider analogs of the DA space for the Siegel domain, the unbounded generalized half-plane biholomorphically equivalent to the ball. We also consider a discrete model of the of the Siegel domain which carries a both a tree and a quotient tree structure. As sometimes happens with passage from function theory on the disk to function theory on a halfplane, the transition to the Siegal domain makes some of the relevant group actions more transparent. In particular this quotient structure, which has no analog on the unit disk (i.e., $n = 0$), has a cleaner presentation in the (discretized) Siegel domain than in the ball.

Along the way, we pose some questions, whose answers might shed more light on the interaction between these new spaces, operator theory and sub-Riemannian geometry.

We start by recalling some basic facts about the space DA. An excellent source of information is the book [2]. The space DA is a reproducing kernel Hilbert space

2000 *Mathematics Subject Classification.* Primary 46E22; Secondary 32A37.

The first author partially supported by the COFIN project Analisi Armonica, funded by the Italian Minister for Research.

The second author's work work supported by the National Science Foundation under Grant No. 0070642.

The third author's work was supported by the National Science and Engineering Council of Canada.

with kernel, for $z, w \in \mathbb{B}_{n+1}$

$$K(z,w) = \frac{1}{1-\overline{w}\cdot z}. \tag{1.1}$$

Elements in DA can be isometrically identified with functions f holomorphic in $\mathbb{B}_{n+1}$, $f(z) = \sum_{m\in\mathbb{N}^{n+1}} a(m)z^m$ (multiindex notation), such that

$$\|f\|_{\mathrm{DA}}^2 = \sum_{m\in\mathbb{N}^{n+1}} \frac{m!}{|m|!}|a(m)|^2 < \infty.$$

When $n = 0$, DA $= H^2$, the classical Hardy space. The multiplier algebra of H^2, the algebra of functions which multiply H^2 boundedly into itself, is H^∞, the algebra of bounded analytic functions. In general the multiplier algebra $M(\mathrm{DA})$ of DA is the space of functions g holomorphic in $\mathbb{B}_{n+1}$ for which the multiplication operator $f \mapsto gf$ from DA to itself has finite operator norm which we denote by $\|g\|_{\mathcal{M}(\mathrm{DA})}$. For $n > 0$, $M(\mathrm{DA})$ is a proper subalgebra of H^∞, however in some ways it plays a role analogous to H^∞. In particular the multiplier norm $\|g\|_{\mathcal{M}(\mathrm{DA})}$ replaces the H^∞ norm in the multivariable version of von Neumann's Inequality [11]. Also, the general theory of Hilbert spaces with the Nevanlinna–Pick property exposes the fact that many operator theoretic results about H^2 and H^∞ are special cases of general results about Hilbert spaces with the Nevanlinna–Pick property, for instance DA, and the associated multiplier algebra.

Given $\{w_j\}_{j=1}^N$ in $\mathbb{B}_{n+1}$ and $\{\lambda_j\}_{j=1}^N$ in $\mathbb{C}$, the interpolation problem of finding g in $M(\mathrm{DA})$ such that. $g(w_j) = \lambda_j$ and $\|g\|_{\mathcal{M}(\mathrm{DA})} \le 1$, has solution if and only if the "Pick matrix" is positive semidefinite,

$$[(1 - w_j\overline{w_h})K(\lambda_j, \lambda_h)]_{j,h=1}^N \ge 0.$$

Agler and McCarthy [1] showed that the (possibly infinite dimensional) DA kernel is universal among the kernels having the *complete* Nevanlinna–Pick property, which is a vector valued analog of the property just mentioned. While for $n = 0$ we have the simple characterization $\|g\|_{\mathcal{M}(\mathrm{DA})} = \|g\|_{\mathcal{M}(H^2)} = \|g\|_{H^\infty}$, no such formula exists in the multidimensional case. However, a sharp, geometric estimate of the multiplier norm was given in [3].

Theorem 1. (A) *A function g, analytic in $\mathbb{B}_{n+1}$, is a multiplier for* DA *if and only if $g \in H^\infty$ and the measure $\mu = \mu_g$, $\mathrm{d}\mu_g := (1-|z|^2)|Rg|^2\,\mathrm{d}A(z)$ is a Carleson measure for* DA,

$$\int_{\mathbb{B}_{n+1}} |f|^2 d\mu \le C(\mu)\|f\|_{\mathrm{DA}}^2. \tag{1.2}$$

Here dA *is the Lebesgue measure in $\mathbb{B}_{n+1}$ and R is the radial differentiation operator. In this case, with $K(\mu)$ denoting the infimum of the possible $C(\mu)$ in the previous inequality,*

$$\|g\|_{\mathcal{M}(\mathrm{DA})} \approx \|g\|_{H^\infty} + K(\mu)^{1/2}$$

(B) *For a in $\mathbb{B}_{n+1}$, let $S(a) = \{w \in \mathbb{B}_{n+1} : |1 - a/|a|\cdot\overline{w}| \le (1-|a|^2)\}$ be the Carleson box with vertex a.*

Given a positive measure μ on $\mathbb{B}_{n+1}$, the following are equivalent:

(a) *μ is a Carleson measure for* DA;

(b) *the inequality*

$$\int_{\mathbb{B}_{n+1}}\int_{\mathbb{B}_{n+1}} \operatorname{Re} K(z,w)\varphi(z)\varphi(w)\,\mathrm{d}\mu(z)\,\mathrm{d}\mu(w) \le C(\mu)\int_{\mathbb{B}_{n+1}} \varphi^2\,\mathrm{d}\mu$$

holds for all nonnegative φ.

(c) *The measure* μ *satisfies both the simple condition*

$$\mu(S(a)) \le C(\mu)(1-|a|^2) \tag{SC}$$

and the split-tree condition, which is obtained by testing (2) *over the characteristic functions of the sets* $S(a)$,

$$\int_{S(a)}\Big(\int_{S(a)} \operatorname{Re} K(z,w)\,\mathrm{d}\mu(z)\Big)\,\mathrm{d}\mu(w) \le C(\mu)\mu(S(a)), \tag{ST}$$

(*with* $C(\mu)$ *independent of* a *in* $\mathbb{B}_{n+1}$).

Here $C(\mu)$ denotes positive constants, possibly with different value at each occurrence.

The conditions (SC) is obtained by testing the boundedness of J, the inclusion of DA into $L^2(\mathrm{d}\mu)$, on a localized bump. The condition (ST) is obtained by testing the boundedness of the adjoint, J^*, on a localized bump. Hence the third statement of the theorem is very similar to the hypotheses in some versions of the $T(1)$ theorem. This viewpoint is developed in [17].

In light of (2) we had used $\operatorname{Re} K(z,w)$ in analyzing Carleson measures. When estimating the size of $\operatorname{Re} K(z,w)$ in the tree model it was useful to split the tree into equivalence classes and use the geometry of the quotient structure. That is the source of the name "split-tree condition" for (ST). Versions of such a quotient structure will be considered in the later part of this paper.

Problem 1. Theorem 1 gives a geometric characterization of the multiplier norm for fixed n, but we do not know how the relationship between the different constants $C(\mu)$, and between them the multiplier norm of g, depend on the dimension. Good control of the dependence of the constants on the dimension would open the possibility of passing to the limit as $n \to \infty$ and providing a characterization of the multiplier norm for the infinite-dimensional DA space.

An alternative approach to the characterization of the Carleson measures is in [17], where Tchoundja exploits the observation made in [3] that, by general Hilbert space theory, the inequality in (2) is equivalent (with a different $C(\mu)$) to

$$\int_{\mathbb{B}_{n+1}}\Big(\int_{\mathbb{B}_{n+1}} \operatorname{Re} K(z,w)\varphi(z)\,\mathrm{d}\mu(z)\Big)^2\,\mathrm{d}\mu(w) \le C(\mu)\int_{\mathbb{B}_{n+1}} \varphi^2\,\mathrm{d}\mu. \tag{1.3}$$

We mention here that (1.3) is never really used in [3], while it is central in [17]. Tchoundja's viewpoint is that (1.3) is the L^2 inequality for the "singular" integral having kernel $\operatorname{Re} K(z,w)$, with respect to the non-doubling measure μ. He uses the fact $\operatorname{Re} K(z,w) \ge 0$ to insure that a generalized "Menger curvature" is positive. With this in hand he adapts some of the methods employed in the solution of the Painlevé problem to obtain his proof. His theorem reads as follows.

Theorem 2. *A measure* μ *on* $\mathbb{B}_{n+1}$ *is Carleson for* DA *if and only if any of the following (hence, all) holds.*

(1) *For some* $1 < p < \infty$,

$$\int_{\mathbb{B}_{n+1}} \left(\int_{\mathbb{B}_{n+1}} \operatorname{Re} K(z,w)\varphi(z)\, \mathrm{d}\mu(z) \right)^p \mathrm{d}\mu(w) \le C(\mu) \int_{\mathbb{B}_{n+1}} \varphi^p \mu$$

(2) *The inequality in* (1) *holds for all* $1 < p < \infty$.

(3) *The measure* μ *satisfies the simple condition* (SC) *and also for some* $1 < p < \infty$ *the inequality*

$$\int_{S(a)} \left(\int_{S(a)} \operatorname{Re} K(z,w)\, \mathrm{d}\mu(w) \right)^p \mathrm{d}\mu(z) \le C(\mu)\mu(S(a)). \tag{1.4}$$

(4) *Condition* (3) *holds for all* $1 < p < \infty$.

(Actually [17] focuses on the $p \ge 2$ but self adjointness and duality then give the expanded range.) Observe that, as a consequence of Theorems 1 and 2 the condition (1.4) equivalently holds for some $1 < p < \infty$ then it holds all $1 < p < \infty$. On the other hand, it is immediate from Jensen's inequality that if the inequality holds for some p then it holds for any smaller p; hence the condition in Theorem 1,(ST), is a priori the weakest such condition.

Problem 2. Which geometric-measure theoretic properties follow from the fact that the Carleson measures for the DA space satisfy such "reverse Hölder inequalities"?

Indeed, the same question might be asked for the Carleson measures for a variety of weighted Dirichlet spaces, to which our and Tchoundja's methods apply. It is interesting to observe that, while our approach is different in the DA case and in other weighted Dirichlet spaces (see [3] and the references quoted there); Tchoundja's method works the same way in both cases. On the other hand, his proof does not encompass (ST) in Theorem 1.

We conclude this introduction with an overview of the article.

Changing coordinates by stereographic projection, we see in Section 2.1 that on the Siegel domain (generalized upper half-plane)

$$\mathcal{U}_{n+1} = \{z = (z', z_{n+1}) \in \mathbb{C}^n \times \mathbb{C} : \operatorname{Im}(z_{n+1}) > |z'|^2\}$$

K is conjugate to a natural kernel H

$$H(z,w) = \frac{1}{\mathrm{i}(\overline{w}_{n+1} - z_{n+1})/2 - z' \cdot \overline{w'}}$$

This is best seen changing to Heisenberg coordinates:

$$[\zeta, t; r] = [z', \operatorname{Re}(z_{n+1}); \operatorname{Im}(z_{n+1}) - |z'|^2].$$

The Heisenberg group $\mathbb{H}^n$ has elements $[\zeta, t] \in \mathbb{C}^n \times \mathbb{R}$ and group law $[\zeta, t] \cdot [\xi, s] = [\zeta + \xi, t + s + 2\operatorname{Im}(\zeta \cdot \overline{\xi})]$. The kernel can now be written as a convolution kernel: writing

$$\varphi_r([\zeta, t]) = \frac{r + |\zeta|^2 - it}{(r + |\zeta|^2)^2 + t^2}$$

we have

$$H([\zeta, t; r], [\xi, s; r]) = 2\varphi_{r+p}([\xi, s]^{-1} \cdot [\zeta, s])$$

Because of the connection with the characterization of the multipliers for DA our main interest is in $\operatorname{Re}\big(H(z,w)\big)$. The numerator and the denominator of $\operatorname{Re}(\varphi_r)$

each have an interpretation on terms of the sub-Riemannian geometry of $\mathbb{H}^n$. The denominator is the Koranyi distance to the origin, at scale $\sqrt{r}$, while the numerator is the Koranyi distance from the center of the group $\mathbb{H}^n$ to its coset passing through $[\zeta, t]$, again at the scale $\sqrt{r}$. We see, then, that the kernel φ_r reflects the two-step stratification of the Lie algebra of $\mathbb{H}^n$.

The Heisenberg group, which has a dilation as well as a translation structure, can be easily discretized, uniformly at each scale; and this is equivalent to a discretization of Whitney type for the Siegel domain $\mathcal{U}_{n+1}$. The dyadic boxes are fractals, but in Section 2.2 we see that they behave sufficiently nicely for us to use them the same way one uses dyadic boxes in real upper-half spaces. The same way the discretization of the upper half space can be thought of in terms of a tree, the discretization of the Siegel domain can be thought of in terms of a quotient structure of trees, which is a discretized version of the two-step structure of the Heisenberg Lie algebra.

In Section 3, we see how the DA kernel (rather, its real part) has a natural discrete analog living on the quotient structure. We show that, although the new kernel is not a complete Nevanlinna – Pick, it is nonetheless a positive definite kernel. In [3], the analysis of a variant of that discrete kernel led to the characterization of the multipliers for DA. We do not know if an analogous fact is true here, if the discrete kernel we introduce contains all the important information about the kernel H.

We conclude by observing, in Section 4, that, as a consequence of its "conformal invariance," a well-known kernel on the tree, which can be seen as the discretization of the kernel for a weighted Dirichlet space in the unit disc, has the complete Nevanlinna – Pick property.

Notation. Given two positive quantities A and B, depending on parameters $\alpha, \beta, \ldots$, we write $A \approx B$ if there are positive $c, C > 0$, independent of $\alpha, \beta, \ldots$, such that $cA \leq B \leq CA$.

2. A flat version of DA_d

2.1. From the ball to Siegel's domain. In this section, we apply stereographic projection to the DA_d kernel and we see that it is conjugate to a natural kernel on the Siegel domain. In this "flat" environment it is easier to see how the DA_d kernel is related to Bergman, and hence also to sub-Riemannian geometry. A discretized version of the kernel, analogous to the dyadic versions of the Hardy space kernel in one complex variable immediately comes to mind.

We follow here the exposition in [15]. As we mentioned, Siegel's domain $\mathcal{U}_{n+1}$ is defined as

$$\mathcal{U}_{n+1} = \{z = (z_1, \ldots, z_{n+1}) = (z', z_{n+1}) \in \mathbb{C}^{n+1} : \operatorname{Im} z_{n+1} > |z'|^2\}.$$

For z, w in $\mathcal{U}_{n+1}$, define

$$r(z, w) = \frac{\mathrm{i}}{2}(\bar{w}_{n+1} - z_{n+1}) - z' \cdot \overline{w'}.$$

Consider the kernel $H : \mathcal{U}_{n+1} \times \mathcal{U}_{n+1} \to \mathbb{C}$,

$$H(z, w) = \frac{1}{r(z, w)} \tag{2.1}$$

Proposition 1. *The kernel H is conjugate to the Drury–Arveson kernel K. Hence, it is a definite positive, (universal) Nevanlinna–Pick kernel.*

In fact, there is a map $\Phi\colon \mathbb{B}_{n+1}\to\mathcal{U}_{n+1}$ such that:

$$K\big(\Phi^{-1}(z),\Phi^{-1}(w)\big)=\frac{(i+z_{n+1})\overline{(\mathrm{i}+w_{n+1})}}{4\cdot r(z,w)} \tag{2.2}$$

PROOF. Let $\mathbb{B}_{n+1}$ be the unit ball of $\mathbb{C}^{n+1}$ and let $\mathcal{U}_{n+1}$ be Siegel's domain. There is a biholomorphic map $z=\Phi(\zeta)$ from $\mathbb{B}_{n+1}$ onto $\mathcal{U}_{n+1}$:

$$\begin{cases} z_{n+1}=\mathrm{i}\left(\dfrac{1-\zeta_{n+1}}{1+\zeta_{n+1}}\right) & \\ z_k=\dfrac{\zeta_k}{1+\zeta_{n+1}}, & \text{if } 1\le k\le n,\end{cases}$$

having inverse

$$\begin{cases} \zeta_{n+1}=\dfrac{\mathrm{i}-z_{n+1}}{\mathrm{i}+z_{n+1}} & \\ \zeta_k=\dfrac{2\mathrm{i}z_k}{\mathrm{i}+z_{n+1}}, & \text{if } 1\le k\le n.\end{cases}$$

Equation (2.2) follows by straightforward calculation. □

Remark 1. The map $f\mapsto\tilde f$, $\tilde f(z)=2/(\mathrm{i}+z_{n+1})f\big(\Phi^{-1}(z)\big)$, is an isometry from the Hilbert space with reproducing kernel K to the Hilbert space with reproducing kernel H. We call the latter $\mathrm{DA}_{\mathcal{U}}$.

Problem 3. Find an interpretation of the $\mathrm{DA}_{\mathcal{U}}$ norm in terms of weighted Dirichlet spaces on $\mathcal{U}_{n+1}$.

Recall (see [3]) that a positive measure μ on $\mathbb{B}_{n+1}$ is a Carleson measure for DA if the inequality

$$\int_{\mathbb{B}_{n+1}}|f|^2\,\mathrm{d}\mu\le C(\mu)\|f\|^2_{\mathrm{DA}} \tag{2.3}$$

holds independently of f. The least constant $\|\mu\|_{CM(\mathrm{DA})}=C(\mu)$ for which (2.3) holds is the Carleson measure norm of μ.

The following proposition is in [3].

Proposition 2. *The Carleson norm of a measure μ on $\mathbb{B}_{n+1}$ is comparable with the least constant $C_1(\mu)$ for which the inequality below hold for all measurable $g\ge 0$ on $\mathbb{B}_{n+1}$,*

$$\int_{\mathbb{B}_{n+1}}\int_{\mathbb{B}_{n+1}}\mathrm{Re}\big(K(z,w)\big)g(z)\,\mathrm{d}\mu(z)g(w)\,\mathrm{d}\mu(w)\le C_1(\mu)\int_{\mathbb{B}_{n+1}}g^2\,\mathrm{d}\mu.$$

As a corollary, we obtain the following.

Theorem 3. *Let $\mu\ge 0$ be a measure on $\mathbb{B}_{n+1}$ and define its normalized pull-back on $\mathcal{U}_{n+1}$,*

$$\mathrm{d}\widetilde{\mu}(z):=|\mathrm{i}+z_{n+1}|^2\,\mathrm{d}\mu\big(\Phi^{-1}(z)\big)$$

Then, $\mu\in CM(\mathrm{DA})$ if and only if $\widetilde{\mu}$ satisfies

$$\int_{\mathcal{U}_{n+1}}\int_{\mathcal{U}_{n+1}}\mathrm{Re}\big(H(z,w)\big)g(z)\,\mathrm{d}\widetilde{\mu}(z)g(w)\,\mathrm{d}\widetilde{\mu}(w)\le C_2(\widetilde{\mu})\int_{\mathcal{U}_{n+1}}g^2\,\mathrm{d}\widetilde{\mu}.$$

Moreover, $C(\mu)=C_2(\widetilde{\mu})$.

Problem 4. Find a natural, operator-theoretic interpretation for H; in analogy with the interpretation of K in [11].

The kernel H is best understood after changing to Heisenberg coordinates which help reveal its algebraic and geometric structure. For z in $\mathcal{U}_{n+1}$, set

$$z = (z', z_{n+1}) \equiv [\zeta, t; r] := [z', \operatorname{Re} z_{n+1}; \operatorname{Im} z_{n+1} - |z'|^2].$$

The map $z \mapsto [\zeta, t; r]$ identifies $\mathcal{U}_{n+1}$ with $\mathbb{R}^{2n+2}$, and its boundary $\partial\mathcal{U}_{n+1}$ with $\mathbb{R}^{2n+1}$. In the new coordinates it is easier to write down the equations of some special families of biholomorphisms of $\mathcal{U}_{n+1}$:

(i) rotations: $R_A\colon [\zeta, t; r] \mapsto [A\zeta, t; r]$, where $A \in \mathrm{SU}(n)$;
(ii) dilations: $D_\rho\colon [\zeta, t; r] \mapsto [\rho\zeta, \rho^2 t; \rho^2 r]$; and
(iii) translations: $\tau_{[\zeta,t]}\colon [\xi, s; p] \mapsto [\zeta + \xi, t + s + 2\operatorname{Im}(\zeta \cdot \overline{\xi}); p]$.

This Lie group of the translation is the Heisenberg group $\mathbb{H}^n \equiv \mathbb{R}^{2n+1}$ which can be identified with $\partial\mathcal{U}_{n+1}$. The group operation is

$$[\zeta, t] \cdot [\xi, s] = [\zeta + \xi, t + s + 2\operatorname{Im}(\zeta \cdot \overline{\xi})]$$

and thus $\tau_{[\zeta,t]}\colon [\xi, s; p] \mapsto [[\zeta, t] \cdot [\xi, s]; p]$.

We can foliate $\mathcal{U}_{n+1} = \bigsqcup_{p>0} \mathbb{H}^n(p)$, where $\mathbb{H}^n(p) = \{[\zeta, t; p] : [\zeta, t] \in \mathbb{H}^n\}$ is the orbit of $[0, 0; p]$ under the action of $\mathbb{H}^n$. The dilations D_ρ on $\mathcal{U}_{n+1}$ induce dilations on the Heisenberg group:

$$\delta_\rho[\zeta, t] := [\rho\zeta, \rho^2 t].$$

The relationship between dilations on $\mathbb{H}^n$ and on $\mathcal{U}_{n+1}$ can be seen as action on the leaves:

$$D_\rho : \mathbb{H}^n(r) \to \mathbb{H}^n(\rho^2 r), D_\rho[\zeta, t; r] = [\delta_\rho[\zeta, t]; \rho^2 r].$$

The zero of the group is $0 = [0, 0]$ and the inverse element of $[\zeta, t]$ is $[-\zeta, -t]$.

The Haar measure on $\mathbb{H}^n$ is $\mathrm{d}\zeta\, \mathrm{d}t$. We let $\mathrm{d}\beta$ to be the measure induced by the Haar measure on $\partial\mathcal{U}_{n+1}$:

$$\mathrm{d}\beta(z) = \mathrm{d}\zeta\, \mathrm{d}t.$$

We also have that $\mathrm{d}z = \mathrm{d}\zeta\, \mathrm{d}t\, \mathrm{d}r$ is the Lebesgue measure in $\mathcal{U}_{n+1}$.

We now change H to Heisenberg coordinates.

Proposition 3. *If* $z = [\zeta, t; r]$ *and* $w = [\xi, s; p]$, *then*

$$H(z, w) = 2 \cdot \frac{r + p + |\xi - \zeta|^2 - \mathrm{i}\big(t - s - 2\operatorname{Im}(\xi \cdot \bar{\zeta})\big)}{(r + p + |\xi - \zeta|^2)^2 + \big(t - s - 2\operatorname{Im}(\xi \cdot \bar{\zeta})\big)^2} \tag{2.4}$$
$$= 2\varphi_{r+p}([\xi, s]^{-1} \cdot [\zeta, t]),$$

where

$$\varphi_r([\zeta, t]) = \frac{r + |\zeta|^2 - \mathrm{i}t}{(r + |\zeta|^2)^2 + t^2}$$

The expression in Proposition 3 is interesting for both algebraic and geometric reasons. Algebraically we see that H can be viewed as a convolution operator. From a geometric viewpoint we note that the quantity $\|[\zeta, t]\| := (t^2 + |\zeta|^4)^{1/4}$ is the Koranyi norm of the point $[\zeta, t]$ in $\mathbb{H}^n$. The distance associated with the norm is

$$d_{\mathbb{H}^n}([\zeta, t], [\xi, s]) := \|[\xi, s]^{-1}[\zeta, t]\|.$$

Hence, the denominator of φ_r might be viewed as the 4th power of the Koranyi norm of $[\zeta, t]$ "at the scale" $r^{1/2}$.

In order to give an intrinsic interpretation of the numerator, consider the center $T = \{[0,t] : t \in \mathbb{R}\}$ of $\mathbb{H}^n$, and the projection $\Pi\colon \mathbb{H}^n \to \mathbb{C}^n \equiv \mathbb{H}^n/\mathcal{T}$: $\Pi([\zeta,t]) = \zeta$. Then, independently of $t \in \mathbb{R}$,

$$|\zeta| = d_{\mathbb{H}^n}(T, [\zeta,t]\cdot T)$$

is the Koranyi distance between the center and its left (hence, right) translate by $[\zeta,t]$. The real part of the DA kernel has a twofold geometric nature: the denominator is purely metric, while the numerator depends on the "quotient structure" induced by the stratification of the Lie algebra of $\mathbb{H}^n$. This duality is ultimately responsible for the difficulty of characterizing the Carleson measures for DA.

The boundary values of $Re(\varphi_r)$,

$$\varphi_0([\zeta,t]) := \frac{|z|^2}{|z|^4+t^2}, \tag{2.5}$$

were considered in [12] (see condition (1.17) on the potential) in connection with the Schrödinger equation and the uncertainty principle in the Heisenberg group.

Problem 5. Explore the connections, if there are any, between the DA space, the uncertainty principle on $\mathbb{H}^n$ and the sub-Riemannian geometry of $\mathbb{H}^n$.

We mention here that, at least when $n=1$, the kernel φ_0 in (2.5) satisfies the following, geometric looking differential equation:

$$\Delta_{\mathbb{H}}\varphi_0([\zeta,t]) = \frac{1}{2}\frac{\partial}{\partial|\zeta|^2}\varphi_0([\zeta,t]),$$

where $\Delta_{\mathbb{H}} = XX+YY$ is the sub-Riemannian Laplacian on $\mathbb{H}$. Here, with $\zeta = x+iy \in \mathbb{C}$, X and Y are the left invariant fields $X = \partial_x + 2y\partial_t$ and $Y = \partial_y - 2x\partial_t$. See [8] for a comprehensive introduction to analysis and PDEs in Lie groups with a sub-Riemannian structure.

2.2. Discretizing Siegel. The space $\mathcal{U}_{n+1}$ admits a dyadic decomposition, which we get from a well-known [16] dyadic multidecomposition of the Heisenberg group, which is well explained in [18]. We might get a similar, less explicit decomposition by means of the general construction in [10].

Theorem 4. *Let $b \geq 2n+1$ be a fixed odd integer. Then, there exists a compact subset T_0 in $\mathbb{H}^n$ such that: T_0 is the closure of its interior and $\Pi(T_0) = [-\frac{1}{2},\frac{1}{2}]^{2n}$;*

1. *$m(\partial T_0) = 0$, the boundary has null measure in $\mathbb{H}^n$;*
2. *there are b^{2n+2} affine maps (compositions of dilations and translations) $\mathring{A}_k$ of $\mathbb{H}^n$ such that: $T_0 = \bigcup_k \mathring{A}_k(T_0)$ and the interiors of the sets $\mathring{A}_k(T_0)$ are disjoint;*
3. *the sets $\Pi(\mathring{A}_k(T_0))$ divide $[-\frac{1}{2},\frac{1}{2}]^{2n}$ into b^{2n} cubes with disjoint interiors, each such cube being the projection of b^2 sets $\mathring{A}_k(T_0)$.*

Consider now $\mathcal{U}_{n+1}$, let b be fixed and let $m \in \mathbb{Z}$. For each $\bar{k} = (k', k_{2n+1}) \in \mathbb{Z}^{2n}\times\mathbb{Z}$, consider the cubes

$$\begin{aligned} Q^m_{\bar{k}} &= \delta_{b^{-m}}\tau_{\bar{k}}(T_0)\times[b^{-2m-2}, b^{-2m}] \\ &= Q^m_{\bar{k}} \times [b^{-2m-2}, b^{-2m}] = Q^m_{\bar{k}} \subset \mathcal{U}_{n+1}, \end{aligned}$$

with $Q^m_{\bar{k}} \subset \mathbb{H}^n$. Let $T^{(m)}$ be the sets of such cubes, $U^{(m)}$ the set of their projections, and $T = \bigcup_{m\in\mathbb{Z}} T^{(m)}$, $U = \bigcup_{m\in\mathbb{Z}} U^{(m)}$. We say that a cube Q' in $T^{(m+1)}$ (respectively $U^{(m+1)}$) is the child of a cube Q in $T^{(m)}$ (respectively $U^{(m)}$), if $Q' \subset Q$.

In order to simplify notation, if Q is a cube in T, we write $[Q] = \Pi(Q)$.
We state some useful consequences Theorem 4.

Proposition 4. (i) *Each cube in* $T^{(m)}$ *has* b^{2n+2} *children in* $T^{(m+1)}$ *and one parent in* $T^{(m-1)}$*; hence,* T *is a (connected) homogeneous tree of degree* b^{2n+2}*.*

(ii) *Each cube in* $U^{(m)}$ *has* b^{2n} *children in* $U^{(m+1)}$ *and one parent in* $U^{(m-1)}$*; hence,* U *is a (connected) homogeneous tree of degree* b^{2n}*.*

(iii) *For each cube* Q *in* $T^{(m)}$*, there are Koranyi balls* $B(z_Q, c_0 b^{-m})$ *and* $B(w_Q, c_1 b^{-m})$ *in* $\mathbb{H}^n$*, such that*

$$B(w_Q, c_1 b^{-m}) \times [b^{-2m-2}, b^{-2m}] \subset Q \subset B(w_Q, c_2 b^{-m}) \times [b^{-2m-2}, b^{-2m}].$$

We say two cubes Q_1, Q_2 in T are *graph related* if they are joined by an edge of the tree T, or if they belong to the same $T^{(m)}$ and there are points $z_1 \in Q_1$, $z_2 \in Q_2$ such that $d_{\mathbb{H}^n}(z_1, z_2) \leq b^{-m}$. An analogous definition is given for the points of U. We consider on T the edge-counting distance: $d(Q_1, Q_2)$ is the minimum number of edges in a path going from Q_1 to Q_2 following the edges of T: the distance is obviously realized by a unique geodesic. We also consider a graph distance, $d_G(Q_1, Q_2) \leq d(Q_1, Q_2)$, in which the paths have to follow edges of the graph G just defined. The edge counting distance on the graph is realized by geodesics, but they are not necessarily unique anymore. Similarly, we define counting distances for the tree and graph structures on U.

Given a cube Q in T, define its predecessor set in T, $P(Q) = \{Q' \in T : Q \subseteq Q'\}$, and its graph-predecessor set $P_G(Q) = \{Q' : d_G(Q', P(Q)) \leq 1\}$. We define the level of the confluent of Q_1 and Q_2 in G as

$$d(Q_1 \tilde{\wedge} Q_2) := \max\{d(Q) : Q \in P_G(Q_1) \cap P_G(Q_2)\}. \tag{2.6}$$

(We don't need, and hence don't define, the confluent $Q_1 \tilde{\wedge} Q_2$ itself.)

Similarly, we define predecessor sets in T and G for the elements of U, and the level of the confluent in the graph structure, using the same notation. Observe that $P([Q]) = [P(Q)] := \{[Q'] : Q' \in P(Q)\}$, $P([Q]) = [P(Q)]$, but that the inequality

$$d(Q_1 \tilde{\wedge} Q_2) \leq d([Q_1] \tilde{\wedge} [Q_2])$$

cannot in general be reversed.

Theorem 5. *Let* $z = [\zeta, t; r]$, $w = [\xi, s; p]$ *be points in the Siegel domain* $\mathcal{U}_{n+1}$*, and let* $Q(z), Q(w)$ *be the cubes in* T *which contain* z, w*, respectively (if* z *is contained in more than one cube, we choose one of them). Then,*

$$b^{d(Q(z)\tilde{\wedge}Q(w))} \approx \left((r + p + |\zeta - \xi|^2)^2 + \left(t - s - 2\operatorname{Im}(\bar{\zeta} \cdot \xi)\right)^2\right)^{1/4} \tag{2.7}$$

is approximately the $\frac{1}{4}$*-power of the denominator of* $H(z, w)$*. On the other hand,*

$$b^{d([Q(z)]\tilde{\wedge}[Q(w)])} \approx (r + p + |\zeta - \xi|^2)^{1/2} \tag{2.8}$$

is approximately the $\frac{1}{2}$*-power of the numerator of* $\operatorname{Re} H(z, w)$*.*

We have then the equivalence of kernels:

$$\operatorname{Re} H(z, w) \approx b^{2d(Q(z)\tilde{\wedge}Q(w)) - d([Q(z)]\tilde{\wedge}[Q(w)])} \tag{2.9}$$

Thus we have modeled the continuous kernel by a discrete kernel. This kernel, however, lives on the graph G, rather than on the tree T.

Theorem 5 allows a discretization of the Carleson measures problem for the DA space on $\mathcal{U}_{n+1}$. Given a measure $\widetilde{\mu}$ on $\mathcal{U}_{n+1}$, define a measure $\mu^\sharp$ on the graph G:

$\mu^\sharp(Q) := \int_{\widetilde{Q}} \mathrm{d}\widetilde{\mu}$. Then, $\widetilde{\mu}$ satisfies the inequality in Theorem 3 if and only if $\mu^\sharp$ is such that the inequality

$$\sum_{q\in G}\sum_{q'\in G} b^{2d(q\tilde{\wedge}q')-d([q]\tilde{\wedge}[q'])}\varphi(q)\mu^\sharp(q)\varphi(q')\mu^\sharp(q') \le C(\mu^\sharp)\sum_G \varphi^2\mu^\sharp \tag{2.10}$$

holds whenever $\varphi \ge 0$ is a positive function on the graph G.

In the Dirichlet case, inequality (2.10) is equivalent to its analog on the tree. Given q, q' in T, let $q\wedge q'$ be the element p contained in $[o,q]\cap[o,q']$ for which $d(p)$ is maximal. An analogous definition can be given for elements in U. The tree-analog of (2.10) is:

$$\sum_{q\in T}\sum_{q'\in T} b^{2d(q\wedge q')-d([q]\wedge[q'])}\varphi(q)\mu^\sharp(q)\varphi(q')\mu^\sharp(q') \le C(\mu^\sharp)\sum_T \varphi^2\mu^\sharp. \tag{2.11}$$

Problem 6. Is it true that the measures $\mu^\sharp$ such that (2.10) holds for all $\varphi\colon T\to[0,+\infty)$, are the same such that (2.11) holds for all $\varphi\colon T\to[0,+\infty)$?

There is a rich literature on the interplay of weighted inequalities, Carleson measures, potential theory and boundary values of holomorphic functions.

Problem 7. Is there a "potential theory" associated with the kernel $\operatorname{Re} H$, giving, e.g., sharp information about the boundary behavior of functions in DA?

Before we proceed, we summarize the zoo of distances usually employed in the study of $\mathcal{U}_{n+1}$ and of $\mathbb{H}^n$ as a guide to defining useful distances on T and U. We have already met the Koranyi distance $\|[\xi,s]^{-1}\cdot[\zeta,t]\|$ between the points $[\zeta,t]$ and $[\xi,s]$ in $\mathbb{H}^n$. The Koranyi distance is bi-Lipschitz equivalent to the Carnot – Carathéodory distance on $\mathbb{H}^n$. We refer the reader to [8] for a thorough treatment of sub-Riemannian distances in Lie groups and their use in analysis. The point we want to stress here is that the Carnot – Carathéodory distance is a length-distance, hence we can talk about approximate geodesics for the Koranyi distance itself.

Although it is not central to our story, for comparison we recall the Bergman metric β on $\mathcal{U}_{n+1}$. It is a Riemannian metric which is invariant under Heisenberg translations, dilations and rotations. Define the 1-form $\omega([\zeta,t]) = \mathrm{d}t - 2\operatorname{Im}(\zeta\cdot\mathrm{d}\bar{\zeta})$. Then,

$$\mathrm{d}\beta([\zeta,t;r])^2 = \frac{|\mathrm{d}\zeta|^2}{r} + \frac{\omega([\zeta,t])^2+\mathrm{d}r^2}{r^2}. \tag{2.12}$$

This can be compared with the familiar Bergman hyperbolic metric in $\mathbb{B}_{n+1}$.

$$\mathrm{d}\beta^2_{\mathbb{B}_{n+1}}(z) = \frac{|\mathrm{d}z|^2}{1-|z|^2} + \frac{|z\cdot\mathrm{d}\bar{z}|^2}{(1-|z|^2)^2}.$$

Lemma 1. (i) *For each $r>0$, consider on $\mathbb{H}^n(r)$ the Riemannian distance $\mathrm{d}\beta_r^2$ obtained by restricting the two-form $\mathrm{d}\beta^2$ to $\mathbb{H}^n(r)$. Then, the following quantities are equivalent for $\|[\zeta,t]\|\ge\sqrt{r}$:*

$$\left((|\zeta|^2+r)^2+t^2\right)^{1/4} \approx \|[\zeta,t]\| \approx \sqrt{r}\beta_r([\zeta,t;r],[0,0;r]).$$

(ii) *A similar relation holds for cosets of the center. Let $[T;r]=T\cdot[0,0;r]$ be the orbit of $[0,0;r]$ under the action of the center T. Then,*

$$(|\zeta|^2+r)^{1/2} \approx d_{\mathbb{H}^n}([\zeta,t]\cdot T, T) \approx \sqrt{r}\beta_r\big([[\zeta,t]\cdot T;r],[T;r]\big)$$

PROOF OF LEMMA 1. The first approximate equality in (i) is obvious. For the second one, using dilation invariance

$$\sqrt{r}\beta([\zeta,t;r],[0,0;r]) = \sqrt{r}\beta\left(D_{\sqrt{r}}\left[\frac{\zeta}{\sqrt{r}},\frac{t}{r};1\right]\right),$$
$$D_{\sqrt{r}}([0,0;1]) = \sqrt{r}\beta\left(\left[\frac{\zeta}{\sqrt{r}},\frac{t}{r};1\right],[0,0;1]\right)$$

Since the metrics β and $d_{\mathbb{H}^n}$ define the same topology on $\mathbb{H}^n(r)$, the last quantity is comparable to $\sqrt{r}d_{\mathbb{H}^n}([\zeta/\sqrt{r},t/r;1],[0,0;1])$ when $1 \le \|\zeta/\sqrt{r},t/r]\| \le 2$, by compactness of the unit ball with respect to the metric and Weierstrass' theorem. Since the metric β_r is a length metric and $d_{\mathbb{H}^n}$ is bi-Lipschitz equivalent to a length metric (the Carnot – Carathéodory distance), then, when $1 \le \|[\zeta/\sqrt{r},t/r]\|$, we have

$$\sqrt{r}\beta\left(\left[\frac{\zeta}{\sqrt{r}},\frac{t}{r};1\right],[0,0;1]\right) \approx \sqrt{r}d_{\mathbb{H}^n}\left(\left[\frac{\zeta}{\sqrt{r}},\frac{t}{r}\right],[0,0]\right) = d_{\mathbb{H}^n}([\zeta,t],[0,0]).$$

The proof of (ii) is analogous. □

PROOF OF THEOREM 5. We prove (2.7), the other case being similar (easier, in fact). Suppose that $d\big(Q(z),Q(w)\big) = m$. Then, $d\big(Q(z)\big), d\big(Q(w)\big) \le m$, hence, $b^{-m} \lesssim \sqrt{r},\sqrt{p}$ and there are Q_1, Q, Q_2 in $T^{(m)}$ such that $Q(z) \ge Q_1 \underset{G}{\sim} Q \underset{G}{\sim} Q_2 \le Q(w)$. We have then that

$$b^{-m} \ge \max\{\sqrt{r},\sqrt{p}, cd_{\mathbb{H}^n}([\zeta,t],[\xi,s])\} \approx \big((r+p+|\zeta-\xi|^2)^2+\big(t-s-2\operatorname{Im}(\bar{\zeta}\cdot\xi)\big)^2\big)^{1/4}:$$

the left-hand side of (2.7) dominates the right-hand side.

To show the opposite inequality, consider two cases. Suppose first that $\sqrt{r} \ge \sqrt{p}, d_{\mathbb{H}^n}([\zeta,t],[\xi,s])$ and that $b^{-m} \ge \sqrt{r} \ge b^{-m-1}$. Then, $Q(z) \underset{G}{\sim} Q(w)$. Hence, $m \le d\big(Q(z) \tilde{\wedge} Q(w)\big) \le m+1$ and

$$b^{-d(Q(z)\tilde{\wedge}Q(w))} \approx b^{-m} \gtrsim \big((r+p+|\zeta-\xi|^2)^2 + \big(t-s-2Im(\bar{\zeta}\cdot\xi)\big)^2\big)^{1/4}.$$

Suppose now that $d_{\mathbb{H}^n}([\zeta,t],[\xi,s]) \ge \sqrt{r},\sqrt{p}$ and choose m with $m \le d\big(Q(z) \tilde{\wedge} Q(w)\big) \le m+1$. Let $Q^m(z)$ and $Q^m(w)$ be the predecessors of $Q(z)$ and $Q(w)$ in $T^{(m)}$ (we use here that $d\big(Q(z)\big), d\big(Q(w)\big) \ge m$). Then, $Q^m(z) \underset{G}{\sim} Q^m(w)$, hence $d_{\mathbb{H}^n}([\zeta,t],[\xi,s]) \lesssim b^{-m}$:

$$b^{-d(Q(z)\tilde{\wedge}Q(w))} \approx b^{-m} \gtrsim d_{\mathbb{H}^n}([\zeta,t],[\xi,s])$$
$$\approx \big((r+p+|\zeta-\xi|^2)^2 + \big(t-s-2\operatorname{Im}(\bar{\zeta}\cdot\xi)\big)^2\big)^{1/4}.$$

The theorem is proved. □

It can be proved that

$$1 + d_G\big(Q(z),Q(w)\big) \approx 1 + \beta(z,w),$$

where β is the Bergman metric and d_G is the edge-counting metric in G.

The expression for the kernel $\operatorname{Re} H$ in Theorem 5 reflects the graph structure of the set of dyadic boxes. We might define a new kernel using the tree structure only as follows. Given cubes Q_1, Q_2 in T, let

$$Q_1 \wedge Q_2 := \max\{Q \in T : Q \le Q_1 \text{ and } Q_2 \le Q_2\}$$

be the element in T such that $[o, Q_1] \cap [o, Q_2] = [o, Q_1 \wedge Q_2]$. Define similarly $[Q_1] \wedge [Q_2]$ in the quotient tree U. Define the kernel:

$$H_T(z, w) := b^{2d\left(\mathcal{Q}(z) \wedge \mathcal{Q}(w)\right) - d([\mathcal{Q}_1] \wedge [\mathcal{Q}_2])}, \qquad z, w \in \mathcal{U}_{n+1}$$

As in Theorem 5, there is a slight ambiguity due to the fact that there are several Q's in T such that $z \in Q$. This ambiguity might be removed altogether by distributing the boundary of the dyadic boxes among the sets sharing it.

Because nearby boxes in a box can be far away in the tree, it is not hard to see that H_T is not pointwise equivalent to $\operatorname{Re} H$. However, when discretizing the reproducing kernel of Dirichlet and related spaces the Carleson measure inequalities are the same for the tree and for the graph structure. We don't know if that holds here. See [6] for a general discussion of this matter.

In the next section, we discuss in greater depth the kernel H_T.

3. The discrete DA kernel

Here, for simplicity, we consider a rooted tree which we informally view as discrete models for the unit ball. The analogous model for the upper half space would have the root "at infinity."

Let $T = \big(V(T), E(T)\big)$ be a tree: $V(T) \equiv T$ is the set of vertices and $E(T)$ is set of edges. We denote by d the natural edge-counting distance on T and,for $x, y \in T$, we write $[x, y]$ for the geodesic joining x and y. Let $o \in T$ be a distinguished element on it, the root. The choice of o induces on T a level structure: $d = d_o \colon T \to \mathbb{N}$, $x \mapsto d(x, o)$. Let (T, o) and (U, p) be rooted trees. We will use the standard notation for trees, $x \wedge y$, $x \geq y$, x^{-1}, $C(x)$ for the parent and children of x, $P(x)$ and $S(x)$ for the predecessor and successor regions. Also recall that for f a function on the tree the operators I and I^* produce the new functions

$$If(x) = \sum_{y \in P(x)} f(y); \qquad I^* f(x) = \sum_{y \in S(x)} f(y).$$

A morphism of trees $\Phi \colon T \to U$ is a couple of maps $\Phi_V : T \to U$, $\Phi_E \colon E(T) \to E(U)$, which preserve the tree structure: if (x, y) is an edge of T, then $\Phi_E(x, y) = \big(\Phi_V(x), \Phi_V(y)\big)$ is an edge of U. A morphism of rooted trees $\Phi \colon (T, o) \to (U, p)$ is a morphism of trees which preserves the level structure:

$$d_p\big(\Phi(x)\big) = d_o(x).$$

The morphism Φ is an epimorphism if Φ_V is surjective: any edge in U is the image of an edge in T.

We adopt the following notation. If $x \in T$, we denote $[x] = \Phi_V(x)$. We use the same symbol $\wedge$ for the confluent in T (with respect to the root o) and in U (with respect to the root $p = [o]$).

A quotient structure on (T, o) is an epimorphism $\Phi \colon (T, o) \to (U, p)$. The rooted tree (U, p) was called the tree of rings in [3].

Recall that $b \geq 2n + 1$ is a fixed odd integer. Fix a positive integer N and let T be a tree with root o, whose elements at level $m \geq 1$ are ordered m-tuples $a = (a_1 a_2 \dots a_m)$, with $a_j \in \mathbb{Z}_{b^{N+1}}$, the cyclic group of order b^{N+1}. The children of a are the $(m+1)$-tuples $(a_1 a_2 \dots a_m \alpha)$, $\alpha \in \mathbb{Z}_{b^{N+1}}$, and the root is identified with a 0-tuple, so that each element in T has b^{N+1} children. The tree U is defined similarly, with b^N instead of b^{N+1}.

Consider now the group homomorphism i from $\mathbb{Z}_{\mathbf{b}}$ to $\mathbb{Z}_{\mathbf{b}^{N+2}}$ given by $i([k]_{\mathrm{mod}\,\mathbf{b}}) = [\mathbf{b}^N k]_{\mathrm{mod}\,\mathbf{b}^{N+1}}$ and the induced short exact sequence

$$0 \hookrightarrow \mathbb{Z}_{\mathbf{b}} \overset{i}{\hookrightarrow} \mathbb{Z}_{\mathbf{b}^{N+1}} \overset{\Pi}{\longrightarrow} \mathbb{Z}_{\mathbf{b}^N} \to 0.$$

The projection Π induces a map $\Phi_V: T \to U$ on the set of vertices,

$$\Phi_V(a_1 a_2 \dots a_m) := \big(\Phi_V(a_1)\Phi_V(a_2)\dots\Phi_V(a_m)\big),$$

which clearly induces a tree epimorphism $\Phi\colon T \to U$. Here a way to picture the map Φ. We think of the elements C of U as "boxes" containing those elements x in T such that $[x] := \Phi(x) = C$. Each box C has $\mathbf{b}^N$ children at the next level, $C_1, \dots, C_{\mathbf{b}^N}$. Now, each x has $\mathbf{b}^{N+1}$ children at the same level, b of them falling in each of the boxes C_j.

We think of the quotient structure (T, U) as a discretization of the Siegel domain $\mathcal{U}_{n+1}$, with $\mathbf{b} = b^2$ and $N = n$.

The discrete DA_{N+1} kernel $K\colon T \times T \to [0, \infty)$ is defined by

$$K(x, y) = \mathbf{b}^{2d(x\wedge y) - d([x]\wedge[y])}$$

Note that it is modeled on the approximate expression in (2.9).

Theorem 6. *The kernel K is positive definite. In fact,*

$$\sum_{x,y\in T} \mathbf{b}^{2d(x\wedge y)-d([x]\wedge[y])}\mu(x)\overline{\mu(y)}$$
$$= |I^*\mu(o)|^2 + \frac{\mathbf{b}-1}{\mathbf{b}}\sum_{z\neq o}|I^*\mu(z)|^2 + \frac{1}{2}\sum_{\substack{z\neq w\in T\\ [z]=[w]}} \mathbf{b}^{2d(z\wedge w)-d([z]\wedge[w])}|I^*\mu(z) - I^*\mu(w)|^2.$$

The theorem will follow from the following lemma and easy counting.

Lemma 2 (Summation by parts). *Let $K\colon T \times T \to \mathbb{C}$ be a kernel on T, having the form $K(x, y) = H(x \wedge y, [x] \wedge [y])$ for some function $H\colon T \times U \to \mathbb{C}$. Then, if $\mu\colon T \to \mathbb{C}$ is a function having finite support,*

$$\sum_{x,y} K(x,y)\mu(x)\overline{\mu(y)} = H(o,[o])|I^*\mu(o)|^2 \tag{3.1}$$
$$+ \sum_{\substack{z,w\in T\setminus\{o\}\\ [z]=[w]}} [H(z\wedge w, [z]\wedge[w]) - H(z^{-1}\wedge w^{-1}, [z^{-1}]\wedge[w^{-1}])]I^*\mu(z)\overline{I^*\mu(w)}.$$

PROOF. Let Q be the left-hand side of (3.1). Then,

$$\begin{aligned} Q &= \sum_{x,y\in T} H(x\wedge y, [x]\wedge[y])\mu(x)\overline{\mu(y)} \\ &= \sum_{C\in U}\sum_{z,w\in T} H(z\wedge w, C) \sum_{\substack{x\geq z, y\geq w\\ [x]\wedge[y]=C}} \mu(x)\overline{\mu(y)} \\ &= \sum_{C\in U}\sum_{z,w\in T} H(z\wedge w, C)A(z,w), \end{aligned}$$

If $z \neq w$,

$$\begin{aligned}
A(z,w) &= \sum_{\substack{x\geq z, y\geq w\\ [x]\wedge[y]=C}} \mu(x)\overline{\mu(y)}\\
&= \sum_{\substack{D\neq F\in U\\ D,F\in\mathcal{C}(C)}} \sum_{\substack{s\in\mathcal{C}(z),[s]=D\\ t\in\mathcal{C}(w),[t]=F}} I^*\mu(s)I^*\mu(t) + \mu(z)\overline{\big(I^*\mu(w)-\mu(w)\big)}\\
&\qquad + \big(I^*\mu(z)-\mu(z)\big)\overline{\mu(w)} + \mu(z)\overline{\mu(w)}.
\end{aligned}$$

On the other hand,

$$\begin{aligned}
I^*\mu(z)\overline{I^*\mu(w)} = \mu(z)\overline{\big(I^*\mu(w)-\mu(w)\big)} + \big(I^*\mu(z)-\mu(z)\big)\overline{\mu(w)} + \mu(z)\overline{\mu(w)}\\
+ \sum_{D,F\in\mathcal{C}(C)} \sum_{\substack{[s]=D,s\in\mathcal{C}(z)\\ [t]=F,t\in\mathcal{C}(w)}} I^*\mu(s)\overline{I^*\mu(w)}.
\end{aligned}$$

Hence, if $z\neq w$,

$$\begin{aligned}
A(z,w) &= I^*\mu(z)\overline{I^*\mu(w)} - \sum_{F\in\mathcal{C}(C)} \sum_{\substack{[s]=D,s\in\mathcal{C}(z)\\ [t]=F,t\in\mathcal{C}(w)}} I^*\mu(s)\overline{I^*\mu(t)}\\
&= I^*\mu(z)\overline{I^*\mu(w)} - \sum_{F\in\mathcal{C}(C)} \sum_{\substack{[s]=D,\\ s\in\mathcal{C}(z)}} I^*\mu(s) \sum_{\substack{[t]=F,\\ t\in\mathcal{C}(w)}} \overline{I^*\mu(t)}.
\end{aligned}$$

In the case of equality,

$$\begin{aligned}
A(z,z) &= \sum_{\substack{x,y\geq z\\ [x]\wedge[y]=C}} \mu(x)\overline{\mu(y)}\\
&= \mu(z)\overline{\big(I^*\mu(z)-\mu(z)\big)} + \overline{\mu(z)}\big(I^*\mu(z)-\mu(z)\big) + |\mu(z)|^2\\
&\qquad + \sum_{\substack{D\neq F\\ D,F\in\mathcal{C}(C)}} \sum_{\substack{[s]=D,s\in\mathcal{C}(z)\\ [t]=F,t\in\mathcal{C}(w)}} I^*\mu(s)\overline{I^*\mu(t)}.
\end{aligned}$$

On the other hand,

$$\begin{aligned}
|I^*\mu(z)|^2 &= \mu(z)\overline{\big(I^*\mu(z)-\mu(z)\big)} + \overline{\mu(z)}\big(I^*\mu(z)-\mu(z)\big) + |\mu(z)|^2\\
&\qquad + \left|\sum_{D\in\mathcal{C}(C)} \sum_{\substack{[s]=D,\\ s\in\mathcal{C}(z)}} I^*\mu(s)\right|^2\\
&= \mu(z)\overline{\big(I^*\mu(z)-\mu(z)\big)} + \overline{\mu(z)}\big(I^*\mu(z)-\mu(z)\big) + |\mu(z)|^2\\
&\qquad + \sum_{D\in\mathcal{C}(D)}\left|\sum_{\substack{[s]=D,\\ s\in\mathcal{C}(z)}} I^*\mu(s)\right|^2 + \sum_{\substack{D\neq F\\ D,E\in\mathcal{C}(C)}} \sum_{\substack{[s]=D,\ s\in\mathcal{C}(z)\\ [t]=F,\ t\in\mathcal{C}(w)}} I^*\mu(s)\overline{I^*\mu(t)}..
\end{aligned}$$

Comparing:

$$A(z,z) = |I^*\mu(z)|^2 - \sum_{D\in\mathcal{C}(D)}\left|\sum_{\substack{[s]=D\\ s\in\mathcal{C}(z)}} I^*\mu(s)\right|^2. \tag{3.2}$$

Then,

$$Q = \sum_{C \in U} \sum_{[z]=[w] \in C} H(z \wedge w, C) \left[I^*\mu(z)\overline{I^*\mu(w)} - \sum_{D \in \mathcal{C}(C)} \left(\sum_{\substack{[s]=D \\ s \in \mathcal{C}(z)}} I^*\mu(s) \sum_{\substack{[t]=F \\ t \in \mathcal{C}(w)}} \overline{I^*\mu(t)} \right) \right]$$

$$= \sum_{\substack{[z]=[w]=C \\ d(z)=d(w)\geq 1}} [H(z \wedge w, C) - H(z^{-1} \wedge w^{-1}, C^{-1})] I^*\mu(z)\overline{I^*\mu(w)} + H(o, [o])|I^*\mu(o)|^2,$$

which is the desired expression.

In the last member of the chain of equalities,we have taken into account that each term $I^*\mu(z)\overline{I^*\mu(w)}$ appears twice in the preceding member (except for the root term). □

Proof of Theorem 6. Let Q be the left-hand side of (3.1). By Lemma 2,

$$Q = |I^*\mu(o)|^2 + \sum_{\substack{z,w \in T \setminus \{o\} \\ [z]=[w]}} \left[\mathbf{b}^{2d(z\wedge w)-d([z]\wedge[w])} - \mathbf{b}^{2d(z^{-1}\wedge w^{-1})-d([z^{-1}]\wedge[w^{-1}])} \right] I^*\mu(z)\overline{I^*\mu(w)}$$

We have two consider two cases. If $z \neq w$, then $z \wedge w = z^{-1} \wedge w^{-1}$, $[z^{-1}] \wedge [w^{-1}] = ([z] \wedge [w])^{-1}$, so that the corresponding part of the sum is

$$Q_1 = -(\mathbf{b} - 1) \sum_{\substack{z \neq w \in T \setminus \{o\} \\ [z]=[w]}} \mathbf{b}^{2d(z\wedge w)-d([z]\wedge[w])} I^*\mu(z)\overline{I^*\mu(w)} \tag{3.3}$$

If $z = w$, then $z^{-1} \wedge z^{-1} = z^{-1}$, hence the remaining summands add up to

$$Q_2 = \frac{\mathbf{b} - 1}{\mathbf{b}} \sum_{z \neq o} \mathbf{b}^{d(z)} |I^*\mu(z)|^2.$$

The term Q_1 in (3.3) contains the mixed products of

$$R = \frac{\mathbf{b} - 1}{2} \sum_{\substack{z \neq w \in T \setminus \{o\} \\ [z]=[w]}} \mathbf{b}^{2d(z\wedge w)-d([z]\wedge[w])} |I^*\mu(z) - I^*\mu(w)|^2$$

$$= Q_1 + (\mathbf{b} - 1) \sum_{z \neq o} |I^*\mu(z)|^2 \sum_{w:[w]=[z]} \mathbf{b}^{2d(z\wedge w)-d([z]\wedge[w])}.$$

The last sum can be computed, taking into account that, for $1 \leq k \leq d(z)$, there are $(\mathbf{b} - 1)\mathbf{b}^{k-1}$ w's for which $[w] = [z]$ and

$$d(z) = d([z] \wedge [w]) = d(z \wedge w) + k,$$

by the special nature of $\Phi \colon T \to U$:

$$\sum_{w:[w]=[z]} \mathbf{b}^{2d(z\wedge w)-d([z]\wedge[w])} = \sum_{k=1}^{d(z)} (\mathbf{b} - 1)\mathbf{b}^{k-1} b^{2(d(z)-k)-d(z)}$$

$$= (\mathbf{b} - 1) b^{d(z)} \sum_{k=1}^{d(z)} 2^{-k-1} = \frac{1}{\mathbf{b}} (\mathbf{b}^{d(z)} - 1)$$

Hence,

$$R = Q_1 + Q_2 - \frac{\mathbf{b}-1}{\mathbf{b}} \sum_{z\neq o} |I^*\mu(z)|^2 = Q - |I^*\mu(o)|^2 - \frac{\mathbf{b}-1}{\mathbf{b}} \sum_{z\neq o} |I^*\mu(z)|^2,$$

as wished. □

Problem 8. The discrete DA kernel in Theorem 6 does not have the complete Nevanlinna – Pick property. This is probably due to the fact that the kernel is a discretization of the real part of the DA kernel on the unit ball, not of the whole kernel. Is there a natural kernel on the quotient structure $\Phi\colon T \to U$ which is complete Nevanlinna – Pick?

In the next section, we exhibit a real valued, complete Nevanlinna – Pick kernel on trees.

4. Complete Nevanlinna – Pick kernels on trees

Let T be a tree: a loopless, connected graph, which we identify with the set of its vertices. Consider a root o in T and define a partial order having o as minimal element: $x \leq y$ if $x \in [o, y]$ belongs to the unique nonintersecting path joining o and y following the edges of T. Given x in T, let $d(x) := \sharp[o, x] - 1$ be the number of edges one needs to cross to go from o to x. Define $x \wedge y =: \max[o, x] \cap [o, y]$ to be the confluent of x and y in T, with respect to o. Given a summable function $\mu\colon T \to \mathbb{C}$, let $I^*\mu(x) = \sum_{y \geq x} \mu(y)$.

Theorem 7. *Let $\Lambda > 1$. The kernel*

$$K(x, y) = \Lambda^{d(x\wedge y)}$$

is a complete Nevanlinna – Pick kernel.

Our primary experience with these kernels is for $1 < \Lambda < 2$. At the level of the metaphors we have been using, $2^{d(x\wedge y)}$ models $|K(x, y)|$ for the kernel K of (1.1). We noted earlier that the real part of that kernel plays an important role in studying Carleson measures. For that particular kernel passage from $\operatorname{Re} K$ to $|K|$ loses a great deal of information. However in the range $1 < \Lambda < 2$ the situation is different. In that range $\Lambda^{d(x\wedge y)}$ models $|K^\alpha|$, $0 < \alpha < 1$ and the K^α are the kernels for Besov spaces between the DA space and Dirichlet spaces. For those kernels we have $|K^\alpha| \approx \operatorname{Re} K^\alpha$ making the model kernels quite useful, for instance in [5].

These kernels also arise in other contexts and the fact that they are positive definite has been noted earlier, [13, Lemma 1.2; 14, (1.4)].

We need two simple lemmas.

Lemma 3 (Summation by parts). *Let $h, \mu\colon T \to \mathbb{C}$ be functions and let $M = I^*\mu$. Then,*

$$\sum_{x,y} h(x\wedge y)\mu(x)\overline{\mu(y)} = h(o)|M(o)|^2 + \sum_{t\in T\setminus\{o\}} [h(t) - h(t^{-1})]|M(t)|^2.$$

PROOF.

$$\begin{aligned}
&\sum_{x,y} h(x \wedge y)\mu(x)\overline{\mu(y)} \\
&\quad = \sum_t h(t) \sum_{x\wedge y=t} \mu(x)\overline{\mu(y)} \\
&\quad = \sum_t h(t) \left[|\mu(t)|^2 + \mu(t)(\overline{M(t) - \mu(t)}) + \overline{\mu(t)}\big(M(t) - \mu(t)\big) + \sum_{\substack{z\neq w; z,w>t;\\ d(w,t)=d(z,t)=1}} M(z)\overline{M(w)} \right] \\
&\quad = \sum_t h(t) \left[|M(t)|^2 - \sum_{\substack{z>t\\ d(z,t)=1}} |M(z)|^2 \right],
\end{aligned}$$

which is the quantity on the right-hand side of the statement. □

Lemma 4. *Fix a new root a in T and let d_a and $\wedge_a$ be the objects related to this new root. Then,*

$$d_a(x \wedge_a y) = d(x \wedge y) + d(a) - d(x \wedge a) - d(a \wedge y).$$

PROOF. The proof is clear after making sketches for the various cases. □

PROOF OF THE THEOREM 7. The kernel K is complete Nevanlinna – Pick if and only if each matrix

$$A = \left[1 - \frac{K(x_i, x_N)K(x_N, x_j)}{K(x_N, x_N)K(x_i, x_j)} \right]_{i,j=1\ldots N-1} \tag{4.1}$$

is positive definite for each choice of $x_1, \ldots, x_N$ in T; see [2].

Let $a = x_N$. The (i,j)th entry of A is, by the second lemma, $A_{ij} = 1 - \Lambda^{-d_a(x_i \wedge x_j)}$. By the first lemma, A is positive definite. □

References

1. J. Agler and J. E. McCarthy, *Complete Nevanlinna – Pick kernels*, J. Funct. Anal. **175** (2000), no. 1, 111 – 124.
2. ———, *Pick interpolation and Hilbert function spaces*, Grad. Stud. Math., vol. 44, Amer. Math. Soc., Providence, RI, 2002.
3. N. Arcozzi, R. Rochberg, and E. Sawyer, *Carleson measures for the Drury – Arveson Hardy space and other Besov – Sobolev spaces on complex balls*, Adv. Math. **218** (2008), no. 4, 1107 – 1180.
4. ———, *The diameter space, a restriction of the Drury – Arveson – Hardy space*, Function Spaces, Contemp. Math., vol. 435, Amer. Math. Soc., Providence, RI, 2007, pp. 21 – 42.
5. ———, *Carleson measures and interpolating sequences for Besov spaces on complex balls*, Mem. Amer. Math. Soc. **182** (2006), no. 859.
6. N. Arcozzi, R. Rochberg, E. Sawyer, and B. Wick, *Potential theory and Carleson measures on trees and graphs*, in preparation.
7. W. Arveson, *Subalgebras of C^*-algebras.* III: *Multivariable operator theory*, Acta Math. **181** (1998), no. 2, 159 – 228.
8. A. Bonfiglioli, E. Lanconelli, and F. Uguzzoni, *Stratified Lie groups and potential theory for their sub-Laplacians*, Springer Monogr. Math., Springer, Berlin, 2007.
9. Z. Chen, *Characterizations of Arveson's Hardy space*, Complex Var. Theory Appl. **48** (2003), no. 5, 453 – 465.
10. M. Christ, *A $T(b)$ theorem with remarks on analytic capacity and the Cauchy integral*, Colloq. Math. **60/61** (1990), no. 2, 601 – 628.
11. S. W. Drury, *A generalization of von Neumann's inequality to the complex ball*, Proc. Amer. Math. Soc. **68** (1978), no. 3, 300 – 304.

12. N. Garofalo and E. Lanconelli, *Frequency functions on the Heisenberg group, the uncertainty principle and unique continuation*, Ann. Inst. Fourier (Grenoble) **40** (1990), no. 2, 313–356.
13. U. Haagerup, *An example of a nonnuclear C^*-algebra, which has the metric approximation property*, Invent. Math. **50** (1978/79), no. 3, 279–293.
14. Yu. A. Neretin, *Groups of hierarchomorphisms of trees and related Hilbert spaces*, J. Funct. Anal. **200** (2003), no. 2, 505–535.
15. E. M. Stein, *Harmonic analysis: real-variable methods, orthogonality, and oscillatory integrals*, Princeton Math. Ser., vol. 43, Princeton Univ. Press, Princeton, NJ, 1993.
16. Robert S. Strichartz, *Self-similarity on nilpotent Lie groups*, Geometric Analysis (Philadelphia, PA, 1991), Contemp. Math., vol. 140, Amer. Math. Soc., Providence, RI, 1992, pp. 123–157.
17. E. Tchoundja, *Carleson measures for the generalized Bergman spaces via a $T(1)$-type theorem*, Ark. Mat. **46** (2008), no. 2, 377–406.
18. J. T. Tyson, *Global conformal Assouad dimension in the Heisenberg group*, Conform. Geom. Dyn. **12** (2008), 32–57.
19. J. von Neumann, *Eine Spektraltheorie für allgemeine Operatoren eines unitären Raumes*, Math. Nachr. **4** (1951), 258–281 (German).

Dipartimento di Matematica, Universita di Bologna, 40127 Bologna, Italy
E-mail address: arcozzi@dm.unibo.it

Department of Mathematics, Washington University, St. Louis, MO 63130, USA
E-mail address: rr@math.wustl.edu

Department of Mathematics & Statistics, McMaster University, Hamilton, ON L8S 4K1, Canada
E-mail address: Saw6453CDN@aol.com

Centre de Recherches Mathématiques
CRM Proceedings and Lecture Notes
Volume **51**, 2010

The Norm of a Truncated Toeplitz Operator

Stephan Ramon Garcia and William T. Ross

ABSTRACT. We prove several lower bounds for the norm of a truncated Toeplitz operator and obtain a curious relationship between the H^2 and H^∞ norms of functions in model spaces.

1. Introduction

In this paper, we continue the discussion initiated in [6] concerning the norm of a truncated Toeplitz operator. In the following, let H^2 denote the classical Hardy space of the open unit disk $\mathbb{D}$ and $K_\Theta := H^2 \cap (\Theta H^2)^\perp$, where Θ is an inner function, denote one of the so-called Jordan model spaces [2, 4, 7]. If H^∞ is the set of all bounded analytic functions on $\mathbb{D}$, the space $K_\Theta^\infty := H^\infty \cap K_\Theta$ is norm dense in K_Θ (see [2, p. 83] or [9, Lemma 2.3]). If P_Θ is the orthogonal projection from $L^2 := L^2(\partial\mathbb{D}, |d\zeta|/2\pi)$ onto K_Θ and $\varphi \in L^2$, then the operator

$$A_\varphi f := P_\Theta(\varphi f), \qquad f \in K_\Theta^\infty,$$

is densely defined on K_Θ and is called a *truncated Toeplitz operator*. Various aspects of these operators were studied in [3, 5, 6, 9, 10].

If $\|\cdot\|$ is the norm on L^2, we let

$$\|A_\varphi\| := \sup\{\|A_\varphi f\| : f \in K_\Theta^\infty, \|f\| = 1\} \tag{1}$$

and note that this quantity is finite if and only if A_φ extends to a bounded operator on K_Θ. When $\varphi \in L^\infty$, the set of bounded measurable functions on $\partial\mathbb{D}$, we have the basic estimates

$$0 \leq \|A_\varphi\| \leq \|\varphi\|_\infty.$$

However, it is known that equality can hold, in nontrivial ways, in either of the inequalities above and hence finding the norm of a truncated Toeplitz operator can be difficult. Furthermore, it turns out that there are many unbounded symbols $\varphi \in L^2$ which yield bounded operators A_φ. Unlike the situation for classical Toeplitz operators on H^2, for a given $\varphi \in L^2$, there many $\psi \in L^2$ for which $A_\varphi = A_\psi$ [9, Theorem 3.1].

2000 *Mathematics Subject Classification.* 47A05, 47B35, 47B99.

Key words and phrases. Toeplitz operator, model space, truncated Toeplitz operator, reproducing kernel, complex symmetric operator, conjugation.

First author partially supported by National Science Foundation Grant DMS-0638789.

For a given symbol $\varphi \in L^2$ and inner function Θ, lower bounds on the quantity (1) are useful in answering the following nontrivial questions:

(1) is A_φ unbounded?
(2) if $\varphi \in L^\infty$, is A_φ norm-attaining (i.e., is $\|A_\varphi\| = \|\varphi\|_\infty$)?

When Θ is a finite Blaschke product and $\varphi \in H^\infty$, the paper [6] computes $\|A_\varphi\|$ and gives necessary and sufficient conditions as to when $\|A_\varphi\| = \|\varphi\|_\infty$. The purpose of this short note is to give a few lower bounds on $\|A_\varphi\|$ for general inner functions Θ and general $\varphi \in L^2$. Along the way, we obtain a curious relationship (Corollary 5) between the H^2 and H^∞ norms of functions in K_Θ^∞.

2. Lower bounds derived from Poisson's formula

For $\varphi \in L^2$, let

$$(\mathfrak{P}\varphi)(z) := \int_{\partial\mathbb{D}} \frac{1-|z|^2}{|\zeta - z|^2}\varphi(\zeta)\frac{|d\zeta|}{2\pi}, \qquad z \in \mathbb{D}, \tag{2}$$

be the standard Poisson extension of φ to $\mathbb{D}$. For $\varphi \in C(\partial\mathbb{D})$, the continuous functions on $\partial\mathbb{D}$, recall that $\mathfrak{P}\varphi$ solves the classical Dirichlet problem with boundary data φ. Also note that

$$k_\lambda(z) := \frac{1-\overline{\Theta(\lambda)}\Theta(z)}{1-\overline{\lambda}z}, \qquad \lambda, z \in \mathbb{D},$$

is the reproducing kernel for K_Θ [9].

Our first result provides a general lower bound for $\|A_\varphi\|$ which yields a number of useful corollaries:

Theorem 1. *If $\varphi \in L^2$, then*

$$\sup_{\lambda\in\mathbb{D}} \frac{1-|\lambda|^2}{1-|\Theta(\lambda)|^2}\left|\int_{\partial\mathbb{D}} \varphi(z)\left|\frac{\Theta(z)-\Theta(\lambda)}{z-\lambda}\right|^2\frac{|dz|}{2\pi}\right| \leq \|A_\varphi\|. \tag{3}$$

In other words,

$$\sup_{\lambda\in\mathbb{D}}\left|\int_{\partial\mathbb{D}} \varphi(z)\, d\nu_\lambda(z)\right| \leq \|A_\varphi\|$$

where

$$d\nu_\lambda(z) := \frac{1-|\lambda|^2}{1-|\Theta(\lambda)|^2}\left|\frac{\Theta(z)-\Theta(\lambda)}{z-\lambda}\right|^2\frac{|dz|}{2\pi}$$

is a family of probability measures on $\partial\mathbb{D}$ indexed by $\lambda \in \mathbb{D}$.

PROOF. For $\lambda \in \mathbb{D}$ we have

$$\|k_\lambda\| = \sqrt{\frac{1-|\Theta(\lambda)|^2}{1-|\lambda|^2}}, \tag{4}$$

from which it follows that

$$\begin{aligned}\|A_\varphi\| \geq \frac{1-|\lambda|^2}{1-|\Theta(\lambda)|^2}|\langle A_\varphi k_\lambda, k_\lambda\rangle| &= \frac{1-|\lambda|^2}{1-|\Theta(\lambda)|^2}|\langle P_\Theta \varphi k_\lambda, k_\lambda\rangle| \\ &= \frac{1-|\lambda|^2}{1-|\Theta(\lambda)|^2}|\langle \varphi k_\lambda, k_\lambda\rangle| \\ &= \frac{1-|\lambda|^2}{1-|\Theta(\lambda)|^2}\left|\int_{\partial\mathbb{D}} \varphi(z)\left|\frac{\Theta(z)-\Theta(\lambda)}{z-\lambda}\right|^2\frac{|dz|}{2\pi}\right|.\end{aligned}$$

That the measures $d\nu_\lambda$ are indeed probability measures follows from (4). □

Now observe that if $\Theta(\lambda) = 0$, the argument in the supremum on the left hand side of (3) becomes the absolute value of the expression in (2). This immediately yields the following corollary:

Corollary 1. *If $\varphi \in L^2$, then*

$$\sup_{\lambda \in \Theta^{-1}(\{0\})} |(\mathfrak{P}\varphi)(\lambda)| \leq \|A_\varphi\|, \tag{5}$$

where the supremum is to be regarded as 0 if $\Theta^{-1}(\{0\}) = \varnothing$.

Under the right circumstances, the preceding corollary can be used to prove that certain truncated Toeplitz operators are norm-attaining:

Corollary 2. *Let Θ be an inner function having zeros which accumulate at every point of $\partial\mathbb{D}$. If $\varphi \in C(\partial\mathbb{D})$ then $\|A_\varphi\| = \|\varphi\|_\infty$.*

PROOF. Let $\zeta \in \partial\mathbb{D}$ be such that $|\varphi(\zeta)| = \|\varphi\|_\infty$. By hypothesis, there exists a sequence λ_n of zeros of Θ which converge to ζ. By continuity, we conclude that

$$\|\varphi\|_\infty = \lim_{n\to\infty} |(\mathfrak{P}\varphi)(\lambda_n)| \leq \|A_\varphi\| \leq \|\varphi\|_\infty$$

whence $\|A_\varphi\| = \|\varphi\|_\infty$. □

The preceding corollary stands in contrast to the finite Blaschke product setting. Indeed, if Θ is a finite Blaschke product and $\varphi \in H^\infty$, then it is known that $\|A_\varphi\| = \|\varphi\|_\infty$ if and only if φ is the scalar multiple of the inner factor of some function from K_Θ [6]Theorem 2.

At the expense of wordiness, the hypothesis of Corollary 2 can be considerably weakened. A cursory examination of the proof indicates that we only need ζ to be a limit point of the zeros of Θ, $\varphi \in L^\infty$ to be continuous on an open arc containing ζ, and $|\varphi(\zeta)| = \|\varphi\|_\infty$.

Theorem 1 yields yet another lower bound for $\|A_\varphi\|$. Recall that an inner function Θ has a finite angular derivative at $\zeta \in \partial\mathbb{D}$ if Θ has a nontangential limit $\Theta(\zeta)$ of modulus one at ζ and Θ' has a finite nontangential limit $\Theta'(\zeta)$ at ζ. This is equivalent to asserting that

$$\frac{\Theta(z) - \Theta(\zeta)}{z - \zeta} \tag{6}$$

has the nontangential limit $\Theta'(\zeta)$ at ζ. If Θ has a finite angular derivative at ζ, then the function in (6) belongs to H^2 and

$$\lim_{r\to 1^-} \int_{\partial\mathbb{D}} \left| \frac{\Theta(z) - \Theta(r\zeta)}{z - r\zeta} \right|^2 \frac{|dz|}{2\pi} = \int_{\partial\mathbb{D}} \left| \frac{\Theta(z) - \Theta(\zeta)}{z - \zeta} \right|^2 \frac{|dz|}{2\pi}.$$

Furthermore, the above is equal to

$$\lim_{r\to 1^-} \frac{1 - |\Theta(r\zeta)|^2}{1 - r^2} = |\Theta'(\zeta)| > 0.$$

See [1, 8] for further details on angular derivatives. Theorem 1 along with the preceding discussion and Fatou's lemma yield the following lower estimate for $\|A_\varphi\|$.

Corollary 3. *For an inner function Θ, let D_Θ be the set of $\zeta \in \partial\mathbb{D}$ for which Θ has a finite angular derivative $\Theta'(\zeta)$ at ζ. If $\varphi \in L^\infty$ or if $\varphi \in L^2$ with $\varphi \geq 0$, then*

$$\sup_{\zeta \in D_\Theta} \frac{1}{|\Theta'(\zeta)|}\left|\int_{\partial\mathbb{D}} \varphi(z)\left|\frac{\Theta(z)-\Theta(\zeta)}{z-\zeta}\right|^2 \frac{|dz|}{2\pi}\right| \leq \|A_\varphi\|.$$

In other words,

$$\sup_{\zeta \in D_\Theta}\left|\int_{\partial\mathbb{D}} \varphi(z)\, d\nu_\lambda(z)\right| \leq \|A_\varphi\|,$$

where

$$d\nu_\lambda(z) := \frac{1}{|\Theta'(\zeta)|}\left|\frac{\Theta(z)-\Theta(\zeta)}{z-\zeta}\right|^2 \frac{|dz|}{2\pi}$$

is a family of probability measures on $\partial\mathbb{D}$ indexed by $\zeta \in D_\Theta$.

3. Lower bounds and projections

Our next several results concern lower bounds on $\|A_\varphi\|$ involving the orthogonal projection $P_\Theta : L^2 \to K_\Theta$.

Theorem 2. *If Θ is an inner function and $\varphi \in L^2$, then*

$$\frac{\|P_\Theta(\varphi) - \overline{\Theta(0)}P_\Theta(\Theta\varphi)\|}{(1-|\Theta(0)|^2)^{1/2}} \leq \|A_\varphi\|.$$

PROOF. First observe that $\|k_0\| = (1-|\Theta(0)|^2)^{1/2}$. Next we see that if $\varphi \in L^2$ and $g \in K_\Theta$ is any unit vector, then

$$(1-|\Theta(0)|^2)^{1/2}\|A_\varphi\| \geq |\langle A_\varphi k_0, g\rangle| = |\langle P_\Theta(\varphi k_0), g\rangle| = |\langle P_\Theta(\varphi) - \overline{\Theta(0)}P_\Theta(\Theta\varphi), g\rangle|.$$

Setting

$$g = \frac{P_\Theta(\varphi) - \overline{\Theta(0)}P_\Theta(\Theta\varphi)}{\|P_\Theta(\varphi) - \overline{\Theta(0)}P_\Theta(\Theta\varphi)\|}$$

yields the desired inequality. □

In light of the fact that $P_\Theta(\Theta\varphi) = 0$ whenever $\varphi \in H^2$, Theorem 2 leads us immediately to the following corollary:

Corollary 4. *If Θ is inner and $\varphi \in H^2$, then*

$$\frac{\|P_\Theta(\varphi)\|}{(1-|\Theta(0)|^2)^{1/2}} \leq \|A_\varphi\|. \tag{7}$$

It turns out that (7) has a rather interesting function-theoretic implication. Let us first note that for $\varphi \in H^\infty$, we can expect no better inequality than

$$\|\varphi\| \leq \|\varphi\|_\infty$$

(with equality holding if and only if φ is a scalar multiple of an inner function). However, if φ belongs to K_Θ^∞, then a stronger inequality holds.

Corollary 5. *If Θ is an inner function, then*

$$\|\varphi\| \leq (1-|\Theta(0)|^2)^{1/2}\|\varphi\|_\infty \tag{8}$$

holds for all $\varphi \in K_\Theta^\infty$. If Θ is a finite Blaschke product, then equality holds if and only if φ is a scalar multiple of an inner function from K_Θ.

PROOF. First observe that the inequality

$$\|\varphi\| \leq (1 - |\Theta(0)|^2)^{\frac{1}{2}} \|\varphi\|_\infty$$

follows from Corollary 4 and the fact that $P_\Theta \varphi = \varphi$ whenever $\varphi \in K_\Theta$. Now suppose that Θ is a finite Blaschke product and assume that equality holds in the preceding for some $\varphi \in K_\Theta^\infty$. In light of (7), it follows that $\|A_\varphi\| = \|\varphi\|_\infty$. From [6, Theorem 2] we see that φ must be a scalar multiple of the inner *part* of a function from K_Θ. But since $\varphi \in K_\Theta^\infty$, then φ must be a scalar multiple of an inner function from K_Θ. □

When Θ is a finite Blaschke product, then K_Θ is a finite dimensional subspace of H^2 consisting of bounded functions [3, 5, 9]. By elementary functional analysis, there are $c_1, c_2 > 0$ so that

$$c_1\|\varphi\| \leq \|\varphi\|_\infty \leq c_2\|\varphi\|$$

for all $\varphi \in K_\Theta$. This prompts the following question:

Question. What are the optimal constants c_1, c_2 in the above inequality?

4. Lower bounds from the decomposition of K_Θ

A result of Sarason [9, [Theorem 3.1]] says, for $\varphi \in L^2$, that

$$A_\varphi \equiv 0 \iff \varphi \in \Theta H^2 + \overline{\Theta H^2}. \tag{9}$$

It follows that the most general truncated Toeplitz operator on K_Θ is of the form $A_{\psi+\overline{\chi}}$ where $\psi, \chi \in K_\Theta$. We can refine this observation a bit further and provide another canonical decomposition for the symbol of a truncated Toeplitz operator.

Lemma 1. *Each bounded truncated Toeplitz operator on K_Θ is generated by a symbol of the form*

$$\varphi = \underbrace{\psi}_{\in H^2} + \underbrace{\chi\overline{\Theta}}_{\in \overline{zH^2}} \tag{10}$$

where $\psi, \chi \in K_\Theta$.

Before getting to the proof, we should remind the reader of a technical detail. It follows from the identity $K_\Theta = H^2 \cap \Theta\overline{zH^2}$ (see [2, p. 82]) that

$$C : K_\Theta \to K_\Theta, \qquad Cf := \overline{zf}\Theta,$$

is an isometric, conjugate-linear, involution. In fact, when A_φ is a bounded operator we have the identity $CA_\varphi C = A_\varphi^*$ [9, Lemma 2.1].

PROOF OF LEMMA 1. If T is a bounded truncated Toeplitz operator on K_Θ, then there exists some $\varphi \in L^2$ such that $T = A_\varphi$. We claim that this φ can be chosen to have the special form (10). First let us write $\varphi = f + \overline{zg}$ where $f, g \in H^2$. Using the orthogonal decomposition $H^2 = K_\Theta \oplus \Theta H^2$, it follows that φ may be further decomposed as

$$\varphi = (f_1 + \Theta f_2) + \overline{z(g_1 + \Theta g_2)}$$

where $f_1, g_1 \in K_\Theta$ and $f_2, g_2 \in H^2$. By (9), the symbols Θf_2 and $\overline{\Theta(zg_2)}$ yield the zero truncated Toeplitz operator on K_Θ. Therefore we may assume that

$$\varphi = f + \overline{zg}$$

for some $f, g \in K_\Theta$. Since $Cg = \overline{gz}\Theta$, we have $\overline{zg} = (Cg)\overline{\Theta}$ and hence (10) holds with $\psi = f$ and $\chi = Cg$. □

Corollary 6. *Let Θ be an inner function. If $\psi_1, \psi_2 \in K_\Theta$ and $\varphi = \psi_1 + \psi_2\overline{\Theta}$, then*

$$\frac{\|\psi_1 - \overline{\Theta(0)}\psi_2\|}{(1 - |\Theta(0)|^2)^{1/2}} \leq \|A_\varphi\|.$$

PROOF. If $\varphi = \psi_1 + \psi_2\overline{\Theta}$, then, since $\psi_1, \psi_2 \in K_\Theta$ and $\psi_2\overline{\Theta} \in \overline{zH^2}$, we have

$$P_\Theta(\varphi) - \overline{\Theta(0)}P_\Theta(\Theta\varphi) = \psi_1 - \overline{\Theta(0)}\psi_2.$$

The result now follows from Theorem 2. □

References

1. J. A. Cima, A. L. Matheson, and W. T. Ross, *The Cauchy transform*, Math. Surveys Monogr., vol. 125, Amer. Math. Soc., Providence, RI, 2006.
2. J. A. Cima and W. T. Ross, *The backward shift on the Hardy space*, Math. Surveys Monogr., vol. 79, Amer. Math. Soc., Providence, RI, 2000.
3. J. A. Cima, W. T. Ross, and W. R. Wogen, *Truncated Toeplitz operators on finite dimensional spaces*, Oper. Matrices **2** (2008), no. 3, 357–369.
4. R. G. Douglas, H. S. Shapiro, and A. L. Shields, *Cyclic vectors and invariant subspaces for the backward shift operator*, Ann. Inst. Fourier (Grenoble) **20** (1970), no. 1, 37–76.
5. S. R. Garcia, *Conjugation and Clark operators*, Recent Advances in Operator-Related Function Theory, Contemp. Math., vol. 393, Amer. Math. Soc., Providence, RI, 2006, pp. 67–111.
6. S. R. Garcia and W. T. Ross, *A nonlinear extremal problem on the Hardy space*, Comput. Methods Funct. Theory **9** (2009), no. 2, 485–524.
7. N. K. Nikol′skiĭ, *Treatise on the shift operator*, Grundlehren Math. Wiss., vol. 273, Springer, Berlin, 1986.
8. D. Sarason, *Sub-Hardy Hilbert spaces in the unit disk*, Univ. Arkansas Lecture Notes Math. Sci., vol. 10, Wiley, New York, 1994.
9. ———, *Algebraic properties of truncated Toeplitz operators*, Oper. Matrices **1** (2007), no. 4, 491–526.
10. ———, *Unbounded Toeplitz operators*, Integral Equations Operator Theory **61** (2008), no. 2, 281–298.

DEPARTMENT OF MATHEMATICS, POMONA COLLEGE, CLAREMONT, CA 91711, USA
E-mail address: `Stephan.Garcia@pomona.edu`
URL: `http://pages.pomona.edu/~sg064747`

DEPARTMENT OF MATHEMATICS AND COMPUTER SCIENCE, UNIVERSITY OF RICHMOND, RICHMOND, VA 23173, USA
E-mail address: `wross@richmond.edu`
URL: `http://facultystaff.richmond.edu/~wross`

Centre de Recherches Mathématiques
CRM Proceedings and Lecture Notes
Volume **51**, 2010

Approximation in Weighted Hardy Spaces for the Unit Disc

André Boivin and Changzhong Zhu

ABSTRACT. In this paper we study polynomial and rational approximation in the weighted Hardy spaces for the unit disc with the weight function satisfying Muckenhoupt's (A^q) condition.

1. Introduction

In [4], some basic properties of the weighted Hardy spaces for the unit disc D with the weight function satisfying Muckenhoupt's (A^q) condition were obtained, including series expansions of functions in these spaces with respect to the systems $\{(2\pi i(1-\bar{a}_k z))^{-1}\}$, with $a_k \in D$, $k = 1, 2\dots.$ In this paper, we continue our study of approximation properties in these spaces. In particular, we obtain some results on the rate of convergence of approximation by polynomial and rational functions. Let us first recall some definitions and known properties.

Assume that w is a nonnegative (with $0 < w < \infty$ a.e.), 2π periodic measurable function defined on $(-\infty, \infty)$. For $1 < q < \infty$, we say w satisfies Muckenhoupt's (A^q) condition or $w \in (A^q)$ (we also call w an (A^q) weight), if there is a constant C such that for every interval I with $|I| \le 2\pi$,

$$\left(\frac{1}{|I|}\int_I w(\theta)\,\mathrm{d}\theta\right)\left(\frac{1}{|I|}\int_I w(\theta)^{-1/(q-1)}\,\mathrm{d}\theta\right)^{q-1} \le C,$$

where $|I|$ denotes the length of I. We say $w \in (A^1)$ if

$$\frac{1}{|I|}\int_I w(\theta)\,\mathrm{d}\theta \le C\|w\|_I,$$

for every interval I with $|I| \le 2\pi$, where $\|w\|_I$ denotes the essential infimum of w over I.

(A^q) weights were introduced in [12]. In the general definition, w is not necessarily 2π periodic and I is not restricted by $|I| \le 2\pi$ (see, [8, Chapter VI; 12]). But in [4] and in the current paper, as in [12, Theorem 10] and [9, Theorem 1], w is additionally assumed to be 2π periodic since the weighted Hardy spaces we consider are over the unit disc and integration takes place over the unit circle. For

2000 *Mathematics Subject Classification.* 30D55.
Key words and phrases. weighted Hardy spaces, Muckenhoupt condition.
Research supported by NSERC (Canada).
This is the final form of the paper.

2π-periodic weights, as shown in [4], the imposition or not of the condition $|I| \leq 2\pi$ does not change the class of (A^q) weights (the value of the constant C appearing in the above definition may change).

Some well-known properties of (A^q) weights include: (i) if $w(\theta) \in (A^q)$ with $1 < q < \infty$, then $w(\theta)$, $[w(\theta)]^{-1/(q-1)}$ and $\log w(\theta)$ are integrable on $[-\pi, \pi]$; (ii) $w \in (A^q)$ if and only if $w^{1-q'} \in (A^{q'})$ where $1 < q < \infty$ and $1/q + 1/q' = 1$; (iii) if $w \in (A^q)$ and $q_0 > q$, then $w \in (A^{q_0})$, and (iv) if $w \in (A^q)$ with $1 < q < \infty$, then $w \in (A^{q_1})$ for some q_1 with $1 < q_1 < q$. Given $w \in (A^q)$ for some q with $1 < q < \infty$, we denote by q_w the *critical exponent* for w, that is, the infimum of all r's such that $w \in (A^r)$. We have $q_w \geq 1$, and $w \in (A^r)$ for all $r > q_w$.

Example 1.1. Let $1 < q < \infty$, $-1 < s < q - 1$, and

$$w(\theta) = |e^{i\theta} - e^{i\pi}|^s \qquad \text{(i.e., } |1+t|^s,\ |t| = 1). \tag{1.1}$$

By [16, p. 236], we have $w(\theta) \in (A^q)$.

For $w(\theta) \in (A^q)$, $1 \leq q < \infty$ and $0 < p < \infty$, the weighted Hardy space $H^p_w(D)$ for the unit disc $D = \{z \in \mathbb{C} : |z| < 1\}$ (see [7]) is the collection of functions $f(z)$ which are holomorphic in D and satisfy

$$\|f\|^p_{H^p_w(D)} := \sup_{r<1} \frac{1}{2\pi} \int_{-\pi}^{\pi} |f(re^{i\theta})|^p w(\theta)\, d\theta < \infty.$$

The classical Hardy space $H^p(D)$ is obtained by taking $w \equiv 1$. The space $L^p_w(T)$ is the collection of measurable functions $f(t)$ on $T = \{t \in \mathbb{C} : |t| = 1\}$ which satisfy

$$\|f\|^p_{L^p_w(T)} := \frac{1}{2\pi} \int_{-\pi}^{\pi} |f(e^{i\theta})|^p w(\theta)\, d\theta < +\infty.$$

For $1 \leq p < \infty, H^p_w(D)$ and $L^p_w(T)$ are Banach spaces. From now on in this paper, we assume that w is an (A^q) weight for some q with $1 < q < \infty$ and with critical exponent q_w (for simplicity of writing, in lemmas and theorems involving w, we will not repeat this assumption), and in most cases, we assume that $q_w < p < \infty$. Under these conditions, $H^p_w(D)$ and $L^p_w(T)$ are Banach spaces since $p > 1$, and by the properties of (A^q) weights mentioned above, we have $w \in (A^p)$.

By [4] and [12, Theorem 10], we have

Lemma 1.2. *Assume that $q_w < p < \infty$, then $H^p_w(D) \subset H^{p_0}(D)$ and $L^p_w(T) \subset L^{p_0}(T)$, for some p_0 with $1 < p_0 < p$, that is for any $f(z) \in H^p_w(D)$,*

$$\|f\|_{H^{p_0}(D)} \leq C'\|f\|_{H^p_w(D)}, \tag{1.2}$$

where C' is a positive constant independent of f.

Lemma 1.3. *Assume that $q_w < p < \infty$. If $f(z) \in H^p_w(D)$, then $f(z)$ has nontangential limits $f(t)$ a.e. on T ($f(t)$ is called the boundary function of $f(z)$), and $f(t)$ belongs to $L^p_w(T)$ and satisfies*

$$\lim_{r\to 1^-} \int_{-\pi}^{\pi} |f(re^{i\theta}) - f(e^{i\theta})|^p w(\theta)\, d\theta = 0, \tag{1.3}$$

$$\lim_{r\to 1^-} \int_{-\pi}^{\pi} |f(re^{i\theta})|^p w(\theta)\, d\theta = \int_{-\pi}^{\pi} |f(e^{i\theta})|^p w(\theta)\, d\theta, \tag{1.4}$$

and

$$\|f(t)\|_{L^p_w(T)} \leq \|f(z)\|_{H^p_w(D)} \leq C_p\|f(t)\|_{L^p_w(T)}, \tag{1.5}$$

where C_p is a constant depending only on p.

Lemma 1.4. *Assume that $q_w < p < \infty$. If $f(z) \in H^p_w(D)$, then*

$$\frac{1}{2\pi \mathrm{i}} \int_{|t|=1} \frac{f(t)}{t-z}\,\mathrm{d}t = \begin{cases} f(z), & z \in D; \\ 0, & z \in \mathbb{C} \setminus \overline{D}. \end{cases} \tag{1.6}$$

Lemma 1.5. *Assume that $q_w < p < \infty$. Let $1/p + 1/p' = 1$. Then, for every bounded linear functional $l \in \left(H^p_w(D)\right)^*$, there is a function $\Phi(z) \in H^{p'}_{w^{1-p'}}(D)$ such that*

$$l(f) = \frac{1}{2\pi} \int_{-\pi}^{\pi} f(\mathrm{e}^{\mathrm{i}\theta}) \overline{\Phi(\mathrm{e}^{\mathrm{i}\theta})}\,\mathrm{d}\theta$$

for $f(z) \in H^p_w(D)$.

2. Smirnov's theorem and examples

It is known (see [10, Chapter IV]) that if $f(z) \in H^p(D)$ and its boundary function $f(t)$ belongs to $L^{p_1}(T)$ for some $p_1 > p$, then $f(z) \in H^{p_1}$ (Smirnov's theorem). A similar theorem also holds for the case with (A^q) weight.

Theorem 2.1. *Assume that $w \in (A^q)$ for some q with $1 < q < \infty$. Moreover assume that $0 < p < \infty$ and that $p_1 > p$. If $f(z) \in H^p(D)$ and if its boundary function $f(t) \in L^{p_1}_w(T)$, then $f(z) \in H^{p_1}_w(D)$.*

Before giving the proof, let us recall that for $f(t) \in L^1(T)$, the Hardy–Littlewood maximal operator is defined by

$$Mf(\mathrm{e}^{\mathrm{i}\theta}) := \sup_{0<\phi<\pi} \frac{1}{2\phi} \int_{\theta-\phi}^{\theta+\phi} |f(\mathrm{e}^{\mathrm{i}s})|\,\mathrm{d}s.$$

By [2, p. 113], the Poisson integral

$$j(z) = j(r\mathrm{e}^{\mathrm{i}\theta}) := \frac{1}{2\pi} \int_{-\pi}^{\pi} P_r(\theta - t) f(\mathrm{e}^{\mathrm{i}t})\,\mathrm{d}t, \qquad |z| = r < 1$$

satisfies

$$|j(r\mathrm{e}^{\mathrm{i}\theta})| \le Mf(\mathrm{e}^{\mathrm{i}\theta}), \qquad 0 \le r < 1, \tag{2.1}$$

where $P_r(\phi)$ is the Poisson kernel:

$$P_r(\phi) = \frac{1 - r^2}{1 - 2r\cos\phi + r^2}.$$

If $w \in (A^q)$ for some q with $1 < q < \infty$, and $1 < p < \infty$, by [12], the operator Mf is bounded from $L^p_w(T)$ into itself, that is, there is a constant C_p such that for every $f(t) \in L^p_w(T)$,

$$\int_{-\pi}^{\pi} |Mf(\mathrm{e}^{\mathrm{i}\theta})|^p w(\theta)\,\mathrm{d}\theta \le C_p \int_{-\pi}^{\pi} |f(\mathrm{e}^{\mathrm{i}\theta})|^p w(\theta)\,\mathrm{d}\theta.$$

We are now ready to prove Theorem 2.1.

Proof. If $f(z) \in H^p(D)$, by [6, Theorem 2.7], and multiplication by p/q, we have

$$\log|f(r\mathrm{e}^{\mathrm{i}\theta})|^{p/q} \le \frac{1}{2\pi} \int_{-\pi}^{\pi} P_r(\theta - t) \log|f(\mathrm{e}^{\mathrm{i}t})|^{p/q}\,\mathrm{d}t.$$

Exponentiating and using the arithmetic-geometric mean inequality [6, p. 29], we have

$$\begin{aligned}|f(re^{i\theta})|^{p/q} &\leq \exp\left\{\frac{1}{2\pi}\int_{-\pi}^{\pi} P_r(\theta-t)\log|f(e^{it})|^{p/q}\,dt\right\}\\ &\leq \frac{1}{2\pi}\int_{-\pi}^{\pi} P_r(\theta-t)|f(e^{it})|^{p/q}\,dt.\end{aligned}$$

But, since $|f(e^{i\theta})| \in L_w^{p_1}(-\pi,\pi)$, we have $|f(e^{i\theta})|^{p/q} \in L_w^{qp_1/p}(T)$, and since $qp_1/p > q > 1$ and $w \in (A^q)$, by property (iii) of the (A^q) weight mentioned in the introduction, it follows that $w \in (A^{qp_1/p})$. Thus, by Lemma 1.2, $|f(e^{i\theta})|^{p/q} \in L^1(T)$. Hence, by (2.1), we have

$$\sup_{r<1}|f(re^{i\theta})|^{p/q} \leq \sup_{r<1}\frac{1}{2\pi}\int_{-\pi}^{\pi} P_r(\theta-t)|f(e^{it})|^{p/q}\,dt \leq M(|f(e^{i\theta})|^{p/q}),$$

where M is the Hardy–Littlewood maximal operator. Thus, for $r < 1$,

$$|f(re^{i\theta})| \leq [M(|f(e^{i\theta})|^{p/q})]^{q/p},$$

and

$$\begin{aligned}\int_{-\pi}^{\pi}|f(re^{i\theta})|^{p_1}w(\theta)\,d\theta &\leq \int_{-\pi}^{\pi}|[M(|f(e^{i\theta})|^{p/q})]^{q/p}|^{p_1}w(\theta)\,d\theta\\ &= \int_{-\pi}^{\pi}|M(|f(e^{i\theta})|^{p/q})|^{qp_1/p}w(\theta)\,d\theta\\ &\leq C\int_{-\pi}^{\pi}(|f(e^{i\theta})|^{p/q})^{qp_1/p}w(\theta)d\theta\\ &= C\int_{-\pi}^{\pi}|f(e^{i\theta})|^{p_1}w(\theta)\,d\theta < \infty,\end{aligned}$$

here in the last 2 steps, we use the facts that the operator M is bounded from $L_w^{qp_1/p}(T)$ into itself, and $f(e^{i\theta}) \in L_w^{p_1}(T)$. Hence $f(z) \in H_w^{p_1}(D)$. □

Note. Another proof can be obtained using [13, Theorem A4.4.5] and an important result (see [4, Lemma 2.3] or [7, p. 6]): If $f(z) \in H_w^p(D)$ then $f(z)W_p(z) \in H^p(D)$ where

$$W_p(z) = \exp\left(\frac{1}{2p\pi}\int_{-\pi}^{\pi}\frac{e^{it}+z}{e^{it}-z}\log w(t)\,dt\right), \qquad z \in D.$$

As an application of Theorem 2.1, let us give an example of a function belonging to $H_w^p(D)$:

Example 2.2. Let $w(\theta) = |e^{i\theta} - e^{i\pi}|^{-1/2}$. We know (see Example 1.1), $w(\theta) \in (A^q)$ for q with $1 < q < \infty$. Fix ϕ_0 with $0 < \phi_0 < \pi$, and define

$$f(z) = \left(z - e^{i(\pi-\phi_0)}\right)^{-1/2}, \qquad z \in D.$$

It is known that $f(z) \in H^p(D)$ for $0 < p < 2$ (see [6, p. 13, Exercise 1])[1], hence $f(z) \in H^1(D)$. Meanwhile, its boundary function $f(e^{i\theta}) \in L_w^{p_1}(T)$ for $1 < p_1 < 2$

[1]As noted in [6], $g(z) = (1-z)^{-1}$ is in $H^p(D)$ for every $p < 1$. From this it follows that $g_q(z) = (1-z)^{-1/q}$ is in $H^p(D)$ for every $p < q$. Thus, with a change of variable, we have $f(z) \in H^p(D)$ for $0 < p < 2$.

because

$$\int_{-\pi}^{\pi} \frac{1}{|e^{i\theta} - e^{i(\pi-\phi_0)})|^{p_1/2}} \cdot \frac{1}{|e^{i\theta} - e^{i\pi}|^{1/2}} \, d\theta < \infty, \qquad 1 < p_1 < 2$$

(here we note that the singularity of the first factor after the integral sign is at $\pi - \phi_0$ with power $p_1/2 < 1$, and the singularity of the second factor, i.e., $w(\theta)$, is at π with power $1/2$). Thus, by Theorem 2.1, $f(z) \in H_w^{p_1}(D)$ for $1 < p_1 < 2$.

Given a function $f(z)$ and an angle ϕ, we will use $f \circ R_\phi(z)$ to denote the function obtained from $f(z)$ by rotating z by ϕ, that is $f \circ R_\phi(z) = f(ze^{i\phi})$.

The following examples show that if $f(z) \in H_w^p(D)$ then $f \circ R_\phi(z)$ may or may not belong to $H_w^p(D)$.

Example 2.3. Let f be given as in Example 2.2, and consider the function

$$g(z) = f \circ R_{-\phi_0}(z) = (ze^{-i\phi_0} - e^{i(\pi-\phi_0)})^{-1/2}, \qquad z \in D,$$

The arguments presented in Example 2.2 show that $g(z) \in H^p(D)$ for $1 < p < 2$. But since

$$\int_{-\pi}^{\pi} \frac{1}{|e^{i(\theta-\phi_0)} - e^{i(\pi-\phi_0)}|^{p/2}} \cdot \frac{1}{|e^{i\theta} - e^{i\pi}|^{1/2}} \, d\theta = \infty, \qquad p > 1$$

(note that the singularity of the function after the integral sign is at π with power $p/2 + 1/2 > 1$ when $p > 1$), the boundary function $g(e^{i\theta}) \notin L_w^p(T)$ for $p > 1$, so by Lemma 1.3, it follows that $g(z) \notin H_w^p(D)$ for $1 < p < 2$.

If instead we consider the function

$$h(z) = f \circ R_{-\phi}(z) = (ze^{-i\phi} - e^{i(\pi-\phi_0)})^{-1/2}, \quad z \in D,$$

then when $0 < \phi < \phi_0$ (or $\phi > \phi_0$ but $\phi - \phi_0 < \pi$), as in Example 2.2, we have $h(z) \in H_w^p(D)$ for $1 < p < 2$.

Example 2.4. Let $w(\theta)$ be given as in Example 2.2, and $\{\phi_n\}$ $(n = 1, 2, \dots)$ be a given sequence with the properties that $0 < \phi_n < \pi$, $\phi_i \neq \phi_j$ for $i \neq j$, and $\phi_n \to 0$ as $n \to \infty$. We will construct a function $h_1(z)$ which is in $H_w^p(D)$ but $h_1 \circ R_{-\phi_n}(z) \notin H_w^p(D)$, $1 < p < 2$, for all $n = 1, 2, \dots$. Let

$$f_n(z) = \left(z - e^{i(\pi-\phi_n)}\right)^{-1/2} \qquad (n = 1, 2, \dots)\ z \in D.$$

By Example 2.2, $f_n(z) \in H_w^p(D)$ for $1 < p < 2$, $n = 1, 2, \dots$. And as shown in Example 2.3, the functions

$$f_n \circ R_{-\phi_n}(z) = \left(ze^{-i\phi_n} - e^{i(\pi-\phi_n)}\right)^{-1/2} \qquad (n = 1, 2, \dots)\ z \in D$$

are in $H^p(D)$ but not in $H_w^p(D)$ for $1 < p < 2$.

Define the function

$$h_1(z) = \sum_{k=1}^{\infty} \frac{f_k(z)}{2^k \|f_k\|_{H_w^p(D)}}, \qquad z \in D. \tag{2.2}$$

It is seen that $h_1(z) \in H_w^p(D)$ for $1 < p < 2$.

For any fixed n, we have

$$h_1 \circ R_{-\phi_n}(z) = \sum_{k=1}^{\infty} \frac{f_k \circ R_{-\phi_n}(z)}{2^k \|f_k\|_{H_w^p(D)}} = \frac{f_n \circ R_{-\phi_n}(z)}{2^n \|f_n\|_{H_w^p(D)}} + \sum_{k \neq n} \frac{f_k \circ R_{-\phi_n}(z)}{2^k \|f_k\|_{H_w^p(D)}}.$$

Thus, for $1 < p < 2$, $h_1 \circ R_{-\phi_n}(z) \notin H_w^p(D)$ since $f_n \circ R_{-\phi_n}(z) \notin H_w^p(D)$ and $f_k \circ R_{-\phi_n}(z) \in H_w^p(D)$ for all $k \neq n$ (see Example 2.3).

3. Approximation in $H^p_w(D)$

Recall that a system of functions is called complete in $H^p_w(D)$ if the closed linear span of elements of the system is the space $H^p_w(D)$; otherwise, it is called incomplete.

3.1. Approximation by polynomials.

As in the classical case, we have:

Lemma 3.1. *Assume that $q_w < p < \infty$, then the system of polynomials is complete in $H^p_w(D)$.*

PROOF. Suppose that $f(z) \in H^p_w(D)$. By (1.3), given $\varepsilon > 0$, for $r < 1$ sufficiently close to 1, we have

$$\|f(z) - f(rz)\|_{H^p_w(D)} < \varepsilon.$$

Since the Taylor series of $f(rz)$ converges uniformly for $|z| \le 1$, it also converges in the topology of $H^p_w(D)$. Choosing sufficiently many terms of the series, we get a polynomial $P(z)$ which satisfies $\|f(z) - P(z)\|_{H^p_w(D)} < 2\varepsilon$. □

Definition. Assume that $q_w < p < \infty$ and $f(z) \in H^p_w(D)$. For $\delta > 0$, let

$$\beta_{f,w}(\delta) = \sup_{|\phi_1 - \phi_2| < \delta} \|f \circ R_{\phi_1} - f \circ R_{\phi_2}\|_{L^p_w(T)}.$$

Now assume that there exists a $\delta_0 > 0$ with $\beta_{f,w}(\delta_0) < \infty$ and define

$$\omega_{h,w}(\delta) = \sup_{j=0,1,2,\dots} \sup_{|\phi| \le \delta} \left(\frac{1}{2\pi} \int_{-\pi}^{\pi} \left| h(\mathrm{e}^{\mathrm{i}[\theta+(j+1)\phi]}) - h(\mathrm{e}^{\mathrm{i}(\theta+j\phi)}) \right|^p w(\theta) d\theta \right)^{1/p}.$$

We call $\omega_{h,w}$ the *generalized modulus of continuity* of h. Note that $\delta_1 \le \delta_2$ implies $\omega_{h,w}(\delta_1) \le \omega_{h,w}(\delta_2)$.

Noting that $\omega_{f,w}(\delta) \le \beta_{f,w}(\delta)$, and when $\delta \le \delta_0$, $\beta_{f,w}(\delta) \le \beta_{f,w}(\delta_0) < \infty$, we have, for $\delta \le \delta_0$, $\omega_{f,w}(\delta) < \infty$.

It may be expected, like in the case $w = 1$, that $\omega_{f,w}(\delta) \to 0$ as $\delta \to 0$, but unfortunately, in general, this is not guaranteed. Indeed, as shown by the following example due to G. Sinnamon, there exists a function f such that the functions $f \circ R_\psi(z)$ are in $H^p_w(D)$ for all $\psi \in \mathbb{R}$, with norms in $H^p_w(D)$ uniformly bounded, and hence the same holds for the norms of their boundary functions in $L^p_w(T)$ by Lemma 1.3, but such that for any $\delta > 0$ there exist angles ψ_1 and ψ_2 with $|\psi_1 - \psi_2| < \delta$ and

$$\|f \circ R_{\psi_1} - f \circ R_{\psi_2}\|_{H^p_w(D)} \ge \frac{1}{2},$$

and hence, by Lemma 1.3, $\|f \circ R_{\psi_1} - f \circ R_{\psi_2}\|_{L^p_w(T)} \ge 1/(2C_p)$, where C_p is a positive constant depending only on p.

Example 3.2 (G. Sinnamon). Fix p with $1 < p < 2$ and define $w(\theta) = |\mathrm{e}^{\mathrm{i}\theta} - 1|^{-1/2}$. Then $w \in A_p$ as in Example 2.2. For each $s \ge 1$ set $f_s(z) = (z-s)^{-1/2}$, where the branch cut of the square root is $[0,\infty)$ so that $f_s(z)$ is analytic in D. Define

$$g(s,\theta) = \|f_s \circ R_\theta\|_{L^p_w(T)}.$$

It is not hard to verify that

(1) $g(1,\theta) < \infty$ for $\theta \ne 0$ (as in Example 2.2);
(2) $g(1,0) = \infty$ (as in Example 2.3);
(3) g is continuous on $[1,2] \times [-\pi,\pi]$ except at $(1,0)$;

(4) $g(s,0) \to \infty$ as $s \to 1^+$;
(5) $g(s,\theta) \leq g(1,\theta)$ for all s, θ;
(6) $g(s,\theta) \leq g(s,0)$ for all s, θ.

Note that Property 5 does not depend on the weight w, but just on the geometric observation that for $s \geq 1$ and $|z| = 1$, $|s - z| \geq |1 - z|$. Thus $|f_s(z)| \leq |f_1(z)|$ for each z with $|z| = 1$ and hence $g(s,\theta) \leq g(1,\theta)$.

To get Property 6, note that for any fixed $s > 1$, the maximum value of the function

$$(g(s,\theta))^p = \frac{1}{2\pi}\int_{-\pi}^{\pi} \left(1 + s^2 - 2s\cos(t+\theta)\right)^{-p/4} \left(2 - 2\cos(t)\right)^{-1/4} dt$$

occurs at $\theta = 0$.

Now let θ_n, ϕ_n and δ_n be three sequences in $(0,\pi)$ that decrease to zero and satisfy

$$\theta_{n+1} + \delta_{n+1} < \phi_n < \theta_n - \delta_n < \theta_n < \theta_n + \delta_n$$

for $n = 1, 2, \ldots$. Notice that the intervals $(\theta_n - \delta_n, \theta_n + \delta_n)$ are all disjoint and contain none of the points ϕ_n.

Let $M = 4\sum_{n=1}^{\infty} 1/n^2$ and choose a decreasing sequence s_n, with each $s_n > 1$, such that

$$g(s_n, 0) \geq Mn^2 \sup\{g(1,\theta) : \delta_n \leq |\theta| \leq \pi\}.$$

Define

$$f(z) = \sum_{n=1}^{\infty} \frac{f_{s_n} \circ R_{\theta_n}(z)}{g(s_n,0)} \qquad z \in D.$$

For each n, $\delta_n < \theta_n$ so $g(s_n,0) \geq Mn^2 g(1,\theta_n)$. Thus,

$$\sum_{n=1}^{\infty} \left\| \frac{f_{s_n} \circ R_{\theta_n}}{g(s_n,0)} \right\|_{H^p_w(D)} = \sum_{n=1}^{\infty} \frac{g(s_n,\theta_n)}{g(s_n,0)} \leq \sum_{n=1}^{\infty} \frac{g(1,\theta_n)}{Mn^2 g(1,\theta_n)} = \frac{1}{4}.$$

Since $H^p_w(D)$ is a Banach space this shows that $f(z) \in H^p_w(D)$.

Now let ψ be an arbitrary angle.

$$\sum_{n=1}^{\infty} \left\| \frac{f_{s_n} \circ R_{\theta_n} \circ R_\psi}{g(s_n,0)} \right\|_{H^p_w(D)} = \sum_{n=1}^{\infty} \frac{g(s_n,\theta_n + \psi)}{g(s_n,0)}.$$

The inequality $|\theta_n + \psi| < \delta_n$ means that $-\psi$ is in the interval $(\theta_n - \delta_n, \theta_n + \delta_n)$ and so it can hold for at most one n. For such an n, we use the estimate, $g(s_n, \theta_n + \psi) \leq g(s_n, 0)$ and for the other values of n we estimate as above to see that the sum is bounded by $1 + \sum_{n=1}^{\infty} 1/(Mn^2) = \frac{5}{4}$. This shows that for any ψ, $f \circ R_\psi(z) \in H^p_w(D$, and $\|f \circ R_\psi\|_{H^p_w(D)} \leq \frac{5}{4}$.

Now we show that for any $\delta > 0$ there exist angles ψ_1 and ψ_2 such that $|\psi_1 - \psi_2| < \delta$ and

$$\|f \circ R_{\psi_1} - f \circ R_{\psi_2}\|_{H^p_w(D)} \geq \frac{1}{2}.$$

Given $\delta > 0$ choose N so large that $|\theta_N - \phi_N| < \delta$. Set $\psi_1 = -\theta_N$ and $\psi_2 = -\phi_N$.

$$\|f \circ R_{\psi_1} - f \circ R_{\psi_2}\|_{H^p_w(D)} \geq \frac{\|f_{s_N} \circ R_{\theta_N} \circ R_{-\theta_N}\|_{H^p_w(D)}}{g(s_N,0)} - \sum_{n \neq N} \frac{\|f_{s_n} \circ R_{\theta_n} \circ R_{-\theta_N}\|_{H^p_w(D)}}{g(s_n,0)} - \sum_{n=1}^{\infty} \frac{\|f_{s_n} \circ R_{\theta_n} \circ R_{-\phi_N}\|_{H^p_w(D)}}{g(s_n,0)}.$$

The first term is equal to 1 and, arguing as above, each of the two sums is at most $\frac{1}{4}$ so the result is at least $\frac{1}{2}$ as required.

Example 3.3. Assume that $q_w < p < \infty$ and w is bounded (say, $w(\theta) < K$ where K is a positive constant), and $f(t) \in L^p(T)$. Clearly, $f(t) \in L^p_w(T)$ (a particular case is $f(t) \in L^p_w(T)$ with $w = 1$). In this case, we can prove that $\omega_{f,w}(\delta) \to 0$ as $\delta \to 0$: Given an $\varepsilon > 0$, there is a polynomial $p(z)$ such that

$$\|f - p\|_{L^p(T)} < \varepsilon/(3K),$$

hence,

$$\|f - p\|_{L^p_w(T)} < \varepsilon/3.$$

For any ϕ_1 and ϕ_2, we have

$$\|f \circ R_{\phi_1} - p \circ R_{\phi_1}\|_{L^p_w(T)} < K\|f \circ R_{\phi_1} - p \circ R_{\phi_1}\|_{L^p(T)} = K\|f - p\|_{L^p(T)} < \varepsilon/3,$$

and similarly,

$$\|f \circ R_{\phi_2} - p \circ R_{\phi_2}\|_{L^p_w(T)} < \varepsilon/3.$$

Meanwhile, by the uniform continuity of $p(z)$ on T, when $|\phi_1 - \phi_2|$ is sufficiently small, we have

$$\|p \circ R_{\phi_1} - p \circ R_{\phi_2}\|_{L^p_w(T)} < \varepsilon/3.$$

Thus, combining the above 3 inequalities, by Minkowski's inequality, when $|\phi_1 - \phi_2|$ sufficiently small, we have

$$\|f \circ R_{\phi_1} - f \circ R_{\phi_2}\|_{L^p_w(T)} < \varepsilon,$$

hence

$$\limsup_{|\phi_1 - \phi_2| \to 0} \|f \circ R_{\phi_1} - f \circ R_{\phi_2}\|_{L^p_w(T)} \leq \varepsilon,$$

and the required result follows as ε can be arbitrarily small.

This is a very strong condition on the weight w, but it guarantees that $\omega_{f,w}(\delta) \to 0$ as $\delta \to 0$ for *all* $f \in L^p(T)$. Simple conditions involving not only w but also f can easily be given to get $\omega_{f,w}(\delta) \to 0$ as $\delta \to 0$. It would be interesting to obtain a set of necessary and sufficient conditions (on w and f) for $\omega_{f,w}(\delta) \to 0$ as $\delta \to 0$ to hold, as it would have implications on our next two theorems. See below.

Lemma 3.4. *For any positive integer k,*

$$\omega_{h,w}(k\delta) \leq k \cdot \omega_{h,w}(\delta). \tag{3.1}$$

PROOF. By definition,

$$\begin{aligned}
&\omega_{h,w}(k\delta)\\
&= \sup_{j=0,1,2,\dots}\sup_{|\phi|\le\delta}\left(\frac{1}{2\pi}\int_{-\pi}^{\pi}\left|h(\mathrm{e}^{\mathrm{i}[\theta+(j+1)k\phi]})-h(\mathrm{e}^{\mathrm{i}[\theta+jk\phi]})\right|^p w(\theta)\,\mathrm{d}\theta\right)^{1/p}\\
&= \sup_{j=0,1,2,\dots}\sup_{|\phi|\le\delta}\left(\frac{1}{2\pi}\int_{-\pi}^{\pi}\left|h(\mathrm{e}^{\mathrm{i}[\theta+jk\phi+k\phi]})-h(\mathrm{e}^{\mathrm{i}[\theta+jk\phi]})\right|^p w(\theta\,\mathrm{d}\theta\right)^{1/p}\\
&= \sup_{j=0,1,2,\dots}\sup_{|\phi|\le\delta}\left(\frac{1}{2\pi}\int_{-\pi}^{\pi}\left|\sum_{l=0}^{k-1}\left[h(\mathrm{e}^{\mathrm{i}[\theta+jk\phi+(l+1)\phi]})-h(\mathrm{e}^{\mathrm{i}[\theta+jk\phi+l\phi]})\right]\right|^p w(\theta)\,\mathrm{d}\theta\right)^{1/p}
\end{aligned}$$

which, by Minkowski's inequality,

$$\begin{aligned}
&\le \sup_{j=0,1,2,\dots}\sup_{|\phi|\le\delta}\sum_{l=0}^{k-1}\left(\frac{1}{2\pi}\int_{-\pi}^{\pi}\left|h(\mathrm{e}^{\mathrm{i}[\theta+(jk+l+1)\phi]})-h(\mathrm{e}^{\mathrm{i}[\theta+(jk+l)\phi]})\right|^p w(\theta)\,\mathrm{d}\theta\right)^{1/p}\\
&\le \sum_{l=0}^{k-1}\sup_{j=0,1,2,\dots} sup_{|\phi|\le\delta}\left(\frac{1}{2\pi}\int_{-\pi}^{\pi}\left|h(\mathrm{e}^{\mathrm{i}[\theta+(jk+l+1)\phi]})-h(\mathrm{e}^{\mathrm{i}[\theta+(jk+l)\phi]})\right|^p w(\theta)\,\mathrm{d}\theta\right)^{1/p}\\
&\le k\cdot\omega_{h,w}(\delta).
\end{aligned}$$

□

Using a classical method similar to that in [14], we obtain

Theorem 3.5. *Assume that $q_w < p < \infty$, $h(z) \in H_w^p(D)$, and there exists a $\delta_0 > 0$ with $\beta_{h,w}(\delta_0) < \infty$. Then for any positive integer m, there is a polynomial $p_m(z)$ of order $\le m$ such that*

$$\|h(t)-p_m(t)\|_{L_w^p(T)} \le C(h)\cdot\omega_{h,w}\left(\frac{1}{m}\right), \tag{3.2}$$

where $C(h)$ is a constant depending on h but not m.

PROOF. Note that $h(t) \in L^1(T)$, as in [14, Chapter III], define

$$I_m(\theta) = \frac{2}{l_m^{1/p}}\int_{-\pi/2}^{\pi/2} h(\mathrm{e}^{\mathrm{i}(\theta+2t)})\left(\frac{\sin mt}{m\sin t}\right)^4 \mathrm{d}t,$$

where

$$l_m^{1/p} = 2\int_{-\pi/2}^{\pi/2}\left(\frac{\sin mt}{m\sin t}\right)^4 \mathrm{d}t.$$

Then

$$h(\mathrm{e}^{\mathrm{i}\theta}) - I_m(\theta) = \frac{2}{l_m^{1/p}}\int_{-\pi/2}^{\pi/2}[h(\mathrm{e}^{\mathrm{i}\theta})-h(\mathrm{e}^{\mathrm{i}(\theta+2t)})]\left(\frac{\sin mt}{m\sin t}\right)^4 \mathrm{d}t,$$

and

$$|h(\mathrm{e}^{\mathrm{i}\theta}) - I_m(\theta)| \le \frac{2}{l_m^{1/p}}\int_{-\pi/2}^{\pi/2}|h(\mathrm{e}^{\mathrm{i}(\theta+2t)})-h(\mathrm{e}^{\mathrm{i}\theta})|\left(\frac{\sin mt}{m\sin t}\right)^4 \mathrm{d}t.$$

Using Hölder's inequality, we have

$$|h(\mathrm{e}^{\mathrm{i}\theta}) - I_m(\theta)| \le \frac{2c_1}{l_m^{1/p}}\left(\int_{-\pi/2}^{\pi/2}|h(\mathrm{e}^{\mathrm{i}(\theta+2t)})-h(\mathrm{e}^{\mathrm{i}\theta})|^p\left(\frac{\sin mt}{m\sin t}\right)^{4p} \mathrm{d}t\right)^{1/p}$$

where $c_1 = \pi^{1/p'}$ with $1/p + 1/p' = 1$. Thus, using Fubini's theorem, we obtain

$$\begin{aligned}\int_{-\pi}^{\pi} &|h(\mathrm{e}^{\mathrm{i}\theta}) - I_m(\theta)|^p w(\theta)\,\mathrm{d}\theta \\ &\leq \frac{(2c_1)^p}{l_m}\int_{-\pi/2}^{\pi/2}\left[\int_{-\pi}^{\pi}|h(\mathrm{e}^{\mathrm{i}(\theta+2t)}) - h(\mathrm{e}^{\mathrm{i}\theta})|^p w(\theta)\,\mathrm{d}\theta\right]\left(\frac{\sin mt}{m\sin t}\right)^{4p}\mathrm{d}t \\ &\leq \frac{(2c_1)^p}{l_m}\int_{-\pi/2}^{\pi/2}[\omega_{h,w}(|2t|)]^p\left(\frac{\sin mt}{m\sin t}\right)^{4p}\mathrm{d}t.\end{aligned}$$

By (3.1), $\omega_{h,w}(k\delta) \leq k\omega_{h,w}(\delta)$, it follows that

$$\|h(\mathrm{e}^{\mathrm{i}\theta}) - I_m(\theta)\|_{L^p_w(-\pi,\pi)} \leq \frac{8c_1}{l_m^{1/p}}\left(\int_0^{\pi/2}[\omega_{h,w}(t)]^p\left(\frac{\sin mt}{m\sin t}\right)^{4p}\right)^{1/p}.$$

Noting that, for $t > 0$,

$$\begin{aligned}\omega_{h,w}(t) = \omega_{h,w}\left(\frac{mt}{m}\right) \leq \omega_{h,w}\left(\frac{[mt]+1}{m}\right) &\leq ([mt]+1)\omega_{h,w}\left(\frac{1}{m}\right) \\ &\leq (mt+1)\omega_{h,w}\left(\frac{1}{m}\right)\end{aligned}$$

(note that, here, and in the following, as usual, for a real number s, we use $[s]$ to denote the greatest integer not over s), we have

$$\begin{aligned}\|h(\mathrm{e}^{\mathrm{i}\theta}) - I_m(\theta)\|_{L^p_w(-\pi,\pi)} &\leq \frac{c_2}{l_m^{1/p}}\cdot\omega_{h,w}\left(\frac{1}{m}\right)\cdot\left(\int_0^{\pi/2}(mt+1)^p\left(\frac{\sin mt}{m\sin t}\right)^{4p}\mathrm{d}t\right)^{1/p} \\ &\leq \frac{2c_2}{l_m^{1/p}}\cdot\omega_{h,w}\left(\frac{1}{m}\right)\cdot\left(\int_0^{\pi/2}((mt)^p+1)\left(\frac{\sin mt}{m\sin t}\right)^{4p}\mathrm{d}t\right)^{1/p},\end{aligned}$$

where we used the inequality

$$(a+b)^p \leq 2^p(a^p+b^p), \qquad a \geq 0,\ b \geq 0,\ 1 < p < \infty.$$

But we have the estimates (see [14, pp. 84–85]):

$$l_m \geq \frac{c_3}{m}, \qquad \int_0^{\pi/2}\left(\frac{\sin mt}{m\sin t}\right)^{4p}\mathrm{d}t \leq \frac{c_4}{m},$$

and

$$\int_0^{\pi/2} t^p\left(\frac{\sin mt}{m\sin t}\right)^{4p}\mathrm{d}t \leq \frac{c_5}{m^{p+1}},$$

where c_i $(i = 1, \ldots, 5)$ are constants independent of m. Thus, we obtain

$$\|h(\mathrm{e}^{\mathrm{i}\theta}) - I_m(\theta)\|_{L^p_w(-\pi,\pi)} \leq C(h)\cdot\omega_{h,w}\left(\frac{1}{m}\right).$$

Noting that, by [14, Chapter III, p. 91], $I_m(\theta)$ can be re-written as a polynomial $p_m(z)$ of $z = \mathrm{e}^{\mathrm{i}\theta}$ with order $m-1$. The lemma is proved. □

Remark 3.6. If $\omega_{h,w}(\delta) \to 0$ as $\delta \to 0$, this is a Jackson-type theorem.

We will need the following fact in the next section:

Lemma 3.7. *Let $q_w < p < \infty$, and $p_m(z)$ be a polynomial of order m. Denote for any $\rho > 0$,*

$$\|p_m\|^p_{L^p_w(|t|=\rho)} = \frac{1}{2\pi}\int_{-\pi}^{\pi} |p_m(\rho e^{i\theta})|^p w(\theta)\,d\theta. \tag{3.3}$$

Then, for $r > 1$,

$$\|p_m\|_{L^p_w(|t|=r)} \le \frac{cr^{m+1}}{r-1}\|p_m\|_{L^p_w(T)}, \tag{3.4}$$

where c is a constant independent of m.

PROOF. Consider the function

$$f(z) := \frac{p_m(z)}{z^{m+1}}, \qquad z \neq 0. \tag{3.5}$$

Noting that $f(\infty) = 0$, by the Cauchy formula, for any $t \in \mathbb{C}$ with $|t| = r > 1$,

$$f(t) = -\frac{1}{2\pi i}\int_{|\xi|=1} \frac{f(\xi)}{\xi - t}\,d\xi.$$

Using Hölder's inequality with $1/p + 1/p' = 1$, we have

$$\begin{aligned}|f(t)| &\le \frac{1}{2\pi}\int_{|\xi|=1}\frac{|f(\xi)|}{|t-\xi|}|d\xi| \le \frac{1}{2\pi}\int_{|\xi|=1}\frac{|f(\xi)|}{|t|-|\xi|}|d\xi| \\ &\le \frac{1}{r-1}\left(\frac{1}{2\pi}\int_{-\pi}^{\pi}|f(e^{i\theta})|^p w(\theta)\right)^{1/p}\left(\frac{1}{2\pi}\int_{-\pi}^{\pi}[w(\theta)]^{1-p'}\,d\theta\right)^{1/p'} \\ &= \frac{c_1}{r-1}\|f\|_{L^p_w(T)},\end{aligned}$$

where $c_1 = \left((1/2\pi)\int_{-\pi}^{\pi}[w(\theta)]^{1-p'}\,d\theta\right)^{1/p'}$. Thus, by (3.5), for $|t| = r > 1$,

$$|p_m(t)| = r^{m+1}|f(t)| \le \frac{c_1 r^{m+1}}{r-1}\cdot\|f\|_{L^p_w(T)} = \frac{c_1 r^{m+1}}{r-1}\cdot\|p_m\|_{L^p_w(T)}$$

since $\|f\|_{L^p_w(T)} = \|p_m\|_{L^p_w(T)}$. Hence, by (3.3),

$$\|p_m\|_{L^p_w(|t|=r)} \le \frac{c_1c_2r^{m+1}}{r-1}\|p_m\|_{L^p_w(T)},$$

where $c_2 = \left((1/2\pi)\int_{-\pi}^{\pi} w(\theta)\,d\theta\right)^{1/p}$. Letting $c = c_1c_2$, (3.4) follows. □

3.2. Approximation by rational functions. For a sequence $\{a_k\} \subset D$, consider the system of rational functions

$$e_k(z) = \frac{1}{2\pi i}\cdot\frac{1}{1-\bar{a}_k z}, \qquad k = 1, 2, \ldots. \tag{3.6}$$

In [4] we studied the system $\{e_k(z)\}$ under the assumption that the sequence $\{a_k\}$ satisfies the Blaschke condition

$$\sum_{k=1}^{\infty}(1-|a_k|) < +\infty.$$

In particular, it was shown that, under this condition, the system $\{e_k(z)\}$ is incomplete in $H^p_w(D)$, for $q_w < p < \infty$. See [4, Lemma 3.3]. We then proceeded to study the subspace generated by the system $\{e_k(z)\}$.

In the current paper, we consider the case when the sequence $\{a_k\}$ does not satisfy the Blaschke condition.

Lemma 3.8. *Assume that $q_w < p < \infty$. If*

$$\sum_{k=1}^{\infty}(1-|a_k|) = +\infty, \tag{3.7}$$

then the above system $\{e_k(z)\}(k = 1, 2, \dots)$ is complete in $H_w^p(D)$.

PROOF. By the Hahn – Banach theorem, and Lemma 1.5, we need only to prove that if

$$\int_{-\pi}^{\pi} e_k(\mathrm{e}^{\mathrm{i}\theta})\overline{\Phi(\mathrm{e}^{\mathrm{i}\theta})}\,\mathrm{d}\theta = \int_{-\pi}^{\pi} \frac{1}{2\pi\mathrm{i}} \cdot \frac{1}{1-\bar{a}_k \mathrm{e}^{\mathrm{i}\theta}} \cdot \overline{\Phi(\mathrm{e}^{\mathrm{i}\theta})}\,\mathrm{d}\theta = 0, \qquad k = 1, 2, \dots, \tag{3.8}$$

where $\Phi(z) \in H^{p'}_{w^{1-p'}}(D)$ with $1/p + 1/p' = 1$, then $\Phi(\mathrm{e}^{\mathrm{i}\theta}) = 0$ a.e. in $(-\pi, \pi)$. But (3.8) is equivalent to

$$\frac{1}{2\pi\mathrm{i}} \int_{|t|=1} \frac{\Phi(t)}{t-a_k}\,\mathrm{d}t = 0,$$

that is, by Lemma 1.4, $\Phi(a_k) = 0$ $(k = 1, 2, \dots)$. We note that, by Lemma 1.2, $\Phi(z) \in H^s$ for some $1 < s < p'$. Thus, by (3.7) and Corollary of [6, Theorem 2.3], we have $\Phi(z) \equiv 0$. The lemma is proved. □

It follows that, under the assumptions of the above lemma, we can use linear combinations of the system $\{e_k(z)\}$ $(k = 1, 2, \dots)$ to approximate any function in $H_w^p(D)$. Assume that $\{a_k\}$ contains the point zero. Without loss of generality, by re-indexing, we assume that $a_0 = 0$ and all $a_k \neq 0$ $(k = 1, 2, \dots)$. Thus, for a fixed positive integer n, a linear combination of the system becomes

$$r_n(z) = c_0 + \sum_{k=1}^{n} \frac{c_k}{1-\bar{a}_k z}. \tag{3.9}$$

The poles of $r_n(z)$ are $b_k = 1/\bar{a}_k$ with $|b_k| > 1$ $(k = 1, 2, \dots, n)$. Denote by $\mathcal{R}_n$ the collection of all rational functions $r_n(z)$ of the above form, and, for $h(z) \in H_w^p(D)$, denote the *best approximation value* of h by r_n in $\mathcal{R}_n$ by

$$E_n(h) := \inf_{r_n \in \mathcal{R}_n} \|h - r_n\|_{H_w^p(D)}.$$

By Lemma 3.8, we have $E_n(h) \to 0$ as $n \to \infty$. An interesting question is how to estimate the speed of $E_n(h) \to 0$. Similar problems were studied for uniform approximation on T or $\overline{D}$ (see [1]), and for the approximation in H^p (see [15]).

Theorem 3.9. *Assume that* (i) *$q_w < p < \infty$, $h(z) \in H_w^p(D)$, and there exists a $\delta_0 > 0$ with $\beta_{h,w}(\delta_0) < \infty$;* (ii) *$\{a_k\}$ satisfies* (3.7) *and $|a_k| \le \rho$ $(k = 1, 2, \dots)$ with $0 < \rho < 1$. If n is a positive integer satisfying*

$$s_n := \frac{1}{3}\sum_{k=1}^{n}(1-|a_k|) > 2, \tag{3.10}$$

and

$$n\left(\frac{2}{\mathrm{e}}\right)^{(1-\rho)n/3} \le \omega_{h,w}(1) \tag{3.11}$$

(here, assume that $\omega_{h,w}(1) < \infty$), then there exists a rational function $r_n(z) \in \mathcal{R}_n$ such that

$$\|h - r_n\|_{H^p_w(D)} < C \cdot \omega_{h,w}\left(\frac{1}{n}\right), \tag{3.12}$$

where C is a positive constant depending on h but not n, and $\omega_{h,w}(\delta)$ is the generalized modulus of continuity of h. Hence, we have

$$E_n(h) < C \cdot \omega_{h,w}\left(\frac{1}{n}\right). \tag{3.13}$$

PROOF. Choose a positive integer $m = [s_n/2]$. By (3.10), $m \geq 1$, and by Theorem 3.5, there is a polynomial $p_m(z)$ of degree $\leq m$ such that

$$\|h(t) - p_m(t)\|_{L^p_w(T)} \leq C_1 \cdot \omega_{h,w}\left(\frac{1}{m}\right), \tag{3.14}$$

where C_1 is a constant depending on h but independent of m.

Noting that $\omega_{h,w}(1/m) \leq \omega_{h,w}(1)$ (since $1/m \leq 1$), by (3.14), we have

$$\|p_m\|_{L^p_w(T)} \leq \|h\|_{L^p_w(T)} + \|h - p_m\|_{L^p_w(T)} \leq \|h\|_{L^p_w(T)} + C_1 \cdot \omega_{h,w}(1) = C_2, \tag{3.15}$$

where C_2 is a constant depending on h but not m and n.

Assume that a rational function $r_n(z) \in \mathcal{R}_n$ interpolates $p_m(z)$ at a_k ($k = 1, 2, \ldots, n+1$). By [17, Chapter VIII, Theorem 2], the error $p_m(z) - r_n(z)$ has the following integral representation: for $|z| \leq 1$,

$$\begin{aligned} &p_m(z) - r_n(z) \\ &\quad = \frac{1}{2\pi \mathrm{i}} \int_{|t|=r} \frac{(z-a_1)\cdots(z-a_{n+1})(t-b_1)\cdots(t-b_n)}{(t-a_1)\cdots(t-a_{n+1})(z-b_1)\cdots(z-b_n)} \cdot \frac{p_m(t)\,\mathrm{d}t}{t-z}, \end{aligned} \tag{3.16}$$

where $r > 1$ and $r \neq |b_k|$ ($k = 1, 2, \ldots$).

Using (3.16), now we estimate the error $|p_m(z) - r_n(z)|$ in $|z| \leq 1$:

$$\begin{aligned} |p_m(z) - r_n(z)| &\leq \frac{1}{2\pi} \int_{|t|=r} \left(\left|\frac{z-a_{n+1}}{t-a_{n+1}}\right| \cdot \prod_{k=1}^n \left|\frac{z-a_k}{z-b_k}\right| \cdot \prod_{k=1}^n \left|\frac{t-b_k}{t-a_k}\right| \right) \frac{|p_m(t)|}{|t-z|} |\mathrm{d}t| \\ &= \frac{1}{2\pi} \int_{|t|=r} \left(\left|\frac{z-a_{n+1}}{t-a_{n+1}}\right| \left|\prod_{k=1}^n \frac{\bar{b}_k z - 1}{z-b_k}\right| \left|\prod_{k=1}^n \frac{t-b_k}{\bar{b}_k t - 1}\right| \right) \frac{|p_m(t)|}{|t-z|} |\mathrm{d}t|. \end{aligned}$$

Clearly, for $|z| \leq 1$,

$$\begin{aligned} |z - a_{n+1}| \left|\prod_{k=1}^n \frac{\bar{b}_k z - 1}{z - b_k}\right| &= |z - a_{n+1}| \left|\prod_{k=1}^n \frac{a_k - z}{1 - \bar{a}_k z}\right| \\ &= |z - a_{n+1}||B_n(z)| \leq (|z| + |a_{n+1}|)|B_n(z)| \leq 2. \end{aligned}$$

And, by [17, Chapter IX, Section 2, Lemma, p. 229], we have for $|t| = r > 1$,

$$\frac{1}{|t - a_{n+1}|} \left|\prod_{k=1}^n \frac{t - b_k}{\bar{b}_k t - 1}\right| \leq \frac{1}{r-1} \prod_{k=1}^n \frac{|b_k| + r}{1 + |b_k| r}.$$

So, for $|z| \le 1$, by Hölder's inequality,

$$(3.17)\qquad |p_m(z) - r_n(z)| \le \frac{2}{(r-1)^2} \cdot \prod_{k=1}^{n} \frac{|b_k| + r}{1 + |b_k| r} \cdot \frac{1}{2\pi} \int_{|t|=r} |p_m(t)|\,|\mathrm{d}t|$$
$$\le \frac{2c_1 r}{(r-1)^2} \cdot \prod_{k=1}^{n} \frac{|b_k| + r}{1 + |b_k| r} \cdot \|p_m\|_{L^p_w(|t|=r)},$$

where $c_1 = \left((1/2\pi) \int_{-\pi}^{\pi} [w(\theta)]^{1-p'}\,\mathrm{d}\theta\right)^{1/p'}$ with $1/p + 1/p' = 1$.

By Lemma 3.7, and taking $r = 2$ (note that if $|b_k| = 2$ for some k, we can choose $r = 2 + \varepsilon$ with a sufficiently small positive number ε), we have for $|z| \le 1$,

$$(3.18)\qquad |p_m(z) - r_n(z)| \le c_2 2^m \cdot \prod_{k=1}^{n} \frac{|b_k| + 2}{1 + 2|b_k|} \cdot \|p_m\|_{L^p_w(T)},$$

where c_2 is a constant independent of m and n. Therefore, by (3.18) and (3.15), we have

$$(3.19)\qquad \|p_m - r_n\|_{L^p_w(T)} \le C_3 \cdot 2^m \prod_{k=1}^{n} \frac{|b_k| + 2}{1 + 2|b_k|},$$

where C_3 is a constant depending on h but not m and n. And by (3.19) and (3.14), noting that $2m = 2[s_n/2] \le s_n$, we obtain:

$$(3.20)\qquad \|h - r_n\|_{L^p_w(T)} \le C_1 \cdot \omega_{h,w}\left(\frac{1}{[s_n/2]}\right) + C_3 \cdot 2^{s_n} \prod_{k=1}^{n} \frac{|b_k| + 2}{1 + 2|b_k|}.$$

Let us estimate the product in the right-hand side of the above inequality. Since

$$0 < \frac{|b_k| + 2}{1 + 2|b_k|} < 2,$$

we have[2]

$$\frac{|b_k| + 2}{1 + 2|b_k|} < \exp\left(\frac{|b_k| + 2}{1 + 2|b_k|} - 1\right) = \exp\left(-\frac{|b_k| - 1}{1 + 2|b_k|}\right).$$

Since $|b_k| > 1$, noting (3.10), we have

$$\prod_{k=1}^{n} \frac{|b_k| + 2}{1 + 2|b_k|} \le \exp\left(-\sum_{k=1}^{n} \frac{|b_k| - 1}{1 + 2|b_k|}\right) < \exp\left(-\sum_{k=1}^{n} \frac{|b_k| - 1}{3|b_k|}\right)$$
$$= \exp\left(-\frac{1}{3}\sum_{k=1}^{n}\left(1 - \frac{1}{|b_k|}\right)\right) = \exp\left(-\frac{1}{3}\sum_{k=1}^{n}(1 - |a_k|)\right) = \mathrm{e}^{-s_n}.$$

Thus, by (3.20), it follows that

$$(3.21)\qquad \|h - r_n\|_{L^p_w(T)} \le C_1 \omega_{h,w}\left(\frac{1}{[s_n/2]}\right) + C_3 \left(\frac{2}{\mathrm{e}}\right)^{s_n}.$$

Since $|a_k| \le \rho$ ($k = 1, 2, \dots$), we have

$$s_n = \frac{1}{3}\sum_{k=1}^{n}(1 - |a_k|) \ge \frac{1}{3}\sum_{k=1}^{n}(1 - \rho) = \frac{1}{3} n(1 - \rho).$$

[2]For $0 < x \le 2$, $\log x = (x-1) - \frac{1}{2}(x-1)^2 + \frac{1}{3}(x-1)^3 - \cdots < x - 1$, and hence $x = \mathrm{e}^{\log x} < \mathrm{e}^{x-1}$.

Hence

$$\left[\frac{s_n}{2}\right] \geq \left[\frac{1}{6}(1-\rho)\right] n.$$

Since $\omega_{h,w}(\delta)$ is nondecreasing as δ increasing, we have by the inequality $\omega_{h,w}(a\delta) \leq (a+1)\omega_{h,w}(\delta)$ for $a > 0$,

$$\omega_{h,w}\left(\frac{1}{[s_n/2]}\right) \leq C_4 \cdot \omega_{h,w}\left(\frac{1}{n}\right),$$

where

$$C_4 = \frac{1}{[\frac{1}{6}(1-\rho)]}.$$

Meanwhile, noting that $2/\mathrm{e} < 1$, we have

$$\left(\frac{2}{\mathrm{e}}\right)^{s_n} \leq \left(\frac{2}{\mathrm{e}}\right)^{(1-\rho)n/3}.$$

Let

$$\lambda = \left(\frac{2}{\mathrm{e}}\right)^{(1-\rho)/3},$$

then clearly we have $0 < \lambda < 1$. Thus, by (3.21), it follows that

$$\|h - r_n\|_{L^p_w(T)} \leq C_5\left(\omega_{h,w}\left(\frac{1}{n}\right) + \cdot\lambda^n\right), \tag{3.22}$$

where C_5 is a positive constant depending only on f and ρ. But

$$\omega_{h,w}(1) = \omega_{h,w}\left(n \cdot \frac{1}{n}\right) \leq n\omega_{h,w}\left(\frac{1}{n}\right).$$

Hence, by (3.11),

$$\omega_{h,w}\left(\frac{1}{n}\right) \geq \frac{\omega_{h,w}(1)}{n} \geq \left(\frac{2}{\mathrm{e}}\right)^{(1-\rho)n/3} = \lambda^n.$$

Thus, (3.22) implies

$$\|f - r_n\|_{L^p_w(T)} \leq C \cdot \omega_{h,w}\left(\frac{1}{n}\right), \tag{3.23}$$

and we have (3.12) since $h - r_n \in H^p_w(D)$ and, by Lemma 1.3, the two norms of $\|h - r_n\|_{H^p_w(D)}$ and $\|h - r_n\|_{L^p_w(T)}$ are equivalent. The proof is complete. □

Remark 3.10. If $\omega_{h,w}(\delta) \to 0$ as $\delta \to 0$, (3.13) gives an estimate of the speed of $E_n(h) \to 0$.

Acknowledgement. We are indebted to the referee for his/her careful reading and valuable suggestions and corrections which greatly improved this paper, and to G. Sinnamon for helpful discussions and providing us in particular with Example 3.2.

References

1. J.-E. Andersson and T. Ganelius, *The degree of approximation by rational functions with fixed poles*, Math. Z. **153** (1977), no. 2, 161–166.
2. S. Axler, P. Bourdon, and W. Ramey, *Harmonic function theory*, Grad. Texts in Math., vol. 137, Springer, New York, 1992.
3. E. F. Beckenbach and R. Bellman, *Inequalities*, Ergeb. Math. Grenzgeb., vol. 30, Springer, New York, 1983.
4. A. Boivin, P. M. Gauthier, and C. Zhu, *Weighted Hardy spaces for the unit disc: approximation properties*, Complex and Harmonic Analysis, DEStech Publ., Lancaster, PA, 2007, pp. 129–155.
5. J. A. Cima, A. L. Matheson, and W. T. Ross, *The Cauchy transform*, Math. Surveys Monogr., vol. 125, Amer. Math. Soc., Providence, RI, 2006.
6. P. L. Duren, *Theory of H^p spaces*, Pure Appl. Math., vol. 38, Academic Press, New York, 1970; Dover, Mineola, NY, 2000.
7. J. García-Cuerva, *Weighted H^p spaces*, Dissertationes Math. (Rozprawy Mat.) **162** (1979).
8. J. B. Garnett, *Bounded analytic functions*, Pure Appl. Math., vol. 96, Academic Press, New York, 1981.
9. R. Hunt, B. Muckenhoupt, and R. Wheeden, *Weighted norm inequalities for the conjugate function and Hilbert transform*, Trans. Amer. Math. Soc. **176** (1973), 227–251.
10. P. Koosis, *Introduction to H_p spaces*, 2nd ed., Cambridge Tracts in Math., vol. 115, Cambridge Univ. Press, Cambridge, 1998.
11. J. Lee and K. S. Rim, *Weighted norm inequalities for pluriharmonic conjugate functions*, J. Math. Anal. Appl. **268** (2002), no. 2, 707–717.
12. B. Muckenhoupt, *Weighted norm inequalities for the Hardy maximal function*, Trans. Amer. Math. Soc. **165** (1972), 207–226.
13. N. K. Nikolski, *Operators, functions, and systems: an easy reading.* Vol. 1: *Hardy, Hankel, and Toeplitz*, Math. Surveys Monogr., vol. 92, Amer. Math. Soc., Providence, RI, 2002; Vol. 2: *Model operators and systems*, vol. 93.
14. W. E. Sewell, *Degree of approximation by polynomials in the complex domain*, Ann. of Math. Stud., vol. 9, Princeton Univ. Press, Princeton, NJ, 1942.
15. X. C. Shen and Y. R. Lou, *Best approximation by rational functions in H_p spaces ($p \geq 1$)*, Beijing Daxue Xuebao **1** (1979), 58–72 (Chinese).
16. A. Torchinsky, *Real-variable methods in harmonic analysis*, Pure Appl. Math., vol. 123, Academic Press, Orlando, FL, 1986.
17. J. L. Walsh, *Interpolation and approximation by rational functions in the complex domain*, 4th ed., Amer. Math. Soc. Colloq. Publ., vol. 20, Amer. Math. Soc., Providence, R.I., 1965.

Department of Mathematics, University of Western Ontario, London, ON N6A 5B7, Canada

E-mail address, A. Boivin: `boivin@uwo.ca`

E-mail address, C. Zhu: `czhu28@uwo.ca`

Centre de Recherches Mathématiques
CRM Proceedings and Lecture Notes
Volume **51**, 2010

Some Remarks on the Toeplitz Corona Problem

Ronald G. Douglas and Jaydeb Sarkar

ABSTRACT. In a recent paper, Trent and Wick [23] establish a strong relation between the corona problem and the Toeplitz corona problem for a family of spaces over the ball and the polydisk. Their work is based on earlier work of Amar [3]. In this note, several of their lemmas are reinterpreted in the language of Hilbert modules, revealing some interesting facts and raising some questions about quasi-free Hilbert modules. Moreover, a modest generalization of their result is obtained.

1. Introduction

While isomorphic Banach algebras of continuous complex-valued functions with the supremum norm can be defined on distinct topological spaces, the results of Gelfand (cf. [12]) showed that for an algebra $A \subseteq C(X)$, there is a canonical choice of domain, the maximal space of the algebra. If the algebra A contains the function 1, then its maximal ideal space, M_A, is compact. Determining M_A for a concrete algebra is not always straightforward. New points can appear, even when the original space X is compact, as the disk algebra, defined on the unit circle T, demonstrates. If A separates the points of X, then one can identify X as a subset of M_A with a point x_0 in X corresponding to the maximal ideal of all functions in A vanishing at x_0. When X is not compact, new points must be present but there is still the question of whether the closure of X in M_A is all of M_A or does there exist a "corona" $M_A \setminus X \neq \varnothing$.

The celebrated theorem of Carleson states that the algebra $H^\infty(\mathbb{D})$ of bounded holomorphic functions on the unit disk $\mathbb{D}$ has no corona. There is a corona problem for $H^\infty(\Omega)$ for every domain Ω in $\mathbb{C}^m$ but a positive solution exists only for the case $m = 1$ with Ω a finitely connected domain in $\mathbb{C}$.

2000 *Mathematics Subject Classification.* 46E50, 46E22, 47B32, 50H05.

Key words and phrases. Corona problem, Toeplitz corona problem, Hilbert modules, Kernel Hilbert spaces.

This research was supported in part by a grant from the National Science Foundation.

This is the final version of the paper.

One can show with little difficulty that the absence of a corona for an algebra A means that for $\{\varphi_i\}_{i=1}^n$ in A, the statement that

$$(1) \qquad \sum_{i=1}^{n} |\varphi_i(x)|^2 \geq \varepsilon^2 > 0 \qquad \text{for all } x \text{ in } X$$

is equivalent to

(2) the existence of functions $\{\psi_i\}_{i=1}^n$ in A such that

$$\sum_{i=1}^{n} \varphi_i(x)\psi_i(x) = 1 \qquad \text{for } x \text{ in } X.$$

The original proof of Carleson [8] for $H^\infty(\mathbb{D})$ has been simplified over the years but the original ideas remain vital and important. One attempt at an alternate approach, pioneered by Arveson [6] and Schubert [20], and extended by Agler – McCarthy [2], Amar [3], and finally Trent – Wick [23] for the ball and polydisk, involves an analogous question about Toeplitz operators. In particular, for $\{\varphi_i\}_{i=1}^n$ in $H^\infty(\Omega)$ for $\Omega = \mathbb{B}^m$ or $\mathbb{D}^m$, one considers the Toeplitz operator $T_\Phi\colon\ H^2(\Omega)^n \to H^2(\Omega)$ defined $T_\Phi \boldsymbol{f} = \sum_{i=1}^n \varphi_i f_i$ for $\boldsymbol{f}$ in $H^2(\Omega)$, where $\boldsymbol{f} = f_1 \oplus \cdots \oplus f_n$ and $\mathcal{X}^n = \mathcal{X} \oplus \cdots \oplus \mathcal{X}$ for any space $\mathcal{X}$. One considers the relation between the operator inequality

$$(3) \qquad T_\Phi T_\Phi^* \geq \varepsilon^2 I \qquad \text{for some } \varepsilon > 0$$

and statement (1). One can readily show that (3) implies that one can solve (2) where the functions $\{\psi_i\}_{n=1}^n$ are in $H^2(\Omega)$. We will call the existence of such functions, statement (4). The original hope was that one would be able to modify the method or the functions obtained to achieve $\{\psi_i\}_{i=1}^n$ in $H^\infty(\Omega)$. That (1) implies (3) follows from earlier work of Andersson – Carlsson [5] for the unit ball and of Varopoulos [24], Li [17], Lin [18], Trent [22] and Treil – Wick [21] for the polydisk.

In the Trent – Wick paper [23] this goal was at least partially accomplished with the use of (3) to obtain a solution to (4) for the case $m = 1$ and for the case $m > 1$ if one assumes (3) for a family of weighted Hardy spaces. Their method was based on that of Amar [3].

In this note we provide a modest generalization of the result of Trent – Wick in which weighted Hardy spaces are replaced by cyclic submodules or cyclic invariant subspaces of the Hardy space and reinterpretations are given in the language of Hilbert modules for some of their other results. It is believed that this reformulation clarifies the situation and raises several interesting questions about the corona problem and Hilbert modules. Moreover, it shows various ways the Corona Theorem could be established for the ball and polydisk algebras. However, most of our effort is directed at analyzing the proof in [23] and identifying key hypotheses.

2. Hilbert modules

A *Hilbert module* over the algebra $A(\Omega)$, for Ω a bounded domain in $\mathbb{C}^m$, is a Hilbert space $\mathcal{H}$ which is a unital module over $A(\Omega)$ for which there exists $C \geq 1$ so that $\|\varphi \cdot f\|_{\mathcal{H}} \leq C\|\varphi\|_{A(\Omega)}\|f\|_{\mathcal{H}}$ for φ in $A(\Omega)$ and f in $\mathcal{H}$. Here $A(\Omega)$ is the closure in the supremum norm over Ω of all functions holomorphic in a neighborhood of the closure of Ω.

We consider Hilbert modules with more structure which better imitate the classical examples of the Hardy and Bergman spaces.

The Hilbert module $\mathcal{R}$ over $A(\Omega)$ is said to be *quasi-free of multiplicity one* if it has a canonical identification as a Hilbert space closure of $A(\Omega)$ such that:

(1) Evaluation at a point z in Ω has a continuous extension to $\mathcal{R}$ for which the norm is locally uniformly bounded.
(2) Multiplication by a φ in $A(\Omega)$ extends to a bounded operator T_φ in $\mathcal{L}(\mathcal{R})$.
(3) For a sequence $\{\varphi_k\}$ in $A(\Omega)$ which is Cauchy in $\mathcal{R}, \varphi_k(z) \to 0$ for all z in Ω if and only if $\|\varphi_k\|_{\mathcal{R}} \to 0$.

We normalize the norm on $\mathcal{R}$ so that $\|1\|_{\mathcal{R}} = 1$.

We are interested in establishing a connection between the corona problem for $\mathcal{M}(\mathcal{R})$ and the Toeplitz corona problem on $\mathcal{R}$. Here $\mathcal{M}(\mathcal{R})$ denotes the multiplier algebra for $\mathcal{R}$; that is, $\mathcal{M}$ $(= \mathcal{M}(\mathcal{R}))$ consists of the functions ψ on Ω for which $\psi\mathcal{R} \subset \mathcal{R}$. Since 1 is in $\mathcal{R}$, we see that $\mathcal{M}$ is a subspace of $\mathcal{R}$ and hence consists of holomorphic functions on Ω. Moreover, a standard argument shows that ψ is bounded (cf. [10]) and hence $\mathcal{M} \subset H^\infty(\Omega)$. In general, $\mathcal{M} \neq H^\infty(\Omega)$.

For ψ in $\mathcal{M}$ we let T_ψ denote the analytic Toeplitz operator in $\mathcal{L}(\mathcal{R})$ defined by module multiplication by ψ. Given functions $\{\varphi_i\}_{i=1}^n$ in $\mathcal{M}$, the set is said to

(1) satisfy the corona condition if $\sum_{i=1}^n |\varphi_i(z)|^2 \geq \varepsilon^2$ for some $\varepsilon > 0$ and all z in Ω;

(2) have a corona solution if there exist $\{\psi_i\}_{i=1}^n$ in $\mathcal{M}$ such that $\sum_{i=1}^n \varphi_i(z)\psi_i(z) = 1$ for z in Ω;

(3) satisfy the Toeplitz corona condition if $\sum_{i=1}^n T_{\varphi_i}T_{\varphi_i}^* \geq \varepsilon^2 I_{\mathcal{R}}$ for some $\varepsilon > 0$; and

(4) satisfy the $\mathcal{R}$-corona problem if there exist $\{f_i\}_{i=1}^n$ in $\mathcal{R}$ such that $\sum_{i=1}^n T_{\varphi_i} f_i = 1$ or $\sum_{i=1}^n \varphi_i(z) f(z_i) = 1$ for z in Ω with $\sum_{i=1}^n \|f_i\|^2 \leq 1/\varepsilon^2$.

3. Basic implications

It is easy to show that (2) $\implies$ (1), (4) $\implies$ (3) and (2) $\implies$ (4). As mentioned in the introduction, it has been shown that (1) $\implies$ (3) in case Ω is the unit ball $\mathbb{B}^m$ or the polydisk $\mathbb{D}^m$ and (1) $\implies$ (2) for $\Omega = \mathbb{D}$ is Carleson's Theorem. For a class of reproducing kernel Hilbert spaces with complete Nevanlinna–Pick kernels one knows that (2) and (3) are equivalent [7] (cf. [4, 15]). These results are closely related to generalizations of the commutant lifting theorem [19]. Finally, (3) $\implies$ (4) results from the range inclusion theorem of the first author as follows (cf. [11]).

Lemma 1. *If $\{\varphi_i\}_{i=1}^n$ in $\mathcal{M}$ satisfy $\sum_{i=1}^n T_{\varphi_i}T_{\varphi_i}^* \geq \varepsilon^2 I_{\mathcal{R}}$ for some $\varepsilon > 0$, then there exist $\{f_i\}_{i=1}^n$ in $\mathcal{R}$ such that $\sum_{i=1}^n \varphi_i(z) f_i(z) = 1$ for z in Ω and $\sum_{i=1}^n \|f_i\|_{\mathcal{R}}^2 \leq 1/\varepsilon^2$.*

PROOF. The assumption that $\sum_{i=1}^n T_{\varphi_i}T_{\varphi_i}^* \geq \varepsilon^2 I$ implies that the operator $X \colon \mathcal{R}^n \to \mathcal{R}$ defined by $X\boldsymbol{f} = \sum_{i=1}^n T_{\varphi_1} f_i$ satisfies $XX^* = \sum_{i=1}^n T_{\varphi_i}T_{\varphi_i}^* \geq \varepsilon^2 I_{\mathcal{R}}$ and hence by [11] there exists $Y \colon \mathcal{R} \to \mathcal{R}^n$ such that $XY = I_{\mathcal{R}}$ with $\|Y\| \leq \frac{1}{\varepsilon}$. Therefore, with $Y1 = f_1 \oplus \cdots \oplus f_n$, we have $\sum_{i=1}^n \varphi_i(z) f_i(z) = \sum_{i=1}^n T_{\varphi_i} f_i = XY1 = 1$ and $\sum_{i=1}^n \|f_i\|_{\mathcal{R}}^2 = \|Y1\|^2 \leq \|Y\|^2 \|1\|_{\mathcal{R}}^2 \leq 1/\varepsilon^2$. Thus the result is proved. □

To compare our results to those in [23], we need the following observations.

Lemma 2. *Let $\mathcal{R}$ be the Hilbert module $L^2_a(\mu)$ over $A(\Omega)$ defined to be the closure of $A(\Omega)$ in $L^2(\mu)$ for some probability measure μ on* clos Ω. *For f in $L^2_a(\mu)$, the Hilbert modules $L^2_a(|f|^2\,\mathrm{d}\mu)$ and $[f]$, the cyclic submodule of $\mathcal{R}$ generated by f, are isomorphic such that $1 \to f$.*

PROOF. Note that $\|\varphi \cdot 1\|_{L^2(|f|^2\,\mathrm{d}\mu)} = \|\varphi f\|_{L^2(\mu)}$ for φ in $A(\Omega)$ and the closure of this map sets up the desired isomorphism. □

Lemma 3. *If $\{f_i\}_{i=1}^n$ are functions in $L^2_a(\mu)$ and $g(z) = \sum_{i=1}^n |f_i(z)|^2$, then $L^2_a(g\,\mathrm{d}\mu)$ is isomorphic to the cyclic submodule $[f_1 \oplus \cdots \oplus f_n]$ of $L^2_a(\mu)^n$ with $1 \to f_1 \oplus \cdots \oplus f_n$.*

PROOF. The same proof as before works. □

In [23], Trent – Wick prove this result and use it to replace the L^2_a spaces used by Amar [3] by weighted Hardy spaces. However, before proceding we want to explore the meaning of this result from the Hilbert module point of view.

Lemma 4. *For $\mathcal{R} = H^2(\mathbb{B}^m)$ (or $H^2(\mathbb{D}^m)$) the cyclic submodule of $\mathcal{R}^n$ generated by $\varphi_1 \oplus \cdots \oplus \varphi_n$ with $\{\varphi_i\}_{i=1}^n$ in $A(\mathbb{B}^m)$ (or $A(\mathbb{D}^m)$) is isomorphic to a cyclic submodule of $H^2(\mathbb{B}^m)$ (or $H^2(\mathbb{D}^m)$).*

PROOF. Combining Lemma 3 in [23] with the observations made in Lemmas 2 and 3 above yields the result. □

There are several remarks and questions that arise at this point. First, does this result hold for arbitrary cyclic submodules in $H^2(\mathbb{B}^m)^n$ or $H^2(\mathbb{D}^m)^n$, which would require an extension of Lemma 3 in [23] to arbitrary $\boldsymbol{f}$ in $H^2(\mathbb{B}^m)^n$ or $H^2(\mathbb{D}^m)^n$? (This equivalence follows from the fact that a converse to Lemma 2 is valid.) It is easy to see that the lemma can be extended to an n-tuple of the form $f_1 h \oplus \cdots \oplus f_n h$, where the $\{f_i\}_{i=1}^n$ are in $A(\Omega)$ and h is in $\mathcal{R}$. Thus one need only assume that the quotients $\{f_i/f_j\}_{i,j=1}^n$ are in $A(\Omega)$ or even only equal a.e. to some continuous functions on $\partial\Omega$.

Second, the argument works for cyclic submodules in $H^2(\mathbb{B}^m) \otimes l^2$ or $H^2(\mathbb{D}^m) \otimes l^2$ as long as the generating vectors are in $A(\Omega)$ since Lemma 3 in [23] holds in this case also.

Note that since every cyclic submodule of $H^2(\mathbb{D}) \otimes l^2$ is isomorphic to $H^2(\mathbb{D})$, the classical Hardy space has the property that all cyclic submodules for the case of infinite multiplicity already occur, up to isomorphism, in the multiplicity one case. Although less trivial to verify, the same is true for the bundle shift Hardy spaces of multiplicity one over a finitely connected domain in $\mathbb{C}$ [1].

Third, one can ask if there are other Hilbert modules $\mathcal{R}$ that possess the property that every cyclic submodule of $\mathcal{R} \otimes \mathbb{C}^n$ or $\mathcal{R} \otimes l^2$ is isomorphic to a submodule of $\mathcal{R}$? The Bergman module $L^2_a(\mathbb{D})$ does not have this property since the cyclic submodule of $L^2_a(\mathbb{D}) \oplus L^2_a(\mathbb{D})$ generated by $1 \oplus z$ is not isomorphic to a submodule of $L^2_a(\mathbb{D})$. If it were, we could write the function $1 + |z|^2 = |f(z)|^2$ for some f in $L^2_a(\mathbb{D})$ which a simple calculation using a Fourier expansion in terms of $\{z^n \bar{z}^m\}$ shows is not possible.

We now abstract some other properties of the Hardy modules over the ball and polydisk.

We say that the Hilbert module $\mathcal{R}$ over $A(\Omega)$ has the *modulus approximation property* (*MAP*) if for vectors $\{f_i\}_{i=1}^N$ in $\mathcal{M} \subseteq \mathcal{R}$, there is a vector k in $\mathcal{R}$ such that

$\|\theta k\|_{\mathcal{R}}^2 = \sum_{j=1}^N \|\theta f_j\|^2$ for θ in $\mathcal{M}$. The map $\theta k \to \theta f_i \oplus \cdots \oplus \theta f_N$ thus extends to a module isomorphism of $[k] \subset \mathcal{R}$ and $[f_1 \oplus \cdots \oplus f_N] \subset \mathcal{R}^N$.

For z_0 in Ω, let I_{z_0} denote the maximal ideal in $A(\Omega)$ of all functions that vanish at z_0. The quasi-free Hilbert module $\mathcal{R}$ over $A(\Omega)$ of multiplicity one is said to satisfy the *weak modulus approximation property* (*WMAP*) if

(1) A nonzero vector k_{z_0} in $\mathcal{R} \ominus I_{z_0} \cdot \mathcal{R}$ can be written in the form $k_{z_0} \cdot 1$, where k_{z_0} is in $\mathcal{M}$, and $T_{k_{z_0}}$ has closed range acting on $\mathcal{R}$. In this case $\mathcal{R}$ is said to have a *good kernel function.*

(2) Property MAP holds for $f_i = \lambda_i k_{z_i}$, $i = 1, \ldots, N$ with $0 \leq \lambda_i \leq 1$ and $\sum_{i=1}^N \lambda_i^2 = 1$.

4. Main result

Our main result relating properties (2) and (3) is the following one which generalizes Theorem 1 of [23].

Theorem. *Let $\mathcal{R}$ be a WMAP quasi-free Hilbert module over $A(\Omega)$ of multiplicity one and $\{\varphi_i\}_{i=1}^n$ be functions in $\mathcal{M}$. Then the following are equivalent:*

(a) *There exist functions $\{\psi_i\}_{i=1}^n$ in $H^\infty(\Omega)$ such that $\sum_{i=1}^n \varphi_i(z)\psi_i(z) = 1$ and $\sum|\psi_i(z)| \leq 1/\varepsilon^2$ for some $\varepsilon > 0$ and all z in Ω, and*

(b) *there exists $\varepsilon > 0$ such that for every cyclic submodule $\mathcal{S}$ of $\mathcal{R}$, $\sum_{i=1}^n T_{\varphi_i}^{\mathcal{S}} T_{\varphi_i}^{\mathcal{S}^*} \geq \varepsilon^2 I_{\mathcal{S}}$, where $T_\varphi^{\mathcal{S}} = T_\varphi|_{\mathcal{S}}$ for φ in $\mathcal{M}$.*

PROOF. We follow the proof in [23] making a few changes. Fix a dense set $\{z_i\}_{i=2}^\infty$ of Ω.

First, we define for each positive integer N, the set $\mathcal{C}_N$ to be the convex hull of the functions $\{|k_{z_i}|^2/\|k_{z_i}\|^2\}_{i=2}^N$ and the function 1 for $i = 1$ with abuse of notation. Since $\mathcal{R}$ being WMAP implies that it has a good kernel function, $\mathcal{C}_N$ consists of nonnegative continuous functions on Ω. For a function g in the convex hull of the set $\{|k_{z_i}|^2/\|k_{z_i}\|^2\}_{i=1}^N$, the vector $\lambda_1 k_{z_1}/\|k_{z_1}\|^2 \oplus \cdots \oplus \lambda_N k_{z_N}/\|k_{z_N}\|^2$ is in $\mathcal{R}^N$. By definition there exists G in $\mathcal{R}$ such that $[G] \cong [\lambda_1 k_{z_1}/\|k_{z_1}\| \oplus \cdots \oplus \lambda_N k_{z_N}/\|k_{z_N}\|]$ by extending the map $\theta G \to \lambda_1 \theta k_{z_1}/\|k_{z_1}\| \oplus \cdots \oplus \lambda_N \theta k_{z_N}/\|k_{z_N}\|$ for θ in $\mathcal{M}$.

Second, let $\{\varphi_1, \ldots, \varphi_n\}$ be in $\mathcal{M}$ and let T_Φ denote the column operator defined from $\mathcal{R}^n$ to $\mathcal{R}$ by $T_\Phi(f_1 \oplus \cdots \oplus f_n) = \sum_{i=1}^n T_{\varphi_i} f_i$ for $\boldsymbol{f} = (f_1 \oplus \cdots \oplus f_n)$ in $\mathcal{R}^n$ and set $\mathcal{K} = \ker T_\Phi \subset \mathcal{R}^n$. Fix $\boldsymbol{f}$ in $\mathcal{R}^n$. Define the function

$$\mathcal{F}_N \colon \mathcal{C}_N \times \mathcal{K} \to [0, \infty)$$

by

$$\mathcal{F}_N(g, \boldsymbol{h}) = \sum_{i=1}^N \lambda_i^2 \left\| \frac{k_{z_i}}{\|k_{z_i}\|}(\boldsymbol{f} - \boldsymbol{h}) \right\|^2 \qquad \text{for } \boldsymbol{h} = h_1 \oplus \cdots \oplus h_n \text{ in } \mathcal{R}^n,$$

where $g = \sum_{i=1}^n \lambda_i^2 |k_{z_i}|^2/\|k_{z_i}\|^2$ and $\sum_{i=1}^n \lambda_i^2 = 1$. We are using the fact that the k_{z_i} are in $\mathcal{M}$ to realize $k_{z_i}(\boldsymbol{f} - \boldsymbol{h})$ in $\mathcal{R}^n$.

Except for the fact we are restricting the domain of $\mathcal{F}_N$ to $\mathcal{C}_N \times \mathcal{K}$ instead of $\mathcal{C}_N \times \mathcal{R}^n$, this definition agrees with that of [23]. Again, as in [23], this function is linear in g for fixed $\boldsymbol{h}$ and convex in $\boldsymbol{h}$ for fixed g. (Here one uses the triangular inequality and the fact that the square function is convex.)

Third, we want to identify $\mathcal{F}_N(g, \boldsymbol{h})$ in terms of the product of Toeplitz operators $(T_\Phi^{\mathcal{S}_g})(T_\Phi^{\mathcal{S}_g})^*$, where $\mathcal{S}_g$ is the cyclic submodule of $\mathcal{R}$ generated by a vector P in $\mathcal{R}$ as given in Lemma 3 such that the map $P \to (\lambda_1 k_{z_1}/\|k_{z_1}\| \oplus \cdots \oplus \lambda_N k_{z_N}/\|k_{z_N}\|$

extends to a module isomorphism with $g = \sum_{i=1}^{N} \lambda_i^2 |k_{z_i}|^2/\|k_{z_i}\|^2$, $0 \le \lambda_j^2 \le 1$, and $\sum_{i=1}^{N} \lambda_j^2 = 1$.

Note for $\boldsymbol{f}$ in $\mathcal{R}^n$, $\inf_{\boldsymbol{h}\in\mathcal{K}} \mathcal{F}_N(g,\boldsymbol{h}) \le 1/\varepsilon^2\|T_\Phi \boldsymbol{f}\|^2$ if $T_\Phi^{\mathcal{S}_g}(T_\Phi^{\mathcal{S}_g})^* \ge \varepsilon^2 I_{\mathcal{S}_g}$. Thus, if $T_\Phi^{\mathcal{S}}(T_\Phi^{\mathcal{S}})^* \ge \varepsilon^2 I_{\mathcal{S}}$ for every cyclic submodule of $\mathcal{R}$, we have $\inf_{\boldsymbol{h}\in\mathcal{K}} \mathcal{F}_N(g,\boldsymbol{h}) \le 1/\varepsilon^2\|T_\Phi \boldsymbol{f}\|^2$. Thus from the von Neumann min-max theorem we obtain $\inf_{\boldsymbol{h}\in\mathcal{K}} \sup_{g\in\mathcal{C}_N} \mathcal{F}_N(g,\boldsymbol{h}) = \sup_{g\in\mathcal{C}_N} \inf_{\boldsymbol{h}\in\mathcal{K}} \mathcal{F}_N(g,\boldsymbol{h}) \le 1/\varepsilon^2\|T_\Phi \boldsymbol{f}\|^2$.

From the inequality $T_\Phi T_\Phi^* \ge \varepsilon^2 I_{\mathcal{R}}$, we know that there exists $\boldsymbol{f}_0$ in $\mathcal{R}^n$ such that $\|\boldsymbol{f}_0\| \le 1/\varepsilon\|1\| = 1/\varepsilon$ and $T_\Phi \boldsymbol{f}_0 = 1$. Moreover, we can find $\boldsymbol{h}_N$ in $\mathcal{K}$ such that $\mathcal{F}_N(g,\boldsymbol{h}_N) \le (1/\varepsilon^2 + 1/N)\|T_\Phi \boldsymbol{f}_0\|^2 = 1/\varepsilon^2 + 1/N$ for all g in $\mathcal{C}_N$. In particular, for $g_i = |k_{z_i}|^2/\|k_{z_i}\|^2$, we have $T_\Phi^{\mathcal{S}_{g_i}}(T_\Phi^{\mathcal{S}_{g_i}})^* \ge \varepsilon^2 I_{\mathcal{S}_{g_i}}$, where $\|k_{z_i}/\|k_{z_i}\|(\boldsymbol{f}_0 - \boldsymbol{h}_N)\|^2 < 1/\varepsilon^2 + 1/N$.

There is one subtle point here in that 1 may not be in the range of $T_\Phi^{\mathcal{S}}$. However, if P is a vector generating the cyclic module $\mathcal{S}_g$, then P is in $\mathcal{M}$ and T_P has closed range. To see this recall that the map

$$\theta P \to \lambda_1 \frac{\theta k_{z_1}}{\|k_{z_1}\|} \oplus \cdots \oplus \lambda_N \frac{\theta k_{z_N}}{\|k_{z_N}\|}$$

for θ in $\mathcal{M}$ is an isometry. Since the functions $\{k_{z_i}/\|k_{z_i}\|\}_{i=1}^N$ are in $\mathcal{M}$ by assumption, it follows that the operator M_P is bounded on $\mathcal{M} \subseteq \mathcal{R}$ and has closed range on $\mathcal{R}$ since the operators $M_{k_{z_i}/\|k_{z_i}\|}$ have closed range, again by assumption. Therefore, find a vector $\boldsymbol{f}$ in $\mathcal{S}_g^n$ so that $T_\Phi \boldsymbol{f} = P$. But if $\boldsymbol{f} = f_1 \oplus \cdots \oplus f_n$, then f_i is in $[P]$ and hence has the form $f_i = P\tilde{f}_i$ for $\tilde{f}_i$ in $\mathcal{R}$. Therefore, $T_\Phi T_P \tilde{\boldsymbol{f}} = P$ or $T_\Phi \tilde{\boldsymbol{f}} = 1$ which is what is needed since in the proof $\boldsymbol{f}_0 - \tilde{\boldsymbol{f}}$ is in $\mathcal{K}$.

To continue the proof we need the following lemma.

Lemma 5. *If z_0 is a point in Ω and $\boldsymbol{h}$ is a vector in $\mathcal{R}^n$, then $\|\boldsymbol{h}(z_0)\|_{\mathbb{C}^n}^2 \le \|k_{z_0}/\|k_{z_0}\|\boldsymbol{h}\|^2$.*

PROOF. Suppose $\boldsymbol{h} = h_1 \oplus \cdots \oplus h_n$ with $\{h_i\}_{i=1}^n$ in $A(\Omega)$. Then $T_{h_i}^* k_{z_0} = \overline{h_i(z_0)} k_{z_0}$ and hence

$$\overline{h_i(z_0)}\|k_{z_0}\|^2 = \langle T_{h_i}^* k_{z_0}, k_{z_0}\rangle = \langle k_{z_0}, T_{h_i} k_{z_0}\rangle$$

since $T_{k_{z_0}} h_i = T_{h_i} k_{z_0}$. (We are using the fact the $k_{z_0} h_i = k_{z_0} h_i \cdot 1 = h_i k_{z_0} \cdot 1 = h_i k_{z_0}$.) Therefore,

$$|\overline{h_i(z_0)}|\|k_{z_0}\|^2 = |\langle k_{z_0}, T_{k_{z_0}} h_i\rangle| \le \|k_{z_0}\|^2 \|T_{\boldsymbol{k_{z_0}}/\|\boldsymbol{k_{z_0}}\|} h_i\|,$$

or,

$$|\overline{h_i(z_0)}| \le \|T_{k_{z_0}/\|k_{z_0}\|} h_i\|.$$

Finally,

$$\|\boldsymbol{h}(z_0)\|_{\mathbb{C}^n}^2 = \sum_{i=1}^{n} |h_i(z_0)|^2 \le \|T_{k_{z_0}/\|k_{z_0}\|}\boldsymbol{h}\|^2,$$

and since both terms of this inequality are continuous in the $\mathcal{R}$-norm, we can eliminate the assumption that $\boldsymbol{h}$ is in $A(\Omega)^n$. □

Returning to the proof of the theorem, we can apply the lemma to conclude that $\|(\boldsymbol{f}_0 - \boldsymbol{h}_0)(z)\|_{\mathbb{C}^n}^2 \le \|k_{z_i}/\|k_{z_i}\|(\boldsymbol{f}_0 - \boldsymbol{h}_0)\|^2 \le 1/\varepsilon^2 + 1/N$. Therefore, we see that the vector $\boldsymbol{f}_N = \boldsymbol{f}_0 - \boldsymbol{h}_N$ in $\mathcal{R}^n$ satisfies

(1) $T_\Phi(\boldsymbol{f}_N - \boldsymbol{h}_N) = 1$,

(2) $\|\boldsymbol{f}_N - \boldsymbol{h}_N\|^2_{\mathcal{R}} \leq 1/\varepsilon^2 + 1/N$ and
(3) $\|(\boldsymbol{f}_N - \boldsymbol{h}_N)(z_i)\|^2_{\mathbb{C}^n} \leq 1/\varepsilon^2 + 1/N$ for $i = 1, \ldots, N$.

Since the sequence $\{\boldsymbol{f}_N\}_{N=1}^{\infty}$ in $\mathcal{R}^n$ is uniformly bounded in norm, there exists a subsequence converging in the weak*-topology to a vector $\boldsymbol{\psi}$ in $\mathcal{R}^n$. Since weak*-convergence implies pointwise convergence, we see that $\sum_{j=1}^n \varphi_j \psi_j = 1$ and $\|\psi_j(z_i)\|^2_{\mathbb{C}^n} \leq 1/\varepsilon^2$ for all z_i. Since $\boldsymbol{\psi}$ is continuous on Ω and the set $\{z_i\}$ is dense in Ω, it follows that $\boldsymbol{\psi}$ is in $H^{\infty}_{\mathbb{C}^n}(\Omega)$ and $\|\boldsymbol{\psi}\| \leq 1/\varepsilon^2$ which concludes the proof. □

Note that we conclude that $\boldsymbol{\psi}$ is in $H^{\infty}(\Omega)$ and not in $\mathcal{M}$ which would be the hoped for result.

One can note that the argument involving the min-max theorem enables one to show that there are vectors $\boldsymbol{h}$ in $\mathcal{K}$ which satisfy

$$\|k_{z_i}(\boldsymbol{f} - \boldsymbol{h})\|^2 \leq \frac{1}{\varepsilon^2} + \frac{1}{N}.$$

Moreover, this shows that there are vectors $\tilde{\boldsymbol{f}}$ so that $T_\Phi \tilde{\boldsymbol{f}} = 1$, $\|\tilde{\boldsymbol{f}}\|^2 \leq 1/\varepsilon^2 + 1/N$, and $\|\tilde{\boldsymbol{f}}(z_i)\|^2 \leq 1/\varepsilon^2 + 1/N$ for $i = 1, \ldots, N$. An easy compactness argument completes the proof since the sets of vectors for each N are convex, compact and nested and hence have a point in common. □

5. Concluding comments

With the definitions given, the question arises of which Hilbert modules are (MAP) or which quasi-free ones are WMAP. Lemma 4 combined with observations in [23] show that both $H^2(\mathbb{B}^m)$ and $H^2(\mathbb{D}^m)$ are WMAP. Indeed any L^2_a space for a measure supported on $\partial\mathbb{B}^m$ or the distinguished boundary of $\mathbb{D}^m$ has these properties. One could also ask for which quasi-free Hilbert modules $\mathcal{R}$ the kernel functions $\{k_z\}_{z\in\Omega}$ are in $\mathcal{M}$ and whether the Toeplitz operators T_{k_z} are invertible operators as they are in the cases of $H^2(\mathbb{B}^m)$ and $H^2(\mathbb{D}^m)$. It seems possible that the kernel functions for all quasi-free Hilbert modules might have these properties when Ω is strongly pseudo-convex, with smooth boundary. In many concrete cases, the k_{z_0} are actually holomorphic on a neighborhood of the closure of Ω for z_0 in Ω, where the neighborhood, of course, depends on z_0.

Note that the formulation of the criteria in terms of a cyclic submodule $\mathcal{S}$ of the quasi-free Hilbert modules makes it obvious that the condition

$$T^{\mathcal{S}}_\Phi (T^{\mathcal{S}}_\Phi)^* \geq \varepsilon^2 I_{\mathcal{S}}$$

is equivalent to

$$T_\Phi T^*_\Phi \geq \varepsilon^2 I_{\mathcal{R}}$$

if the generating vector for $\mathcal{S}$ is a cyclic vector. This is Theorem 2 of [23]. Also it is easy to see that the assumption on the Toeplitz operators for all cyclic submodules is equivalent to assuming it for all submodules. That is because

$$\|(P_{\mathcal{S}} \otimes I_{\mathbb{C}^n}) T^*_\Phi f\| \geq \|(P_{[\boldsymbol{f}]} \otimes I_{\mathbb{C}^n}) T^*_\Phi f\|$$

for f in the submodule $\mathcal{S}$.

If for the ball or polydisk we knew that the function "representing" the modulus of a vector-valued function could be taken to be continuous on $\text{clos}(\Omega)$ or cyclic, the corona problem would be solved for those cases. No such result is known, however, and it seems likely that such a result is false.

Finally, one would also like to reach the conclusion that the function ψ is in the multiplier algebra even if it is smaller than $H^\infty(\Omega)$. In the recent paper [9] of Costea, Sawyer and Wick this goal is achieved for a family of spaces which includes the Drury – Arveson space. It seems possible that one might be able to modify the line of proof discussed here to involve derivatives of the $\{\varphi_i\}_{i=1}^n$ to accomplish this goal in this case, but that would clearly be more difficult.

References

1. M. B. Abrahamse and R. G. Douglas, *A class of subnormal operators related to multiply-connected domains*, Advances in Math. **19** (1976), no. 1, 106 – 148.
2. J. Agler and J. E. McCarthy, *Nevanlinna – Pick interpolation on the bidisk*, J. Reine Angew. Math. **506** (1999), 191 – 204.
3. E. Amar, *On the Toeplitz corona problem*, Publ. Mat. **47** (2003), no. 2, 489 – 496.
4. C.-G. Ambrozie and D. Timotin, *On an intertwining lifting theorem for certain reproducing kernel Hilbert spaces*, Integral Equations Operator Theory **42** (2002), no. 4, 373 – 384.
5. M. Andersson and H. Carlsson, *Estimates of solutions of the H^p and BMOA corona problem*, Math. Ann. **316** (2000), no. 1, 83 – 102.
6. W. Arveson, *Interpolation problems in nest algebras*, J. Functional Analysis **20** (1975), no. 3, 208 – 233.
7. J. A. Ball, T. T. Trent, and V. Vinnikov, *Interpolation and commutant lifting for multipliers on reproducing kernel Hilbert spaces*, Operator Theory and Analysis (Amsterdam, 1997), Oper. Theory Adv. Appl., vol. 122, Birkhäuser, Basel, 2001, pp. 89 – 138.
8. L. Carleson, *Interpolations by bounded analytic functions and the corona problem*, Ann. of Math. (2) **76** (1962), 547 – 559.
9. S. Costea, E. T. Sawyer, and B. D. Wick, *The corona theorem for the Drury – Arveson Hardy space and other holomorphic Besov – Sobolev spaces on the unit ball in $\mathbb{C}^n$*, available at `arXiv: 0811.0627`.
10. K. R. Davidson and R. G. Douglas, *The generalized Berezin transform and commutator ideals*, Pacific J. Math. **222** (2005), no. 1, 29 – 56.
11. R. G. Douglas, *On majorization, factorization, and range inclusion of operators on Hilbert space*, Proc. Amer. Math. Soc. **17** (1966), 413 – 415.
12. ———, *Banach algebra techniques in operator theory*, Vol. 49, Academic Press, New York, 1972. Pure Appl. Math.
13. R. G. Douglas and G. Misra, *On quasi-free Hilbert modules*, New York J. Math. **11** (2005), 547 – 561.
14. R. G. Douglas and V. I. Paulsen, *Hilbert modules over function algebras*, Pitman Res. Notes Math. Ser., vol. 217, Longman Sci. Tech., Harlow, 1989.
15. J. Eschmeier and M. Putinar, *Spherical contractions and interpolation problems on the unit ball*, J. Reine Angew. Math. **542** (2002), 219 – 236.
16. T. W. Gamelin, *Uniform algebras*, Prentice-Hall, Englewood Cliffs, NJ, 1969.
17. S.-Y. Li, *Corona problem of several complex variables*, The Madison Symposium on Complex Analysis (Madison, WI, 1991), Contemp. Math., vol. 137, Amer. Math. Soc., Providence, RI, 1992, pp. 307 – 328.
18. K.-C. Lin, *H^p-solutions for the corona problem on the polydisc in $\mathbf{C}^n$*, Bull. Sci. Math. (2) **110** (1986), no. 1, 69 – 84.
19. B. Sz.-Nagy and C. Foias, *Harmonic analysis of operators on Hilbert space*, North-Holland, Amsterdam, 1970.
20. C. F. Schubert, *The corona theorem as an operator theorem*, Proc. Amer. Math. Soc. **69** (1978), no. 1, 73 – 76.
21. S. Treil and B. D. Wick, *The matrix-valued H^p corona problem in the disk and polydisk*, J. Funct. Anal. **226** (2005), no. 1, 138 – 172.
22. T. T. Trent, *Solutions for the $H^\infty(D^n)$ corona problem belonging to* $\exp(L^{1/(2n-1)})$, Recent Advances in Matrix and Operator Theory, Oper. Theory Adv. Appl., vol. 179, Birkhäuser, Basel, 2008, pp. 309 – 328.
23. T. T. Trent and B. D. Wick, *Toeplitz corona theorems for the polydisk and the unit ball*, Complex Anal. Oper. Theory **3** (2009), no. 3, 729 – 738.

24. N. Th. Varopoulos, *Probabilistic approach to some problems in complex analysis*, Bull. Sci. Math. (2) **105** (1981), no. 2, 181–224.

DEPARTMENT OF MATHEMATICS, TEXAS A&M UNIVERSITY, COLLEGE STATION, TX 77843-3368, USA

E-mail address, R. G. Douglas: `rdouglas@math.tamu.edu`

E-mail address, J. Sarkar: `jsarkar@math.tamu.edu`

Centre de Recherches Mathématiques
CRM Proceedings and Lecture Notes
Volume **51**, 2010

Regularity on the Boundary in Spaces of Holomorphic Functions on the Unit Disk

Emmanuel Fricain and Andreas Hartmann

ABSTRACT. We review some results on regularity on the boundary in spaces of analytic functions on the unit disk connected with backward shift invariant subspaces in H^p.

1. Introduction

Fatou's theorem shows that every function of the Nevanlinna class $\mathcal{N} := \{f \in \operatorname{Hol}(\mathbb{D}) : \sup_{0<r<1} \int_{-\pi}^{\pi} \log_+|f(re^{it})| \, dt < \infty\}$ admits nontangential limits at almost every point ζ of the unit circle $\mathbb{T} = \partial\mathbb{D}$, $\mathbb{D} = \{z \in \mathbb{C} : |z| < 1\}$ being the unit disk. One can easily construct functions (even contained in smaller classes) which do not admit nontangential limits on a dense set of $\mathbb{T}$. The question that arises from such an observation is whether one can gain regularity of the functions at the boundary when restricting the problem to interesting subclasses of $\mathcal{N}$. We will discuss two kinds of subclasses corresponding to two different ways of generalizing the class of standard backward shift invariant subspaces in $H^2 := \{f \in \operatorname{Hol}(\mathbb{D}) : \|f\|_2^2 := \lim_{r\to 1}(1/2\pi) \int_{-\pi}^{\pi} |f(re^{it})|^2 \, dt < \infty\}$. Recall that backward shift invariant subspaces have shown to be of great interest in many domains in complex analysis and operator theory. In H^2, they are given by $K_I^2 := H^2 \ominus IH^2$, where I is an inner function, that is a bounded analytic function in $\mathbb{D}$ the boundary values of which are in modulus equal to 1 a.e. on $\mathbb{T}$. Another way of writing K_I^2 is

$$K_I^2 = H^2 \cap I\overline{H_0^2},$$

where $H_0^2 = zH^2$ is the subspace of functions in H^2 vanishing in 0. The bar sign means complex conjugation here. This second writing $K_I^2 = H^2 \cap I\overline{H_0^2}$ does not appeal to the Hilbert space structure and thus generalizes to H^p (which is defined as H^2 but replacing the integration power 2 by $p \in (0,\infty)$; it should be noted that for $p \in (0,1)$ the expression $\|f\|_p^p$ defines a metric; for $p = \infty$, H^∞ is the Banach space of bounded analytic functions on $\mathbb{D}$ with obvious norm). When $p = 2$, then

2000 *Mathematics Subject Classification.* 30B40, 30D55, 47A15, 47B37.
Key words and phrases. analytic continuation, backward shift, de Branges – Rovnyak spaces, Carleson embedding, connected level set condition, invariant subspace, Muckenhoupt condition, reproducing kernels, rigid functions, spectrum, Toeplitz operators.
This is the final form of the paper.

these spaces are also called model spaces because they arise in the construction of a universal model for Hilbert space contractions developped by Sz.-Nagy – Foias (see [71]). Note that if I is a Blaschke product associated with a sequence $(z_n)_{n\geq 1}$ of points in $\mathbb{D}$, then K_I^p coincides with the closed linear span of simple fractions with poles of corresponding multiplicities at the points $1/\overline{z_n}$.

Many questions concerning regularity on the boundary for functions in standard backward shift invariant subspaces were investigated in the extensive existing literature. In particular, it is natural to ask whether one can find points in the boundary where every function f in K_I^p and its derivatives up to a given order have nontangential limits; or even can one find some arc on the boundary where every function f in K_I^p can be continued analytically? Those questions were investigated by Ahern – Clark, Cohn, Moeller,.... Another interest in backward shift invariant subspaces concerns embedding questions, especially when K_I^p embeds into some $L^p(\mu)$. This question is related to the famous Carleson embedding theorem and was investigated for instance by Aleksandrov, Cohn, Treil, Volberg and many others (see below for some results).

In this survey, we will first review the important results in connection with regularity questions in standard backward shift invariant subspaces. Then we will discuss these matters in the two generalizations we are interested in: de Branges – Rovnyak spaces on the one hand, and weighted backward shift invariant subspaces — which occur naturally in the context of kernels of Toeplitz operators — on the other hand. Results surveyed here are mainly not followed by proofs. However, some of the material presented in Section 4 is new. In particular Theorem 18 for which we provide a proof and Example 4.1 that we will discuss in more detail. The reader will notice that for the de Branges – Rovnyak situation there now exists a quite complete picture analogous to that in the standard K_I^p spaces whereas the weighted situation has not been investigated very much yet. The example 4.1 should convince the reader that the weighted situation is more intricate in that the Ahern – Clark condition even under strong conditions on the weight — that ensure, e.g., analytic continuation off the spectrum of the inner function — is not sufficient.

2. Backward shift invariant subspaces

We will need some notation. Recall that the spectrum of an inner function I is defined as $\sigma(I) = \{\zeta \in \operatorname{clos}\mathbb{D} : \liminf_{z\to\zeta} I(z) = 0\}$. This set corresponds to the zeros in $\mathbb{D}$ and their accumulation points on $\mathbb{T} = \partial\mathbb{D}$, as well as the closed support of the singular measure μ_S of the singular factor of I.

The first important result goes back to Moeller [56] (see also [1] for a several variable version):

Theorem 1 (Moeller, 1962). *Let Γ be an open arc of $\mathbb{T}$. Then every function $f \in K_I^p$ can be continued analytically through Γ if and only if $\Gamma \cap \sigma(I) = \varnothing$.*

Moeller also establishes a link with the spectrum of the compression of the backward shift operator to K_I^p.

It is of course easy to construct inner functions the spectrum of which on $\mathbb{T}$ is equal to $\mathbb{T}$ so that there is no analytic continuation possible. Take for instance for I the Blaschke product associated with the sequence $\Lambda = \{(1 - 1/n^2)e^{in}\}_n$, the zeros of which accumulate at every point on $\mathbb{T}$. So it is natural to ask what happens in points which are in the spectrum, and what kind of regularity can be expected there.

Ahern – Clark and Cohn gave an answer to this question in [2, 28]. Recall that an arbitrary inner function I can be factored into a Blaschke product and a singular inner function: $I = BS$, where $B = \prod_n b_{a_n}$, $b_{a_n}(z) = (|a_n|/a_n)(a_n - z)/(1 - \overline{a_n}z)$, $\sum_n (1 - |a_n|^2) < \infty$, and

$$S(z) = \exp\left(- \int_{\mathbb{T}} \frac{\zeta + z}{\zeta - z} \, \mathrm{d}\mu_S(\zeta) \right),$$

where μ_S is a finite positive measure on $\mathbb{T}$ singular with respect to normalized Lebesgue measure m on $\mathbb{T}$. The regularity of functions in K_I^p is then related with the zero distribution of B and the measure μ_S as indicated in the following result.

Theorem 2 (Ahern – Clark, 1970, Cohn, 1986). *Let I be an inner function and let $1 < p < +\infty$ and q its conjugated exponent. If l is a nonnegative integer and $\zeta \in \mathbb{T}$, then the following are equivalent:*

(i) *for every f in K_I^p, the functions $f^{(j)}$, $0 \le j \le l$, have finite nontangential limits at ζ;*

(ii) *we have $S^I_{q(l+1)}(\zeta) < +\infty$, where*

$$(1) \qquad S_r^I(\zeta) := \sum_{n=1}^{\infty} \frac{(1 - |a_n|^2)}{|1 - \bar{\zeta} a_n|^r} + \int_0^{2\pi} \frac{1}{|1 - \bar{\zeta}\mathrm{e}^{\mathrm{i}t}|^r} \, \mathrm{d}\mu_S(\mathrm{e}^{\mathrm{i}t}) \qquad (1 \le r < \infty).$$

Moreover in that case, the function $(k_\zeta^I)^{l+1}$ belongs to K_I^q and we have

$$(2) \qquad f^{(l)}(\zeta) = l! \int_{\mathbb{T}} \bar{z}^l f(z) \overline{k_\zeta^I(z)}^{l+1} \, \mathrm{d}m(z),$$

for every function $f \in K_I^p$.

Here k_ζ^I is the reproducing kernel of the space K_I^2 corresponding to the point ζ and defined by

$$(3) \qquad k_\zeta^I(z) = \frac{1 - \overline{I(\zeta)} I(z)}{1 - \bar{\zeta} z}.$$

The quantity $S_r^I(\zeta)$ is closely related to the angular derivatives of the inner function I. Recall that a holomorphic selfmap f of the unit disk $\mathbb{D}$ is said to have an angular derivative at $\zeta \in \mathbb{T}$ if f has nontangential limit of modulus 1 in ζ and $f'(\zeta) := \lim_{r \to 1} f'(r\zeta)$ exists and is finite. Now, in the case where $f = I$ is an inner function, if $S_2^I(\zeta) < +\infty$, then I has an angular derivative at ζ and $S_2^I(\zeta) = |I'(\zeta)|$ (see [4, Theorem 2]). Moreover, if $S_{l+1}^I(\zeta) < +\infty$, then I and all its derivatives up to order l have finite radial limits at ζ (see [3, Lemma 4]).

Note that the case $p = 2$ of Theorem 2 is due to Ahern – Clark and Cohn generalizes the result to $p > 1$ (when $l = 0$). Another way to read into the results of Ahern – Clark, Cohn and Moeller is to introduce the representing measure of the inner function I, $\mu_I = \mu_S + \mu_B$, where

$$\mu_B := \sum_{n \ge 1} (1 - |a_n|^2) \delta_{\{a_n\}}.$$

Then Theorems 1 and 2 allow us to formulate the following general principle: if the measure μ_I is "small" near a point $\zeta \in \mathbb{T}$, then the functions f in K_I^p must be smooth near that point.

Another type of regularity questions in backward shift invariant subspaces was studied by A. Aleksandrov, K. Dyakonov and D. Khavinson. It consists in asking if

K_I^p contains a nontrivial smooth function. More precisely, Aleksandrov in [5] proved that the set of functions $f \in K_I^p$ continuous in the closed unit disc is dense in K_I^p. It should be noted nevertheless that the result of Aleksandrov is not constructive and indeed we do not know how to construct explicit examples of functions $f \in K_I^p$ continuous in the closed unit disc. In the same direction, Dyakonov and Khavinson, generalizing a result by Shapiro on the existence of $\mathcal{C}^l$-functions in K_I^p [68], proved in [42] that the space K_I^2 contains a nontrivial function of class $\mathcal{A}^\infty$ if and only if either I has a zero in $\mathbb{D}$ or there is a Carleson set $E \subset \mathbb{T}$ with $\mu_S(E) > 0$; here $\mathcal{A}^\infty$ denotes the space of analytic functions on $\mathbb{D}$ that extend continuously to the closed unit disc and that are $C^\infty(\mathbb{T})$; recall that a set E included in $\mathbb{T}$ is said to be a Carleson set if the following condition holds

$$\int_{\mathbb{T}} \log \operatorname{dist}(\zeta, E)\, dm(\zeta) > -\infty.$$

In [34,36,37,40], Dyakonov studied some norm inequalities in backward shift invariant subspaces of $H^p(\mathbb{C}_+)$; here $H^p(\mathbb{C}_+)$ is the Hardy space of the upper half-plane $\mathbb{C}_+ := \{z \in \mathbb{C} : \operatorname{Im} z > 0\}$ and if Θ is an inner function for the upper half-plane, then the corresponding backward shift invariant subspace of $H^p(\mathbb{C}_+)$ is also denoted by K_Θ^p and defined to be

$$K_\Theta^p = H^p(\mathbb{C}_+) \cap \Theta\overline{H^p(\mathbb{C}_+)}.$$

In the special case where $\Theta(z) = e^{iaz}$ $(a > 0)$, the space K_Θ^p is equal to $PW_a^p \cap H^p(\mathbb{C}_+)$, where PW_a^p is the Paley–Wiener space of entire functions of exponential type at most a that belong to L^p on the real axis. Dyakonov shows that several classical regularity inequalities pertaining to PW_a^p apply also to K_Θ^p provided Θ' is in $H^\infty(\mathbb{C}_+)$ (and only in that case). In particular, he proved the following result.

Theorem 3 (Dyakonov, 2000 and 2002). *Let $1 < p < +\infty$ and let Θ be an inner function in $H^\infty(\mathbb{C}_+)$. The following are equivalent:*

(i) $K_\Theta^p \subset C_0(\mathbb{R})$.
(ii) $K_\Theta^p \subset L^q(\mathbb{R})$, *for some (or all) $q \in (p, +\infty)$.*
(iii) *The differentiation operator is bounded as an operator from K_Θ^p to $L^p(\mathbb{R})$, that is*

$$\|f'\|_p \le C(p,\Theta)\|f\|_p, \qquad f \in K_\Theta^p. \tag{4}$$

(iv) $\Theta' \in H^\infty(\mathbb{C}_+)$.

Notice that in (4) one can take $C(p,\Theta) = C_1(p)\|\Theta'\|_\infty$, where $C_1(p)$ depends only on p but not on Θ. Moreover, Dyakonov also showed that the embeddings in (i), (ii) and the differentiation operator on K_Θ^p are compact if and only if Θ satisfies (iv) and $\Theta'(x) \to 0$ as $|x| \to +\infty$ on the real line. In [38], the author discusses when the differentiation operator is in Schatten – von Neumann ideals. Finally in [40], Dyakonov studied coupled with (4) the reverse inequality. More precisely, he characterized those Θ for which the differentiation operator $f \mapsto f'$ provides an isomorphism between K_Θ^p and a closed subspace of H^p, with $1 < p < +\infty$; namely he showed that such Θ's are precisely the Blaschke products whose zero-set lies in some horizontal strip $\{a < \operatorname{Im} z < b\}$, with $0 < a < b < +\infty$ and splits into finitely many separated sequences.

The inequality (4) corresponds for the case $\Theta(z) = e^{iaz}$ to a well-known inequality of S. Bernstein (see [17, Premier lemme, p. 75] for the case $p = +\infty$ and

[19, Theorem 11.3.3] for the general case). For $p=+\infty$, a beautiful generalization of Bernstein's inequality was obtained by Levin [55]: let $x\in\mathbb{R}$ and $|\Theta'(x)|<+\infty$; then for each $f\in K_\Theta^\infty$, the derivative $f'(x)$ exists in the sense of nontangential boundary values and

$$\left|\frac{f'(x)}{\Theta'(x)}\right|\le\|f\|_\infty,\qquad f\in K_\Theta^\infty.$$

Recently, differentiation in the backward shift invariant subspaces K_Θ^p was studied extensively by A. Baranov. In [11, 13], for a general inner function Θ in $H^\infty(\mathbb{C}_+)$, he proved estimates of the form

$$\|f^{(l)}\omega_{p,l}\|_{L^p(\mu)}\le C\|f\|_p,\qquad f\in K_\Theta^p, \tag{5}$$

where $l\ge 1$, μ is a Carleson measure in the closed upper half-plane and $\omega_{p,l}$ is some weight related to the norm of reproducing kernels of the space K_Θ^2 which compensates for possible growth of the derivative near the boundary. More precisely, put

$$\omega_{p,l}(z)=\|(k_z^\Theta)^{l+1}\|_q^{-p/(p+1)},\qquad \big(z\in\operatorname{clos}(\mathbb{C}_+)\big),$$

where q is the conjugate exponent of $p\in[1,+\infty)$. We assume that $\omega_{p,l}(x)=0$, whenever $S_{q(l+1)}^\Theta(x)=+\infty$, $x\in\mathbb{R}$ (here we omit the exact formula of k_z^Θ and S_r^Θ in the upper half-plane but it is not difficult to imagine what will be the analogue of (1) and (3) in that case).

Theorem 4 (Baranov, 2005). *Let μ be a Carleson measure in* $\operatorname{clos}(\mathbb{C}_+)$, $l\in\mathbb{N}$, $1\le p<+\infty$. *Then the operator*

$$(T_{p,l}f)(z)=f^{(l)}(z)\omega_{p,l}(z)$$

is of weak type (p,p) as an operator from K_Θ^p to $L^p(\mu)$ and is bounded as an operator from K_Θ^r to $L^r(\mu)$ for any $r>p$; moreover there is a constant $C=C(\mu,p,r,l)$ such that

$$\|f^{(l)}\omega_{p,l}\|_{L^r(\mu)}\le C\|f\|_r,\qquad f\in K_\Theta^r.$$

The proof of Baranov's result is based on the integral representation (2) which reduces the study of differentiation operators to the study of certain integral singular operators.

To apply Theorem 4, one should have effective estimates of the considered weights, that is, of the norms of reproducing kernels. Let

$$\Omega(\Theta,\varepsilon):=\{z\in\mathbb{C}_+:|\Theta(z)|<\varepsilon\}$$

be the level sets of the inner function Θ and let $d_\varepsilon(x)=\operatorname{dist}\big(x,\Omega(\Theta,\varepsilon)\big)$, $x\in\mathbb{R}$. Then Baranov showed in [12] the following estimates:

$$d_\varepsilon^l(x)\lesssim\omega_{p,l}(x)\lesssim|\Theta'(x)|^{-l},\qquad x\in\mathbb{R}. \tag{6}$$

Using a result of A. Aleksandrov [8], he also proved that for the special class of inner functions Θ satisfying the connected level set condition (see below for the definition in the framework of the unit disc) and such that $\infty\in\sigma(\Theta)$, we have

$$\omega_{p,l}(x)\asymp|\Theta'(x)|^{-l}\qquad(x\in\mathbb{R}). \tag{7}$$

In fact, the inequalities (6) and (7) are proved in [12, Corallary 1.5 and Lemma 4.5] for $l=1$; but the argument extends to general l in an obvious way. We should mention that Theorem 4 implies Theorem 3 on boundeness of differentiation operator. Indeed if $\Theta'\in L^\infty(\mathbb{R})$, then it is clear (and well known) that $\sup_{x\in\mathbb{R}}\|k_x^2\|_q<+\infty$, for any $q \in (1, \infty)$. Thus the weights $\omega_r = \omega_{r,1}$ are bounded from below and thus inequality

$$\|f'\omega_r\|_p \leq C\|f\|_p \qquad (f \in K^p_\Theta)$$

implies inequality (4).

Another type of results concerning regularity on the boundary for functions in standard backward shift invariant subspaces is related to Carleson's embedding theorem. Recall that Carleson proved (see [21, 22]) that $H^p = H^p(\mathbb{D})$ embeds continuously in $L^p(\mu)$ (where μ is a positive Borel measure on clos($\mathbb{D}$)) if and only if μ is a Carleson measure, that is there is a constant $C = C(\mu) > 0$ such that

$$\mu\big(S(\zeta, h)\big) \leq Ch,$$

for every "square" $S(\zeta, h) = \{z \in \operatorname{clos}\mathbb{D} : 1 - h/2\pi \leq |z| \leq 1, |\arg(z\bar{\zeta})| \leq h/2\}$, $\zeta \in \mathbb{T}$, $h \in (0, 2\pi)$. The motivation of Carleson comes from interpolation problems but his result acquired wide importance in a larger context of singular integrals of Calderon – Zygmund type. In [27], Cohn studied a similar question for model subspaces K^2_I. More precisely, he asked the following question: given an inner function I in $\mathbb{D}$ and $p \geq 1$, can we describe the class of positive Borel measure μ in the closed unit disc such that K^p_I is embedded into $L^p(\mu)$? In spite of a number of beautiful and deep (partial) results, this problem is still open. Of course, due to the closed graph theorem, the embedding $K^p_I \subset L^p(\mu)$ is equivalent to the estimate

$$\|f\|_{L^p(\mu)} \leq C\|f\|_p \qquad (f \in K^p_I). \tag{8}$$

Cohn solved this question for a special class of inner functions. We recall that I is said to satisfy the connected level set condition (and we write $I \in CLS$) if the level set $\Omega(I, \varepsilon)$ is connected for some $\varepsilon \in (0, 1)$.

Theorem 5 (Cohn, 1982). *Let μ be a positive Borel measure on* clos $\mathbb{D}$. *Let I be a an inner function such that $I \in CLS$. The following are equivalent:*

(i) *K^2_I embedds continuously in $L^2(\mu)$.*
(ii) *There is $c > 0$ such that*

$$\int_{\operatorname{clos}\mathbb{D}} \frac{1 - |z|^2}{|1 - \bar{z}\zeta|^2}\, d\mu(\zeta) \leq \frac{C}{1 - |I(z)|^2}, \qquad z \in \mathbb{D}. \tag{9}$$

It is easy to see that if we have inequality (8) for $f = k^I_z$ $z \in \mathbb{D}$, then we have inequality (9). Thus Cohn's theorem can be reformulated in the following way: inequality (8) holds for every function $f \in K^2_I$ if and only if it holds for reproducing kernels $f = k^I_z$, $z \in \mathbb{D}$. Recently, F. Nazarov and A. Volberg [58] showed that this is no longer true in the general case. We should compare this property of the embedding operator $K^2_I \subset L^2(\mu)$ (for CLS inner functions) to the "reproducing kernel thesis," which is shared by Toeplitz or Hankel operators in H^2 for instance. The reproducing kernel thesis says roughly that in order to show the boundeness of an operator on a reproducing kernel Hilbert space, it is sufficient to test its boundeness only on reproducing kernels (see, e.g.. [59, Vol.1, pp. 131, 204, 244, 246] for some discussions of this remarkable property).

A geometric condition on μ sufficient for the embedding of K^p_I is due to Volberg – Treil [73].

Theorem 6 (Volberg – Treil, 1986). *Let μ be a positive Borel measure on* clos $\mathbb{D}$, *let I be a an inner function and let $1 \leq p < +\infty$. Assume that there is*

$C > 0$ such that

$$\mu(S(\zeta, h)) \leq Ch, \tag{10}$$

for every square $S(\zeta, h)$ satisfying $S(\zeta, h) \cap \Omega(I, \varepsilon) \neq \varnothing$. Then K_I^p embeds continuously in $L^p(\mu)$.

Moreover they showed that for the case where I satisfies the connected level set condition, the sufficient condition (10) is also necessary, and they extend Theorem 5 to the Banach setting. In [8], Aleksandrov proved that the condition of Volberg–Treil is necessary if and only if $I \in CLS$. Moreover, if I does not satisfy the connected level set condition, then the class of measures μ such that the inequality (8) is valid depend essentially on the exponent p (in contrast to the classical theorem of Carleson).

Of special interest is the case when $\mu = \sum_{n\in\mathbb{N}} a_n \delta_{\{\lambda_n\}}$ is a discrete measure; then embedding is equivalent to the Bessel property for the system of reproducing kernels $\{k_{\lambda_n}^I\}$. In fact, Carleson's initial motivation to consider embedding properties comes from interpolation problems. These are closely related with the Riesz basis property which itself is linked with the Bessel property. The Riesz basis property of reproducing kernels $\{k_{\lambda_n}^I\}$ has been studied by S. V. Hruščëv, N. K. Nikol′skiĭ and B. S. Pavlov in the famous paper [51], see also the recent papers by A. Baranov [13, 14] and by the first author [23, 43]. It is of great importance in applications such as for instance control theory (see [59, Vol. 2]).

Also the particular case when μ is a measure on the unit circle is of great interest. In contrast to the embeddings of the whole Hardy space H^p (note that Carleson measures on $\mathbb{T}$ are measures with bounded density with respect to Lebesgue measure m), the class of Borel measures μ such that $K_I^p \subset L^p(\mu)$ always contains nontrivial examples of singular measures on $\mathbb{T}$; in particular, for $p = 2$, the Clark measures [26] for which the embeddings $K_I^2 \subset L^2(\mu)$ are isometric. Recall that given $\lambda \in \mathbb{T}$, the Clark measure σ_λ associated with a function b in the ball of H^∞ is defined as the unique positive Borel measure on $\mathbb{T}$ whose Poisson integral is the real part of $(\lambda + b)/(\lambda - b)$. When b is inner, the Clark measures σ_α are singular with respect to the Lebesgue measure on $\mathbb{T}$. The situation concerning embeddings for Clark measures changes for $p \neq 2$ as shown by Aleksandrov [6]: while for $p \geq 2$ this embedding still holds (see [6, Corollary 2, p. 117]), he constructed an example for which the embedding fails when $p < 2$ (see [6, Example, p. 123]). See also the nice survey by Poltoratski and Sarason on Clark measures [60] (which they call Aleksandrov-Clark measures). On the other hand, if $\mu = wm$, $w \in L^2(\mathbb{T})$, then the embedding problem is related to the properties of the Toeplitz operator T_w (see [29]).

In [11, 12], Baranov developped a new approach based on the (weighted norm) Bernstein inequalities and he got some extensions of Cohn and Volberg–Treil results. Compactness of the embedding operator $K_I^p \subset L^p(\mu)$ is also of interest and is considered in [12, 15, 24, 29, 72].

Another important result in connection with K_I^p-spaces is that of Douglas, Shapiro and Shields ([32], see also [25, Theorem 1.0.5; 62]) and concerns pseudocontinuation. Recall that a function holomorphic in $\mathbb{D}_e := \hat{\mathbb{C}} \setminus \operatorname{clos} \mathbb{D}$ — $\operatorname{clos} E$ means the closure of a set E — is a pseudocontinuation of a function f meromorphic in $\mathbb{D}$ if ψ vanishes at ∞ and the outer nontangential limits of ψ on $\mathbb{T}$ coincide with the inner nontangential limits of f on $\mathbb{T}$ in almost every point of $\mathbb{T}$. Note that

$f \in K_I^2 = H^2 \cap I\overline{H_0^2}$ implies that $f = I\overline{\psi}$ with $\psi \in H_0^2$. Then the meromorphic function f/I equals $\overline{\psi}$ a.e. $\mathbb{T}$, and writing $\psi(z) = \sum_{n\geq 1} b_n z^n$, it is clear that $\widetilde{\psi}(z) := \sum_{n\geq 1} \overline{b_n}/z^n$ is a holomorphic function in $\mathbb{D}_e$, vanishing at ∞, and being equal to f/I almost everywhere on $\mathbb{T}$ (in fact, $\widetilde{\psi} \in H^2(\mathbb{D}_e)$). The converse is also true: if f/I has a pseudocontinuation in $\mathbb{D}_e$, where f is a H^p-function and I some inner function I, then f is in K_I^p. This can be resumed this in the following result.

Theorem 7 (Douglas – Shapiro – Shields, 1972). *Let I be an inner function. Then a function $f \in H^p$ is in K_I^p if and only if f/I has a pseudocontinuation to a function in $H^p(\mathbb{D}_e)$ which vanishes at infinity.*

Note that there are functions analytic on $\mathbb{C}$ that do not admit a pseudocontinuation. An example of such a function is $f(z) = e^z$ which has an essential singularity at infinity.

As already mentioned, we will be concerned with two generalizations of the backward shift invariant subspaces. One direction is to consider weighted versions of such spaces. The other direction is to replace the inner function by more general functions. The appropriate definition of K_I^2 in this setting is that of de Branges – Rovnyak spaces (requiring that $p = 2$).

Our aim is to discuss some of the above results in the context of these spaces. For analytic continuation it turns out that the conditions in both cases are quite similar to the original K_I^2-situation. However in the weighted situation some additional condition is needed. For boundary behaviour in points in the spectrum the situation changes. In the de Branges – Rovnyak spaces the Ahern – Clark condition generalizes naturally, whereas in weighted backward shift invariant subspaces the situation is not clear and awaits further investigation. This will be illustrated in Example 4.1.

3. de Branges – Rovnyak spaces

Let us begin with defining de Branges – Rovnyak spaces. We will be essentially concerned with the special case of Toeplitz operators. Recall that for $\varphi \in L^\infty(\mathbb{T})$, the Toeplitz operator T_φ is defined on H^2 by

$$T_\varphi(f) := P_+(\varphi f) \qquad (f \in H^2),$$

where P_+ denotes the orthogonal projection of $L^2(\mathbb{T})$ onto H^2. Then, for $\varphi \in L^\infty(\mathbb{T})$, $\|\varphi\|_\infty \leq 1$, the de Branges – Rovnyak space $\mathcal{H}(\varphi)$, associated with φ, consists of those H^2 functions which are in the range of the operator $(\mathrm{Id} - T_\varphi T_{\overline{\varphi}})^{1/2}$. It is a Hilbert space when equipped with the inner product

$$\langle (\mathrm{Id} - T_\varphi T_{\overline{\varphi}})^{1/2} f, (\mathrm{Id} - T_\varphi T_{\overline{\varphi}})^{1/2} g \rangle_\varphi = \langle f, g \rangle_2,$$

where $f, g \in H^2 \ominus \ker(\mathrm{Id} - T_\varphi T_{\overline{\varphi}})^{1/2}$.

These spaces (and more precisely their general vector-valued version) appeared first in L. de Branges and J. Rovnyak [30, 31] as universal model spaces for Hilbert space contractions. As a special case, when $b = I$ is an inner function (that is $|b| = |I| = 1$ a.e. on $\mathbb{T}$), the operator $(\mathrm{Id} - T_I T_{\overline{I}})$ is an orthogonal projection and $\mathcal{H}(I)$ becomes a closed (ordinary) subspace of H^2 which coincides with the model spaces $K_I = H^2 \ominus IH^2$. Thanks to the pioneering work of Sarason, e.g., [64–67], we know that de Branges – Rovnyak spaces play an important role in numerous questions of complex analysis and operator theory. We mention a recent paper by the second named author and Sarason and Seip [47] who gave a characterization of surjectivity of Toeplitz operator the proof of which involves de Branges-Rovnyak spaces. We also refer to work of J. Shapiro [69, 70] concerning the notion of angular derivative for holomorphic self-maps of the unit disk. See also a paper of J. Anderson and J. Rovnyak [10], where generalized Schwarz – Pick estimates are given and a paper of M. Jury [52], where composition operators are studied by methods based on $\mathcal{H}(b)$ spaces.

In what follows we will assume that b is in the unit ball of H^∞. We recall here that since $\mathcal{H}(b)$ is contained contractively in H^2, it is a reproducing kernel Hilbert space. More precisely, for all function f in $\mathcal{H}(b)$ and every point λ in $\mathbb{D}$, we have

$$f(\lambda) = \langle f, k_\lambda^b \rangle_b, \tag{11}$$

where $k_\lambda^b = (\mathrm{Id} - T_b T_{\bar{b}})k_\lambda$. Thus

$$k_\lambda^b(z) = \frac{1 - \overline{b(\lambda)}b(z)}{1 - \bar{\lambda} z}, \qquad z \in \mathbb{D}.$$

We also recall that $\mathcal{H}(b)$ is invariant under the backward shift operator and in the following, we denote by X the contraction $X := S^*_{|\mathcal{H}(b)}$. Its adjoint satisfies the important formula

$$X^* h = Sh - \langle h, S^* b \rangle_b b, \qquad h \in \mathcal{H}(b).$$

In the case where b is inner, then X coincides with the so-called model operator of Sz.-Nagy – Foias which serves as a model for certain Hilbert space contractions (in fact, those contractions T which are $C_{\cdot 0}$ and with $\partial_T = \partial_{T^*} = 1$; for the general case, the model operator is quite complicated).

Finally, let us recall that a point $\lambda \in \overline{\mathbb{D}}$ is said to be regular (for b) if either $\lambda \in \mathbb{D}$ and $b(\lambda) \neq 0$, or $\lambda \in \mathbb{T}$ and b admits an analytic continuation across a neighbourhood $\mathcal{V}_\lambda = \{z : |z - \lambda| < \varepsilon\}$ of λ with $|b| = 1$ on $\mathcal{V}_\lambda \cap \mathbb{T}$. The spectrum of b, denoted by $\sigma(b)$, is then defined as the complement in $\overline{\mathbb{D}}$ of all regular points of b. For the case where $b = I$ is an inner function, this definition coincides with the definition given before.

In this section we will summarize the results corresponding to Theorems 1 and 2 above in the setting of de Branges – Rovnyak spaces. It turns out that Moeller's result remains valid in the setting of de Branges – Rovnyak spaces. Concerning the result by Ahern-Clark, it turns out that if we replace the inner function I by a general function b in the ball of H^∞, meaning that $b = I b_0$ where b_0 is now outer, then we have to add to condition (ii) in Theorem 2 the term corresponding to the absolutely continuous part of the measure: $\big|\log|b_0|\big|$.

In [44], the first named author and J. Mashreghi studied the continuity and analyticity of functions in the de Branges – Rovnyak spaces $\mathcal{H}(b)$ on an open arc of $\mathbb{T}$. As we will see the theory bifurcates into two opposite cases depending on whether b is an extreme point of the unit ball of H^∞ or not. Let us recall that if X is a linear space and S is a convex subset of X, then an element $x \in S$ is called an extreme point of S if it is not a proper convex combination of any two distinct points in S. Then, it is well known (see [33, p. 125]) that a function f is an extreme point of the unit ball of H^∞ if and only if

$$\int_{\mathbb{T}} \log(1 - |f(\zeta)|)\, dm(\zeta) = -\infty.$$

The following result is a generalization of Theorem 1 of Moeller.

Theorem 8 (Sarason 1995, Fricain – Mashreghi, 2008). *Let b be in the unit ball of H^∞ and let Γ be an open arc of $\mathbb{T}$. Then the following are equivalent:*

(i) *b has an analytic continuation across Γ and $|b| = 1$ on Γ;*
(ii) *Γ is contained in the resolvent set of X^*;*
(iii) *any function f in $\mathcal{H}(b)$ has an analytic continuation across Γ;*
(iv) *any function f in $\mathcal{H}(b)$ has a continuous extension to $\mathbb{D} \cup \Gamma$;*
(v) *b has a continuous extension to $\mathbb{D} \cup \Gamma$ and $|b| = 1$ on Γ.*

The equivalence of (i), (ii) and (iii) were proved in [67, p. 42] under the assumption that b is an extreme point. The contribution of Fricain – Mashreghi concerns the last two points. The mere assumption of continuity implies analyticity and this observation has interesting application as we will see below. Note that this implication is true also in the weighted situation (see Theorem 18).

The proof of Theorem 8 is based on reproducing kernel of $\mathcal{H}(b)$ spaces. More precisely, we use the fact that given $\omega \in \mathbb{D}$, then $k_\omega^b = (\mathrm{Id} - \bar{\omega} X^*)^{-1} k_0^b$ and thus

$$f(\omega) = \langle f, k_\omega^b \rangle_b = \langle f, (\mathrm{Id} - \bar{\omega} X^*)^{-1} k_0^b \rangle_b,$$

for every $f \in \mathcal{H}(b)$. Another key point in the proof of Theorem 8 is the theory of Hilbert spaces contractions developped by Sz.-Nagy – Foias. Indeed, if b is an extreme point of the unit ball of H^∞, then the characteristic function of the contraction X^* is b (see [63]) and then we know that $\sigma(X^*) = \sigma(b)$.

It is easy to see that condition (i) in the previous result implies that b is an extreme point of the unit ball of H^∞. Thus, the continuity (or equivalently, the analytic continuation) of b or of the elements of $\mathcal{H}(b)$ on the boundary completely depends on whether b is an extreme point or not. If b is not an extreme point of the unit ball of H^∞ and if Γ is an open arc of $\mathbb{T}$, then there exists necessarily a function $f \in \mathcal{H}(b)$ such that f has not a continuous extension to $\mathbb{D} \cup \Gamma$. On the opposite case, if b is an extreme point such that b has continuous extension to $\mathbb{D} \cup \Gamma$ with $|b| = 1$ on Γ, then all the functions $f \in \mathcal{H}(b)$ are continuous on Γ (and even can be continued analytically across Γ).

As in the inner case (see Ahern – Clark's result, Theorem 2), it is natural to ask what happens in points which are in the spectrum and what kind of regularity can be expected there. In [44], we gave an answer to this question and this result generalizes the Ahern – Clark result.

Theorem 9 (Fricain – Mashreghi, 2008). *Let b be a point in the unit ball of H^∞ and let*

$$b(z) = \gamma \prod_n \left(\frac{|a_n|}{a_n} \frac{a_n - z}{1 - \bar{a}_n z} \right) \exp\left(- \int_{\mathbb{T}} \frac{\zeta + z}{\zeta - z} \, d\mu(\zeta) \right) \exp\left(\int_{\mathbb{T}} \frac{\zeta + z}{\zeta - z} \log|b(\zeta)| \, dm(\zeta) \right) \tag{12}$$

be its canonical factorization. Let $\zeta_0 \in \mathbb{T}$ and let l be a nonnegative integer. Then the following are equivalent.

(i) *each function in $\mathcal{H}(b)$ and all its derivatives up to order l have (finite) radial limits at ζ_0;*
(ii) *$\|\partial^l k_z^b / \partial \bar{z}^l\|_b$ is bounded as z tends radially to ζ_0;*
(iii) *${X^*}^l k_0^b$ belongs to the range of $(\mathrm{Id} - \overline{\zeta_0} X^*)^{l+1}$;*

(iv) *we have* $S^b_{2l+2}(\zeta_0) < +\infty$, *where*

$$S^b_r(\zeta_0) := \sum_n \frac{1-|a_n|^2}{|\zeta_0 - a_n|^r} + \int_0^{2\pi} \frac{d\mu(e^{it})}{|\zeta_0 - e^{it}|^r} + \int_0^{2\pi} \frac{|\log|b(e^{it})||}{|\zeta_0 - e^{it}|^r}\, dm(e^{it}),$$

$$(1 \le r < +\infty).$$

In the following, we denote by $E_r(b)$ the set of points $\zeta_0 \in \mathbb{T}$ which satisfy $S^b_r(\zeta_0) < +\infty$.

The proof of Theorem 9 is based on a generalization of technics of Ahern – Clark. However, we should mention that the general case is a little bit more complicated than the inner case. Indeed if $b = I$ is an inner function, for the equivalence of (iii) and (iv) (which is the hard part of the proof), Ahern – Clark noticed that the condition (iii) is equivalent to the following interpolation problem: there exists $k, g \in H^2$ such that

$$(1 - \overline{\zeta_0} z)^{l+1} k(z) - l! z^l = I(z) g(z).$$

This reformulation, based on the orthogonal decomposition $H^2 = \mathcal{H}(I) \oplus IH^2$, is crucial in the proof of Ahern – Clark. In the general case, this is no longer true because $\mathcal{H}(b)$ is not a closed subspace of H^2 and we cannot have such an orthogonal decomposition. This induces a real difficulty that we can overcome using other arguments: in particular, we use (in the proof) the fact that if $\zeta_0 \in E_{l+1}(b)$ then, for $0 \le j \le l$, the limits

$$\lim_{r \to 1^-} b^{(j)}(r\zeta_0) \qquad \text{and} \qquad \lim_{R \to 1^+} b^{(j)}(R\zeta_0)$$

exist and are equal (see [3]). Here by reflection we extend the function b outside the unit disk by the formula (12), which represents an analytic function for $|z| > 1$, $z \ne 1/\overline{a_n}$. We denote this function also by b and it is easily verified that it satisfies

$$b(z) = \frac{1}{\overline{b(1/\bar{z})}}, \qquad \forall z \in \mathbb{C}. \tag{13}$$

Maybe we should compare condition (iii) of Theorem 9 and condition (ii) of Theorem 8. For the question of analytic continuation through a neighbourhood $\mathcal{V}_{\zeta_0}$ of a point $\zeta_0 \in \mathbb{T}$, we impose that for every $z \in \mathcal{V}_{\zeta_0} \cap \mathbb{T}$, the operator $\mathrm{Id} - \bar{z}X^*$ is bijective (or onto which is equivalent because it is always one-to-one as noted in [43, Lemma 2.2]) whereas for the question of the existence of radial limits at ζ_0 for the derivative up to a given order l, we impose that the range of the operator $(\mathrm{Id} - \overline{\zeta_0} X^*)^{l+1}$ contains the only function $X^{*l} k^b_0$. We also mention that Sarason has obtained another criterion in terms of the Clark measure σ_λ associated with b (see above for a definition of Clark measures; note that the Clark measures here are not always singular as they are when b is inner).

Theorem 10 (Sarason, 1995). *Let ζ_0 be a point of $\mathbb{T}$ and let l be a nonnegative integer. The following conditions are equivalent.*

(i) *Each function in $\mathcal{H}(b)$ and all its derivatives up to order l have nontangential limits at ζ_0.*

(ii) *There is a point $\lambda \in \mathbb{T}$ such that*

$$\int_{\mathbb{T}} |e^{i\theta} - \zeta_0|^{-2l-2}\, d\sigma_\lambda(e^{i\theta}) < +\infty. \tag{14}$$

(iii) *The last inequality holds for all $\lambda \in \mathbb{T} \setminus \{b(\zeta_0)\}$.*

(iv) *There is a point* $\lambda \in \mathbb{T}$ *such that* μ_λ *has a point mass at* ζ_0 *and*

$$\int_{\mathbb{T}\setminus\{z_0\}} |e^{i\theta} - \zeta_0|^{-2l}\, d\sigma_\lambda(e^{i\theta}) < \infty.$$

Recently, Bolotnikov and Kheifets [20] gave a third criterion (in some sense more algebraic) in terms of the Schwarz – Pick matrix. Recall that if b is a function in the unit ball of H^∞, then the matrix $\mathbf{P}_l^\omega(z)$, which will be refered to as to a Schwarz-Pick matrix and defined by

$$\mathbf{P}_l^b(z) := \left[\frac{1}{i!\,j!}\frac{\partial^{i+j}}{\partial z^i \partial \bar{z}^j}\frac{1-|b(z)|^2}{1-|z|^2}\right]_{i,j=0}^l,$$

is positive semidefinite for every $l \geq 0$ and $z \in \mathbb{D}$. We extend this notion to boundary points as follows: given a point $\zeta_0 \in \mathbb{T}$, the boundary Schwarz-Pick matrix is

$$\mathbf{P}_l^b(\zeta_0) = \lim_{\substack{z\to\zeta_0 \\ \sphericalangle}} \mathbf{P}_l^b(z) \qquad (l \geq 0),$$

provided this nontangential limit exists.

Theorem 11 (Bolotnikov – Kheifets, 2006). *Let b be a point in the unit ball of H^∞, let $\zeta_0 \in \mathbb{T}$ and let l be a nonnegative integer. Assume that the boundary Schwarz – Pick matrix $\mathbf{P}_l^b(\zeta_0)$ exists. Then each function in $\mathcal{H}(b)$ and all its derivatives up to order l have nontangential limits at ζ_0.*

Further it is shown in [20] that the boundary Schwarz – Pick matrix $\mathbf{P}_l^b(\zeta_0)$ exists if and only if

$$\lim_{\substack{z\to\zeta_0 \\ \sphericalangle}} d_{b,l}(z) < +\infty, \tag{15}$$

where

$$d_{b,l}(z) := \frac{1}{(l!)^2}\frac{\partial^{2l}}{\partial z^l \partial \bar{z}^l}\frac{1-|b(z)|^2}{1-|z|^2}.$$

We should mention that it is not clear to show direct connections between conditions (14), (15) and condition (iv) of Theorem 9.

Once we know the points ζ_0 in the unit circle where $f^{(l)}(\zeta_0)$ exists (in a nontangential sense) for every function $f \in \mathcal{H}(b)$, it is natural to ask if we can obtain an integral formula for this derivative similar to (2) for the inner case. However, if one tries to generalize techniques used in the model spaces K_I^2 in order to obtain such a representation for the derivatives of functions in $\mathcal{H}(b)$, some difficulties appear mainly due to the fact that the evaluation functional in $\mathcal{H}(b)$ (contrary to the model space K_I^2) is not a usual integral operator. To overcome this difficulty and nevertheless provide an integral formula similar to (2) for functions in $\mathcal{H}(b)$, the first named author and Mashreghi used in [45] two general facts about the de Branges – Rovnyak spaces that we recall now. The first one concerns the relation between $\mathcal{H}(b)$ and $\mathcal{H}(\bar{b})$. For $f \in H^2$, we have [67, p. 10]

$$f \in \mathcal{H}(b) \iff T_{\bar{b}} f \in \mathcal{H}(\bar{b}).$$

Moreover, if $f_1, f_2 \in \mathcal{H}(b)$, then

$$\langle f_1, f_2\rangle_b = \langle f_1, f_2\rangle_2 + \langle T_{\bar{b}} f_1, T_{\bar{b}} f_2\rangle_{\bar{b}}. \tag{16}$$

We also mention an integral representation for functions in $\mathcal{H}(\bar{b})$ [67, p. 16]. Let $\rho(\zeta) := 1 - |b(\zeta)|^2$, $\zeta \in \mathbb{T}$, and let $L^2(\rho)$ stand for the usual Hilbert space of measurable functions $f\colon \mathbb{T} \to \mathbb{C}$ with $\|f\|_\rho < \infty$, where

$$\|f\|_\rho^2 := \int_{\mathbb{T}} |f(\zeta)|^2 \rho(\zeta)\, dm(\zeta).$$

For each $\lambda \in \mathbb{D}$, the Cauchy kernel k_λ belongs to $L^2(\rho)$. Hence, we define $H^2(\rho)$ to be the (closed) span in $L^2(\rho)$ of the functions k_λ ($\lambda \in \mathbb{D}$). If q is a function in $L^2(\rho)$, then $q\rho$ is in $L^2(\mathbb{T})$, being the product of $q\rho^{1/2} \in L^2(\mathbb{T})$ and the bounded function $\rho^{1/2}$. Finally, we define the operator $C_\rho\colon L^2(\rho) \to H^2$ by

$$C_\rho(q) := P_+(q\rho).$$

Then C_ρ is a partial isometry from $L^2(\rho)$ onto $\mathcal{H}(\bar{b})$ whose initial space equals to $H^2(\rho)$ and it is an isometry if and only if b is an extreme point of the unit ball of H^∞.

Now let $\omega \in \operatorname{clos}\mathbb{D}$ and let l be a nonnegative integer. In order to get an integral representation for the lth derivative of f at point ω for functions in the de Branges – Rovnyak spaces, we need to introduce the following kernels

$$(17) \qquad k_{\omega,l}^b(z) := l!\,\frac{z^l - b(z)\sum_{p=0}^{l} \overline{(b^{(p)}(\omega)/p!)} z^{l-p}(1-\overline{\omega}z)^p}{(1-\overline{\omega}z)^{l+1}}, \qquad (z \in \mathbb{D}),$$

and

$$(18) \qquad k_{\omega,l}^\rho(\zeta) := l!\,\frac{\sum_{p=0}^{l} \overline{(b^{(p)}(\omega)/p!)} \zeta^{l-p}(1-\overline{\omega}\zeta)^p}{(1-\overline{\omega}\zeta)^{l+1}}, \qquad (\zeta \in \mathbb{T}).$$

Of course, for $\omega = \zeta_0 \in \mathbb{T}$, these formulae have a sense only if b has derivatives (in a radial or nontangential sense) up to order l; as we have seen this is the case if $\zeta_0 \in E_{l+1}(b)$ (which obviously contains $E_{2(l+1)}(b)$).

For $l = 0$, we see that $k_{\omega,0}^b = k_\omega^b$ is the reproducing kernel of $\mathcal{H}(b)$ and $k_{\omega,0}^\rho = \overline{b(\omega)} k_\omega$ is (up to a constant) the Cauchy kernel. Moreover (at least formally) the function $k_{\omega,l}^b$ (respectively $k_{\omega,l}^\rho$) is the lth derivative of $k_{\omega,0}^b$ (respectively of $k_{\omega,0}^\rho$) with respect to $\overline{\omega}$.

Theorem 12 (Fricain – Mashreghi, 2008). *Let b be a function in the unit ball of H^∞ and let l be a nonnegative integer. Then for every point $\zeta_0 \in \mathbb{D} \cup E_{2l+2}(b)$ and for every function $f \in \mathcal{H}(b)$, we have $k_{\zeta_0,l}^b \in \mathcal{H}(b)$, $k_{\zeta_0,l}^\rho \in L^2(\rho)$ and*

$$(19) \qquad f^{(l)}(\zeta_0) = \int_{\mathbb{T}} f(\zeta)\overline{k_{\zeta_0,l}^b(\zeta)}\, dm(\zeta) + \int_{\mathbb{T}} g(\zeta)\rho(\zeta)\overline{k_{\zeta_0,l}^\rho(\zeta)}\, dm(\zeta),$$

where $g \in H^2(\rho)$ satisfies $T_{\bar{b}} f = C_\rho g$.

We should say that Theorem 12 (as well as Theorem 13, Proposition 1, Theorem 14, Theorem 15 and Theorem 16 below) are stated and proved in [16, 45] in the framework of the upper half-plane; however it is not difficult to see that the same technics can be adapted to the unit disc and we give the analogue of these results in this context.

We should also mention that in the case where $\zeta_0 \in \mathbb{D}$, the formula (19) follows easily from the formulae (16) and (11). For $\zeta_0 \in E_{2n+2}(b)$, the result is more delicate and the key point of the proof is to show that

$$(20) \qquad f^{(l)}(\zeta_0) = \langle f, k_{\zeta_0,l}^b \rangle_b,$$

for every function $f \in \mathcal{H}(b)$ and then show that $T_{\bar{b}} k^b_{\zeta_0,l} = C_\rho k^\rho_{\zeta_0,l}$ to use once again (16).

A consequence of (20) and Theorem 9 is that if $\zeta_0 \in E_{2l+2}(b)$, then $k^b_{\omega,l}$ tends weakly to $k^b_{\zeta_0,l}$ as ω approaches radially to ζ_0. It is natural to ask if this weak convergence can be replaced by norm convergence. In other words, is it true that $\|k^b_{\omega,l} - k^b_{\zeta_0,l}\|_b \to 0$ as ω tends radially to ζ_0?

In [2], Ahern and Clark claimed that they can prove this result for the case where b is inner and $l = 0$. For general functions b in the unit ball of H^∞, Sarason [67, Chapter V] got this norm convergence for the case $l = 0$. In [45], we answer this question in the general case and get the following result.

Theorem 13 (Fricain – Mashreghi, 2008). *Let b be a point in the unit ball of H^∞, let l be a nonnegative integer and let $\zeta_0 \in E_{2l+2}(b)$. Then*

$$\|k^b_{\omega,l} - k^b_{\zeta_0,l}\|_b \to 0, \qquad \textit{as } \omega \textit{ tends radially to } \zeta_0.$$

The proof is based on explicit computations of $\|k^b_{\omega,l}\|_b$ and $\|k^b_{\zeta_0,l}\|_b$ and we use a nontrivial formula of combinatorics for sums of binomial coefficient. We should mention that we have obtained this formula by hypergeometric series. Let us also mention that Bolotnikov – Kheifets got a similar result in [20] using different techniques and under their condition (15).

We will now discuss the weighted norm inequalities obtained in [16]. The main goal was to get an analogue of Theorem 4 in the setting of the de Branges – Rovnyak spaces. To get these weighted Bernstein type inequalities, we first used a slight modified formula of (19).

Proposition 1 (Baranov – Fricain – Mashreghi, 2009). *Let b be in the unit ball of H^∞. Let $\zeta_0 \in \mathbb{D} \cup E_{2l+2}(b)$, $l \in \mathbb{N}$, and let*

$$\mathfrak{K}^\rho_{\zeta_0,l}(\zeta) := \overline{b(\zeta_0)} \frac{\sum_{j=0}^{l} \binom{l+1}{j+1} (-1)^j \overline{b^j(\zeta_0)} b^j(\zeta)}{(1 - \overline{\zeta_0}\zeta)^{l+1}}, \qquad \zeta \in \mathbb{T}. \tag{21}$$

Then $(k^b_{\zeta_0})^{l+1} \in H^2$ and $\mathfrak{K}^\rho_{\zeta_0,l} \in L^2(\rho)$. Moreover, for every function $f \in \mathcal{H}(b)$, we have

$$f^{(l)}(\zeta_0) = l!\Big(\int_{\mathbb{T}} f(\zeta) \bar{\zeta}^l \overline{(k^b_{\zeta_0})^{l+1}(\zeta)}\, \mathrm{d}m(\zeta) + \int_{\mathbb{T}} g(\zeta)\rho(\zeta) \bar{\zeta}^l \overline{\mathfrak{K}^\rho_{\zeta_0,l}(\zeta)}\, \mathrm{d}m(\zeta) \Big), \tag{22}$$

where $g \in H^2(\rho)$ is such that $T_{\bar{b}} f = C_\rho g$.

We see that if b is inner, then it is clear that the second integral in (19) is zero (because $\rho \equiv 0$) and we obtain the formula (2) of Ahern – Clark.

We now introduce the weight involved in our Bernstein-type inequalities. Let $1 < p \le 2$ and let q be its conjugate exponent. Let $l \in \mathbb{N}$. Then, for $z \in \operatorname{clos} \mathbb{D}$, we define

$$w_{p,l}(z) := \min\big\{ \|(k^b_z)^{l+1}\|_q^{-pl/(pl+1)}, \|\rho^{1/q} \mathfrak{K}^\rho_{z,l}\|_q^{-pl/(pl+1)} \big\};$$

we assume $w_{p,l}(\zeta) = 0$, whenever $\zeta \in \mathbb{T}$ and at least one of the functions $(k^b_\zeta)^{l+1}$ or $\rho^{1/q} \mathfrak{K}^\rho_{\zeta,l}$ is not in $L^q(\mathbb{T})$.

The choice of the weight is motivated by the representation (22) which shows that the quantity $\max\big\{ \|(k^b_z)^{l+1}\|_2, \|\rho^{1/2} \mathfrak{K}^\rho_{z,l}\|_2 \big\}$ is related to the norm of the functional $f \mapsto f^{(l)}(z)$ on $\mathcal{H}(b)$. Moreover, we strongly believe that the norms of reproducing kernels are an important characteristic of the space $\mathcal{H}(b)$ which captures many geometric properties of b. Using similar arguments as in the proof of Proposition 1, it is easy to see that $\rho^{1/q}\mathfrak{K}^{\rho}_{\zeta,l} \in L^q(\mathbb{T})$ if $\zeta \in E_{q(l+1)}(b)$. It is also natural to expect that $(k^b_\zeta)^{l+1} \in L^q(\mathbb{T})$ for $\zeta \in E_{q(l+1)}(b)$. This is true when b is an inner function, by a result of Cohn [28]; for a general function b with $q=2$ it was noticed in [16]. However, it seems that the methods of [16, 28] do not apply in the general case.

If $f \in \mathcal{H}(b)$ and $1 < p \leq 2$, then $(f^{(l)}w_{p,l})(x)$ is well-defined on $\mathbb{T}$. Indeed it follows from [44] that $f^{(l)}(\zeta)$ and $w_{p,l}(\zeta)$ are finite if $\zeta \in E_{2l+2}(b)$. On the contrary if $\zeta \notin E_{2l+2}(b)$. then $\|(k^b_\zeta)^{l+1}\|_2 = +\infty$. Hence, $\|(k^b_\zeta)^{l+1}\|_q = +\infty$ which, by definition, implies $w_{p,l}(\zeta) = 0$, and thus we may assume $(f^{(l)}w_{p,l})(\zeta) = 0$.

In the inner case, we have $\rho(t) \equiv 0$, then the second term in the definition of the weight $w_{p,l}$ disappears and we recover the weights considered in [12]. It should be emphasized that in the general case both terms are essential: in [16] we give an example where the norm $\|\rho^{1/q}\mathfrak{K}^{\rho}_{z,l}\|_q$ cannot be majorized uniformly by the norm $\|(k^b_z)^{l+1}\|_q$.

Theorem 14 (Baranov – Fricain – Mashreghi, 2009). *Let μ be a Carleson measure on* $\operatorname{clos}\mathbb{D}$ *, let $l \in \mathbb{N}$, let $1 < p \leq 2$, and let*

$$(T_{p,l}f)(z) = f^{(l)}(z)w_{p,l}(z), \qquad f \in \mathcal{H}(b).$$

If $1 < p < 2$, then $T_{p,l}$ is a bounded operator from $\mathcal{H}(b)$ into $L^2(\mu)$, that is, there is a constant $C = C(\mu, p, l) > 0$ such that

$$\|f^{(l)}w_{p,l}\|_{L^2(\mu)} \leq C\|f\|_b, \qquad f \in \mathcal{H}(b). \tag{23}$$

If $p = 2$, then $T_{2,l}$ is of weak type $(2,2)$ as an operator from $\mathcal{H}(b)$ into $L^2(\mu)$.

The proof of this result is based on the representation (22) which reduces the problem of Bernstein type inequalities to estimates on singular integrals. In particular, we use the following estimates on the weight: for $1 < p \leq 2$ and $l \in \mathbb{N}$, there exists a constant $A = A(l,p) > 0$ such that

$$w_{p,l}(z) \geq A\frac{(1-|z|)^l}{(1-|b(z)|)^{pl/(q(pl+1))}}, \qquad z \in \mathbb{D}.$$

To apply Theorem 14 one should have effective estimates for the weight $w_{p,l}$, that is, for the norms of the reproducing kernels. In the following, we relate the weight $w_{p,l}$ to the distances to the level sets of $|b|$. We start with some notations. Denote by $\sigma_i(b)$ the boundary spectrum of b, i.e.,

$$\sigma_i(b) := \Big\{\zeta \in \mathbb{T} : \liminf_{\substack{z\to\zeta \\ z\in\mathbb{D}}} |b(z)| < 1\Big\}.$$

Then $\operatorname{clos}\sigma_i(b) = \sigma(b) \cap \mathbb{T}$ where $\sigma(b)$ is the spectrum defined at the begining of this section. For $\varepsilon \in (0,1)$, we put

$$\Omega(b,\varepsilon) := \{z \in \mathbb{D} : |b(z)| < \varepsilon\} \qquad \text{and} \qquad \widetilde{\Omega}(b,\varepsilon) := \sigma_i(b) \cup \Omega(b,\varepsilon).$$

Finally, for $\zeta \in \mathbb{T}$, we introduce the following two distances

$$d_\varepsilon(\zeta) := \operatorname{dist}\big(\zeta, \Omega(b,\varepsilon)\big) \qquad \text{and} \qquad \tilde{d}_\varepsilon(\zeta) := \operatorname{dist}\big(\zeta, \widetilde{\Omega}(b,\varepsilon)\big).$$

Note that whenever $b = I$ is an inner function, for all $\zeta \in \sigma_i(I)$, we have

$$\liminf_{\substack{z\to\zeta\\ z\in\mathbb{D}}} |I(z)| = 0,$$

and thus $d_\varepsilon(\zeta) = \tilde{d}_\varepsilon(\zeta)$, $\zeta \in \mathbb{T}$. However, for an arbitrary function b in the unit ball of H^∞, we have to distinguish between the distance functions d_ε and $\tilde{d}_\varepsilon$.

Using fine estimates on the derivatives $|b'(\zeta)|$, we got in [16] the following result.

Lemma 1. *For each $p > 1$, $l \geq 1$ and $\varepsilon \in (0,1)$, there exists $C = C(\varepsilon, p, l) > 0$ such that*

$$(\tilde{d}_\varepsilon(\zeta))^l \leq C\, w_{p,l}(r\zeta), \tag{24}$$

for all $\zeta \in \mathbb{T}$ and $0 \leq r \leq 1$.

This lemma combined with Theorem 14 imply immediately the following.

Corollary 1 (Baranov – Fricain – Mashreghi, 2009). *For each $\varepsilon \in (0,1)$ and $l \in \mathbb{N}$, there exists $C = C(\varepsilon, l)$ such that*

$$\|f^{(l)}\tilde{d}_\varepsilon^l\|_2 \leq C\|f\|_b, \qquad f \in \mathcal{H}(b).$$

As we have said in Section 2, weighted Bernstein-type inequalities of the form (23) turned out to be an efficient tool for the study of the so-called Carleson-type embedding theorems for backward shift invariant subspaces K_I^p. Notably, methods based on the Bernstein-type inequalities allow to give unified proofs and essentially generalize almost all known results concerning these problems (see [12,15]). Here we obtain an embedding theorem for de Branges – Rovnyak spaces. The first statement generalizes Theorem 6 (of Volberg – Treil) and the second statement generalizes a result of Baranov (see [12]).

Theorem 15 (Baranov – Fricain – Mashreghi, 2009). *Let μ be a positive Borel measure in* $\operatorname{clos}\mathbb{D}$, *and let $\varepsilon \in (0,1)$.*

(a) *Assume that $\mu(S(\zeta,h)) \leq Kh$ for all Carleson squares $S(\zeta,h)$ satisfying*

$$S(\zeta,h) \cap \widetilde{\Omega}(b,\varepsilon) \neq \varnothing.$$

Then $\mathcal{H}(b) \subset L^2(\mu)$, that is, there is a constant $C > 0$ such that

$$\|f\|_{L^2(\mu)} \leq C\|f\|_b, \qquad f \in \mathcal{H}(b).$$

(b) *Assume that μ is a vanishing Carleson measure for $\mathcal{H}(b)$, that is, $\mu(S(\zeta,h))/h \to 0$ whenever $S(\zeta,h) \cap \widetilde{\Omega}(b,\varepsilon) \neq \varnothing$ and $h \to 0$. Then the embedding $\mathcal{H}(b) \subset L^2(\mu)$ is compact.*

Note that whenever $b = I$ is an inner function, the sufficient condition that appears in (a) of Theorem 15 is equivalent to the condition of Volberg – Treil theorem because in that case (as already mentionned) we always have $\sigma_i(I) \subset \operatorname{clos}\Omega(I,\varepsilon)$ for every $\varepsilon > 0$.

In Theorem 15 we need to verify the Carleson condition only on a *special* subclass of squares. Geometrically this means that when we are far from the spectrum $\sigma(b)$, the measure μ in Theorem 15 can be essentially larger than standard Carleson measures. The reason is that functions in $\mathcal{H}(b)$ have much more regularity at the points $\zeta \in \mathbb{T} \setminus \sigma(b)$ (see Theorem 8). On the other hand, if $|b(\zeta)| \leq \delta < 1$, almost everywhere on some arc $\Gamma \subset \mathbb{T}$, then the functions in $\mathcal{H}(b)$ behave on Γ essentially the same as a general element of H^2 on that arc, and for any Carleson measure for $\mathcal{H}(b)$ its restriction to the square $S(\Gamma)$ is a standard Carleson measure.

For a class of functions b the converse to Theorem 15 is also true. As in the inner case, we say that b satisfies the *connected level set condition* if the set $\Omega(b,\varepsilon)$ is connected for some $\varepsilon \in (0,1)$. Our next result generalizes Theorem 5 of Cohn.

Theorem 16 (Baranov – Fricain – Mashreghi, 2009). *Let b satisfy the connected level set condition for some $\varepsilon \in (0,1)$. Assume that $\sigma(b) \subset \operatorname{clos}\Omega(b,\varepsilon)$. Let μ be a positive Borel measure on $\operatorname{clos}\mathbb{D}$. Then the following statements are equivalent:*

(a) $\mathcal{H}(b) \subset L^2(\mu)$.

(b) *There exists $C > 0$ such that $\mu\big(S(\zeta,h)\big) \le Ch$ for all Carleson squares $S(\zeta,h)$ such that $S(\zeta,h) \cap \widetilde{\Omega}(b,\varepsilon) \neq \varnothing$.*

(c) *There exists $C > 0$ such that*

$$\int_{\operatorname{clos}\mathbb{D}} \frac{1-|z|^2}{|1-\bar{z}\zeta|^2}\,\mathrm{d}\mu(\zeta) \le \frac{C}{1-|b(z)|}, \qquad z \in \mathbb{D}. \tag{25}$$

In [16], we also discuss another application of our Bernstein type inequalities to the problem of stability of Riesz bases consisting of reproducing kernels in $\mathcal{H}(b)$.

4. Weighted backward shift invariant subspaces

Let us now turn to weighted backward shift invariant subspaces. As will be explained below, the weighted versions we are interested in appear naturally in the context of kernels of Toeplitz operators. In Section 4.1 we will present an example showing that the generalization of the Ahern – Clark result to this weighted situation is far from being immediate. For this reason we will focus essentially on analytic continuation in this section.

For an outer function g in H^p, we define weighted Hardy spaces in the following way:

$$\begin{aligned} H^p(|g|^p) := \frac{1}{g}H^p = \Big\{ f \in \operatorname{Hol}(\mathbb{D}) : \|f\|^p_{|g|^p} &= \sup_{0<r<1} \frac{1}{2\pi}\int_{-\pi}^{\pi} |f(re^{it})|^p |g(re^{it})|^p\,dt \\ &= \int_{-\pi}^{\pi} |f(e^{it})|^p |g(e^{it})|^p\,dt < \infty \Big\}. \end{aligned}$$

Clearly $f \mapsto fg$ induces an isometry from $H^p(|g|^p)$ onto H^p. Let now I be any inner function.

We shall discuss the situation when $p = 2$. There are at least two ways of generalizing the backward shift invariant subspaces to the weighted situation. We first discuss the simple one. As in the unweighted situation we can consider the orthogonal complement of shift invariant subspaces $IH^2(|g|^2)$, the shift $S\colon H^2(|g|^2) \to H^2(|g|^2)$ being given as usual by $Sf(z) = zf(z)$. The weighted scalar product is defined by

$$\langle f,h\rangle_{|g|^2} = \frac{1}{2\pi}\int_{-\pi}^{\pi} f(e^{it})\overline{h(e^{it})}|g(e^{it})|^2\,dt = \langle fg, hg\rangle.$$

Then

$$\langle Sf,h\rangle_{|g|^2} = \langle zfg, hg\rangle = \langle fg, \bar{z}hg\rangle = \langle fg, P_+(\bar{z}hg)\rangle = \left\langle f, \frac{1}{g}P_+(g\bar{z}h)\right\rangle_{|g|^2}.$$

In other words, with respect to the scalar product $\langle\cdot,\cdot\rangle_{|g|^2}$ the adjoint shift is given by $S_g^* := \frac{1}{g}P_+g\bar{z}$, and

$$\begin{aligned} K_I^{2,g} &:= \big(IH^2(|g|^2)\big)^\perp = \{f\in H^2(|g|^2) : \langle fg, Ihg\rangle = 0, h\in H^2(|g|^2)\}\\ &= \{f\in H^2(|g|^2) : \langle fg, Ih\rangle = 0, h\in H^2\}\\ &= \{f\in H^2(|g|^2) : \langle P_+(\bar{I}fg), h\rangle = 0, h\in H^2\}\\ &= \left\{f\in H^2(|g|^2) : \left\langle \frac{1}{g}P_+\bar{I}fg, h\right\rangle_{|g|^2} = 0, h\in H^2(|g|^2)\right\}. \end{aligned}$$

So, $K_I^{2,g} = \ker\big((1/g)P_+\bar{I}g\big) = \ker(\frac{I}{g}P_+\bar{I}g)$. Setting $P_I^g := (I/g)P_-\bar{I}g$ we get a self-adjoint projection such that

$$K_I^{2,g} = P_I^g H^2(|g|^2) = \frac{1}{g}P_I\big(gH^2(|g|^2)\big) = \frac{1}{g}P_I H^2 = \frac{1}{g}K_I^2,$$

where P_I is the unweighted orthogonal projection onto K_I^2. Hence, in this situation continuation is completely determined by that in K_I^2 and that of $1/g$.

We will thus rather consider the second approach. The spaces to be discussed now appear in the context of kernels of Toeplitz operators. Set

$$K_I^p(|g|^p) = H^p(|g|^p)\cap I\overline{H_0^p(|g|^p)},$$

where now $H_0^p(|g|^p) = zH^p(|g|^p)$.

The connection with Toeplitz operators arises in the following way: if $\varphi = \overline{Ig}/g$ is a unimodular symbol, then $\ker T_\varphi = gK_I^2(|g|^2)$ (see [48]). Conversely, whenever $0\neq f\in\ker T_\varphi$, where φ is unimodular and $f = Jg$ is the inner-outer factorization of f, then there exists an inner function I such that $\varphi = \overline{Ig}/g$ (see also [48]).

Note also that the following simple example shows that in general $K_I^{2,g}$ is different from $K_I^2(|g|^2)$. Let $I(z) = z$ be the simplest Blaschke factor. Then $H^2(|g|^2)\cap I\overline{H_0^2(|g|^2)} = H^2(|g|^2)\cap\overline{H^2(|g|)} = \mathbb{C}$ whenever g is rigid (more on rigidity follows later). On the other hand, $(1/g)K_I^2$ is the one-dimensional space spanned by $1/g$ which is different from $\mathbb{C}$ when g is not a constant.

The representation $\ker T_\varphi = gK_I^p(|g|^p)$ is particularly interesting when g is the extremal function of $\ker T_\varphi$. Then we know from a result by Hitt [50] (see also [66] for a de Branges – Rovnyak spaces approach to Hitt's result) that when $p=2$, $\ker T_\varphi = gK_I^2$, and that g is an isometric divisor on $\ker T_\varphi = gK_I^2$ (or g is an isometric multiplier on K_I^2). In this situation we thus have $K_I^2(|g|^2) = K_I^2$. Note, that for $p\neq 2$, if g is extremal for $gK_I^p(|g|^p)$, then $K_I^p(|g|^p)$ can still be imbedded into K_I^2 when $p>2$ and in K_I^p when $p\in(1,2)$ (see [48], where it is also shown that these imbeddings can be strict). In these situations when considering questions concerning pseudocontinuation and analytic continuation, we can carry over to $K_I^p(|g|^p)$ everything we know about K_I^2 or K_I^p, i.e., Theorems 1 and 7. Concerning the Ahern – Clark and Cohn results however, when $p\neq 2$, we lose information since condition (ii) in Theorem 2 depends on p.

In general the extremal function is not easily detectable (explicit examples of extremal functions were given in [48]), in that we cannot determine it, or for a given g it is not a simple matter to check whether it is extremal or not. So a natural question is to know under which conditions on g and I, we can still say something about analytic continuation of functions in $K_I^p(|g|^p)$. It turns out that Moeller's result is valid under an additional local integrability condition of $1/g$ on a closed arc not meeting the spectrum of I. Concerning the regularity questions in points contained in the spectrum, the situation is more intricate. As mentioned earlier, an example in this direction will be discussed at the end of this section.

Regularity of functions in kernels of Toeplitz operators have been considered by Dyakonov. He in particular establishes global regularity properties of functions in the kernel of a Toeplitz operator — such as being in certain Sobolov and Besov spaces [35] or Lipschitz and Zygmund spaces [41] — depending on the smoothness of the corresponding Toeplitz operator.

The following simple example hints at some difference between this situation and the unweighted situation or the context of de Branges–Rovnyak spaces discussed before. Let I be arbitrary with $-1 \notin \sigma(I)$, and let $g(z) = 1+z$, so that $\sigma(I)$ is far from the only point where g vanishes. We know that $\ker T_{\overline{Ig}/g} = gK_I^p(|g|^p)$. We first observe that $\overline{(1+z)}/(1+z) = \bar{z}$. Hence,

$$1 \in K_{zI}^p = \ker T_{\overline{zI}} = \ker T_{\overline{Ig}/g} = gK_I^p(|g|^p)$$

So, $K_I^p(|g|^p)$ contains the function $1/g$ which is badly behaved in -1, and thus cannot extend analytically through -1.

This observation can be made more generally as stated in the following result [46].

Proposition 2 (Hartmann 2008). *Let g be an outer function in H^p. If $\ker T_{\bar{g}/g} \neq \{0\}$ contains an* inner *function, then $1/g \in K_I^p(|g|^p)$ for* every *inner function I.*

Note that if the inner function J is in $\ker T_{\bar{g}/g}$ then $T_{\overline{Jg}/g}1 = 0$, and hence $1 \in \ker T_{\overline{Jg}/g} = gK_J^p(|g|^2)$ and $1/g \in K_J^p(|g|^2)$, which shows that with this simple argument the proposition holds with the more restrictive condition $I = J$.

Let us comment on the case $p = 2$:

The claim that the kernel of $T_{\bar{g}/g}$ contains an inner function implies in particular that $T_{\bar{g}/g}$ is not injective and so g^2 is not rigid in H^1 (see [67, X-2]), which means that it is not uniquely determined — up to a real multiple — by its argument (or equivalently, its normalized version $g^2/\|g^2\|_1$ is not exposed in the unit ball of H^1).

It is clear that if the kernel of a Toeplitz operator is not reduced to $\{0\}$ — or equivalently (since $p = 2$) g^2 is not rigid — then it contains an *outer* function (just divide out the inner factor of any nonzero function contained in the kernel). However, Toeplitz operators with nontrivial kernels containing no inner functions can be easily constructed. Take for instance $T_{\overline{zg_0}/g_0} = T_{\bar{z}}T_{\overline{g_0}/g_0}$, where $g_0(z) = (1-z)^\alpha$ and $\alpha \in (0, \frac{1}{2})$. The Toeplitz operator $T_{\overline{g_0}/g_0}$ is invertible ($|g_0|^2$ satisfies the Muckenhoupt (A_2) condition) and $(T_{\overline{g_0}/g_0})^{-1} = g_0 P_+ \frac{1}{\overline{g_0}}$ [61] so that the kernel of $T_{\overline{zg_0}/g_0}$ is given by the preimage under $T_{\overline{g_0}/g_0}$ of the constants (which define the kernel of $T_{\bar{z}}$). Since $g_0 P_+(c/\overline{g_0}) = cg_0/\overline{g_0(0)}$, c being any complex number, we have $\ker T_{\overline{zg_0}/g_0} = \mathbb{C}g_0$ which does not contain any inner function.

So, without any condition on g, we cannot hope for reasonable results. In the above example, when $p = 2$, then the function $g^2(z) = (1+z)^2$ is in fact not rigid (for instance the argument of $(1+z)^2$ is the same as that of z). As already pointed out, rigidity of g^2 is also characterized by the fact that $T_{\bar{g}/g}$ is injective (see [67, X-2]). Here $T_{\bar{g}/g} = T_{\bar{z}}$ the kernel of which is $\mathbb{C}$. From this it can also be deduced that g^2 is rigid if and only if $H^p(|g|^p) \cap \overline{H^p(|g|^p)} = \{0\}$ which indicates again that rigidity should be assumed if we want to have $K_I^p(|g|^p)$ reasonably defined. (See [53] for some discussions on the intersection $H^p(|g|^p) \cap \overline{H^p(|g|^p)}$.)

A stronger condition than rigidity (at least when $p = 2$) is that of a Muckenhoupt weight. Let us recall the Muckenhoupt (A_p) condition: for general $1 < p < \infty$ a weight w satisfies the (A_p) condition if

$$B := \sup_{I \text{ subarc of } \mathbb{T}} \left\{ \frac{1}{|I|} \int_I w(x)\,\mathrm{d}x \times \left(\frac{1}{|I|} \int_I w^{-1/(p-1)}(x)\,\mathrm{d}x \right)^{p-1} \right\} < \infty.$$

When $p = 2$, it is known that this condition is equivalent to the so-called Helson–Szegő condition. The Muckenhoupt condition will play some role in the results to come. However, our main theorem on analytic continuation (Theorem 17) works under a weaker local integrability condition.

Another observation can be made now. We have already mentioned that rigidity of g^2 in H^1 is equivalent to injectivity of $T_{\bar{g}/g}$, when g is outer. It is also clear that $T_{g/\bar{g}}$ is *always* injective so that when g^2 is rigid, the operator $T_{\bar{g}/g}$ is injective with dense range. On the other hand, by a result of Devinatz and Widom (see, e.g., [59, Theorem B4.3.1]), the invertibility of $T_{\bar{g}/g}$, where g is outer, is equivalent to $|g|^2$ being (A_2). So the difference between rigidity and (A_2) is the surjectivity (in fact the closedness of the range) of the corresponding Toeplitz operator. A criterion for surjectivity of noninjective Toeplitz operators can be found in [47]. It appeals to a parametrization which was earlier used by Hayashi [49] to characterize kernels of Toeplitz operators among general nearly invariant subspaces. Rigid functions do appear in the characterization of Hayashi.

As a consequence of Theorem 17 below analytic continuation can be expected on arcs not meeting the spectrum of I when $|g|^p$ is (A_p) (see Remark 1). However the (A_p) condition cannot be expected to be necessary since it is a global condition whereas continuation depends on the local behaviour of I and g. We will even give an example of a nonrigid function g (hence not satisfying the (A_p) condition) for which analytic continuation is always possible in certain points of $\mathbb{T}$ where g vanishes essentially.

Closely connected with the continuation problem in backward shift invariant subspaces is the spectrum of the backward shift operator on the space under consideration. The following result follows from [9, Theorem 1.9]: Let B be the backward shift on $H^p(|g|^p)$, defined by $Bf(z) = \big(f - f(0)\big)/z$. Clearly, $K_I^p(|g|^p)$ is invariant with respect to B whenever I is inner. Then, $\sigma\big(B \mid K_I^p(|g|^p)\big) = \sigma_{\mathrm{ap}}(B \mid K_I^p(|g|^p))$, where $\sigma_{\mathrm{ap}}(T) = \{\lambda \in \mathbb{C} : \exists (f_n)_n \text{ with } \|f_n\| = 1 \text{ and } (\lambda - T)f_n \to 0\}$ denotes the approximate point spectrum of T, and this spectrum is equal to

$$\mathbb{T} \setminus \{1/\zeta \in \mathbb{T} : \text{every } f \in K_I^p(|g|^p) \text{ extends analytically in a neighbourhood of } \zeta\}.$$

The aim is to link this set and $\sigma(I)$. Here we will need the Muckenhoupt condition. Then, as in the unweighted situation, the approximate spectrum of $B \mid K_I^p(|g|^p)$ on $\mathbb{T}$ contains the conjugated spectrum of I. We will see later that the inclusion in the following proposition [46] actually is an equality.

Proposition 3 (Hartmann, 2008). *Let g be outer in H^p such that $|g|^p$ is a Muckenhoupt (A_p)-weight. Let I be an inner function with spectrum $\sigma(I)$. Then $\overline{\sigma(I)} \subset \sigma_{\mathrm{ap}}\big(B \mid K_I^p(|g|^p)\big)$.*

We now come to the main result in the weighted situation (see [46]).

Theorem 17 (Hartmann, 2008). *Let g be an outer function in H^p, $1 < p < \infty$ and I an inner function with associated spectrum $\sigma(I)$. Let Γ be a closed arc in $\mathbb{T}$. If there exists $s > q$, $1/p + 1/q = 1$, with $1/g \in L^s(\Gamma)$, then every function $f \in K_I^p(|g|^p)$ extends analytically through Γ if and only if Γ does not meet $\sigma(I)$.*

Note that in [46] only the sufficiency part of the above equivalence was shown. However the condition that Γ must not meet $\sigma(I)$ is also necessary (even under the a priori weaker condition of continuation through Γ) as follows from the proof of Theorem 18 below. A stronger version of Theorem 17 can be deduced from [5, Corollay 1 of Theorem 3]

It turns also out that — like in the de Branges – Rovnyak situation discussed in Theorem 8 — for analytic continuation it is actually sufficient to have continuation. This result is new, and we will state it as a theorem provided with a proof. It is based on ideas closed to the proof of the previous theorem.

Theorem 18. *Let g be an outer function in H^p, $1 < p < \infty$ and I an inner function with associated spectrum $\sigma(I)$. Let Γ be an open arc in $\mathbb{T}$. Suppose that every function $f \in K_I^p(|g|^p)$ extends continuously to Γ then $\Gamma \cap \sigma(I) = \varnothing$, and every function in $K_I^p(|g|^p)$ extends analytically through Γ.*

PROOF. Observe first that obviously $k_\lambda^I \in K_I^2(|g|^2)$. By the Schwarz reflection principle, in order that k_λ^I continues through Γ we need that Γ does not meet $\sigma(I)$ (note that $\operatorname{clos}\Gamma$ could meet $\sigma(I)$).

As in the unweighted situation, every meromorphic function f/I, $f = I\overline{\psi} \in K_I^2(|g|^2)$, admits a pseudocontinuation $\widetilde{\psi}$, defined by $\widetilde{\psi}(z) = \sum_{n\geq 0} \overline{\widehat{\psi}(n)} 1/z^n$ in the exterior disk $\mathbb{D}_e = \widehat{\mathbb{C}} \setminus \operatorname{clos}\mathbb{D}$.

Fix Γ_0 any closed subarc of Γ. Since $\sigma(I)$ is closed, the distance between $\sigma(I)$ and Γ_0 is strictly positive. Then there is a neigbourhood of Γ_0 intersected with $\mathbb{D}$ where $|I(z)| \geq \delta > 0$. It is clear that in this neighbourhood we are far away from the part of the spectrum of I contained in $\mathbb{D}$. Thus I extends analytically through Γ_0. For what follows we will call the endpoints of this arc $\zeta_1 := \mathrm{e}^{\mathrm{i}t_1}$ and $\zeta_2 := \mathrm{e}^{\mathrm{i}t_2}$ (oriented in the positive sense).

The following argument is in the spirit of Moeller [56] and based on Morera's theorem. Let us introduce some notation (see Figure 1).

For suitable $r_0 \in (0,1)$ let $\Omega_0 = \{z = r\mathrm{e}^{\mathrm{i}t} \in \mathbb{D} : t \in [t_1, t_2], r_0 \leq r < 1\}$. and $\widetilde{\Omega}_0 = \{z = \mathrm{e}^{\mathrm{i}t}/r \in \mathbb{D}_e : t \in [t_1, t_2], r_0 \leq r < 1\}$. Define

$$F(z) = \begin{cases} f(z)/I(z) & z \in \Omega_0; \\ \widetilde{\psi}(z) & z \in \widetilde{\Omega}_0. \end{cases}$$

By construction this function is analytic on $\Omega_0 \cup \widetilde{\Omega}_0$ and continuous on $\overline{\Omega_0 \cup \widetilde{\Omega}_0}$. Such a function is analytical on $\overline{\Omega_0 \cup \widetilde{\Omega}_0}$. □

Remark 1. It is known (see, e.g., [57]) that when $|g|^p \in (A_p)$, $1 < p < \infty$, then there exists $r_0 \in (1,p)$ such that $|g|^p \in (A_r)$ for every $r > r_0$. Take $r \in (r_0, p)$. Then in particular $1/g \in L^s$, where $1/r + 1/s = 1$. Since $r < p$ we have $s > q$ which allows to conclude that in this situation $1/g \in L^s(\Gamma)$ for every $\Gamma \subset \mathbb{T}$ (s independant of Γ).

We promised earlier an example of a nonrigid function g for which analytic continuation of K_I^p-functions *is* possible in certain points where g vanishes.

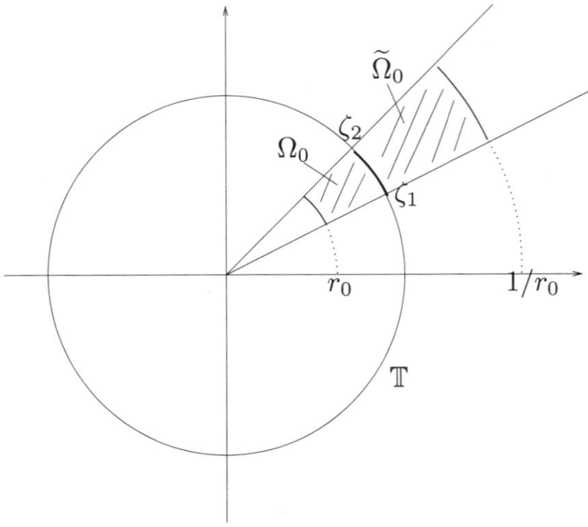

FIGURE 1. The regions Ω_0 and $\widetilde{\Omega}_0$

Example. For $\alpha \in (0, \frac{1}{2})$, let $g(z) = (1+z)(1-z)^\alpha$. Clearly g is an outer function vanishing essentially in 1 and -1. Set $h(z) = z(1-z)^{2\alpha}$, then by similar arguments as those employed in the introducing example to this section one can check that $\arg g^2 = \arg h$ a.e. on $\mathbb{T}$. Hence g is not rigid (it is the "big" zero in -1 which is responsible for nonrigidity). On the other hand, the zero in $+1$ is "small" in the sense that g satisfies the local integrability condition in a neighbourhood of 1 as required in the theorem, so that whenever I has its spectrum far from 1, then every $K_I^2(|g|^2)$-function can be analytically continued through suitable arcs around 1.

This example can be pushed a little bit further. In the spirit of Proposition 2 we check that (even) when the spectrum of an inner function I does not meet -1, there are functions in $K_I^p(|g|^p)$ that are badly behaved in -1. Let again $g_0(z) = (1-z)^\alpha$. Then

$$\frac{\overline{g(z)}}{g(z)} = \frac{\overline{(1+z)(1-z)^\alpha}}{(1+z)(1-z)^\alpha} = \bar{z}\frac{\overline{g_0(z)}}{g_0(z)}.$$

As already explained, for every inner function I, we have $\ker T_{\overline{Ig}/g} = gK_I^p(|g|^p)$, so that we are interested in the kernel $\ker T_{\overline{Ig}/g}$. We have $T_{\overline{Ig}/g} f = 0$ when $f = Iu$ and $u \in \ker T_{\bar{g}/g} = \ker T_{\overline{zg_0}/g_0} = \mathbb{C}g_0$ (see the discussion just before the proof of Proposition 2). Hence the function defined by

$$F(z) = \frac{f(z)}{g(z)} = I(z)\frac{g_0(z)}{g(z)} = \frac{I(z)}{1+z}$$

is in $K_I^p(|g|^p)$ and it is badly behaved in -1 when the spectrum of I does not meet -1 (but not only).

The preceding discussions motivate the following question: does rigidity of g suffice to get analytic continuation for $K_I^p(|g|^2)$-function whenever $\sigma(I)$ is far from zeros of g?

Theorem 17 together with Proposition 3 and Remark 1 allow us to obtain the following result. We should mention that it is easy to check that $H^p(|g|^p)$ satisfies the conditions required of a Banach space of analytic functions in order to apply the results of [9].

Corollary 2 (Hartmann, 2008). *Let g be outer in H^p such that $|g|^p$ is a Muckenhoupt (A_p) weight. Let I be an inner function with spectrum $\sigma(I) = \{\lambda \in \operatorname{clos}\mathbb{D} : \liminf_{z\to\lambda} I(z) = 0\}$. Then $\overline{\sigma(I)} = \sigma_{\mathrm{ap}}(B \mid K_I^p(|g|^p))$.*

Another simple consequence of Theorem 17 concerns embeddings. Contrarily to the situations discussed in Sections 2 and 3, the weight is here on the K_I^p-side.

Corollary 3 (Hartmann, 2008). *Let I be an inner function with spectrum $\sigma(I)$. If $\Gamma \subset \mathbb{T}$ is a closed arc not meeting $\sigma(I)$ and if g is an outer function in H^p such that $|g| \geq \delta$ on $\mathbb{T}\setminus\Gamma$ for some constant $\delta > 0$ and $1/g \in L^s(\Gamma)$, $s > q$, $1/p + 1/q = 1$. Then $K_I^p(|g|^p) \subset K_I^p$. If moreover g is bounded, then the last inclusion is an equality.*

Suppose now $p = 2$. We shall use this corollary to construct an example where $K_I^2(|g|^2) = K_I^2$ without g being extremal for $gK_I^2(|g|^2)$. Recall from Hitt's result [50], that when g is the extremal function of a nearly invariant subspace $M \subset H^2$, then there exists an inner function I such that $M = gK_I^2$, and g is an isometric multiplier on K_I^2 so that $K_I^2 = K_I^2(|g|^2)$. Recall from [48, Lemma 3] that a function g is extremal for $gK_I^2(|g|^2)$ if $\int f|g|^2 dm = f(0)$ for every function $f \in K_I^2(|g|^2)$. Our example is constructed in the spirit of [48, p. 356]. Fix $\alpha \in (0, \frac{1}{2})$. Let $\gamma(z) = (1-z)^\alpha$ and let g be an outer function in H^2 such that $|g|^2 = \operatorname{Re}\gamma$ a.e. on $\mathbb{T}$ (such a function clearly exists). Let now $I = B_\Lambda$ be an infinite Blaschke product with $0 \in \Lambda$. If Λ accumulates to points outside 1, then the corollary shows that $K_I^2 = K_I^2(|g|^2)$. Let us check that g is not extremal. To this end we compute $\int k_\lambda |g|^2 dm$ for $\lambda \in \Lambda$ (recall that for $\lambda \in \Lambda$, $k_\lambda \in K_I^2 = K_I^2(|g|^2)$):

$$\begin{aligned}(26)\qquad \int k_\lambda |g|^2\, dm &= \int k_\lambda \operatorname{Re}\gamma\, dm = \frac{1}{2}\Big(\int k_\lambda \gamma\, dm + \int k_\lambda \overline{\gamma}\, dm\Big)\\ &= \frac{1}{2}k_\lambda(0)\gamma(0) + \frac{1}{2}\langle k_\lambda, \gamma\rangle = \frac{1}{2}(1 + \overline{(1-\lambda)^\alpha})\end{aligned}$$

which is different from $k_\lambda(0) = 1$ (except when $\lambda = 0$). Hence g is not extremal.

We could also have obtained the nonextremality of g from Sarason's result [64, Theorem 2] using the parametrization $g = a/(1-b)$ appearing in Sarason's and Hayashi's work (see [46] for details on this second argument).

It is clear that the corollary is still valid when Γ is replaced by a finite union of intervals. However, we can construct an infinite union of intervals $\Gamma = \bigcup_{n\geq 1}\Gamma_n$ each of which does not meet $\sigma(I)$, an outer function g satisfying the yet weaker integrability condition $1/g \in L^s(\Gamma)$, $s < 2$, and $|g| \geq \delta$ on $\mathbb{T}\setminus\Gamma$, and an inner function I such that $K_I^2(|g|^2) \not\subset K_I^2$. The function g obtained in this construction does not satisfy $|g|^2 \in (A_2)$. (See [46] for details.)

Another simple observation concerning the local integrability condition $1/g \in L^s(\Gamma)$, $s > q$: if it is replaced by the global condition $1/g \in L^s(\mathbb{T})$, then by Hölder's inequality we have an embedding into a bigger backward shift invariant subspace:

Proposition 4 (Hartmann, 2008). *Let $1 < p < \infty$ and $1/p + 1/q = 1$. If there exists $s > q$ such that $1/g \in L^s(\mathbb{T})$, then for r with $1/r = 1/p + 1/s$ we have $L^p(|g|^p) \subset L^r$.*

So in this situation we of course also have $K_I^p(|g|^p) \subset K_I^r$. In particular, every function $f \in K_I^p(|g|^p)$ admits a pseudocontinuation and extends analytically outside $\sigma(I)$. Again the Ahern–Clark condition does not give complete information for the points located in the spectrum of I since (ii) of Theorem 2 depends on p.

When one allows g to vanish in points contained in $\sigma(I)$, then it is possible to construct examples with $|g|^p \in (A_p)$ and $K_I^p(|g|^p) \not\subset K_I^p$: take for instance $I = B_\Lambda$ the Blasche product vanishing exactly in $\Lambda = \{1-1/2^n\}_n$ and $g(z) = (1-z)^\alpha$, where $\alpha \in (0, \frac{1}{2})$ and $p = 2$ (see [46] for details; the condition $|g|^2 \in (A_2)$ is required in the proof to show that $K_I^2(|g|^2) = P_+\big((1/\bar{g})K_I^2\big)$ — see Lemma 2 below — which gives an explicit description of K_I^2 in terms of coefficients with respect to an unconditional basis). The following crucial example is in the spirit of this observation.

4.1. An example. In the spirit of the example given in [46, Proposition 4] we shall now discuss the condition (ii) of Theorem 2 in the context of weighted backward shift invariant subspaces.

We first have to recall Lemma 1 from [46]:

Lemma 2 (Hartmann, 2008). *Suppose $|g|^p$ is an (A_p) weight and I an inner function. Then $A_0 = P_+1/\bar{g}\colon H^p \to H^p(|g|^p)$ is an isomorphism of K_I^p onto $K_I^p(|g|^p)$. Also, for every $\lambda \in \mathbb{D}$ we have*

$$A_0 k_\lambda = \frac{k_\lambda(\mu)}{\overline{g(\lambda)}}. \tag{27}$$

We return to the situation $p = 2$. Take $g(z) = (1-z)^\alpha$ with $\alpha \in (0, 1/2)$. Then $|g|^2$ is (A_2). Let

$$r_n = 1 - \frac{1}{2^n}, \qquad \theta_n = (1-r_n)^s = \frac{1}{2^{ns}}, \qquad \lambda_n = r_n e^{i\theta_n},$$

where $s \in (0, \frac{1}{2})$. Hence the sequence $\Lambda = \{\lambda_n\}_n$ tends tangentially to 1. Set $I = B_\Lambda$. We check the Ahern–Clark condition in $\zeta = 1$ for $l = 0$ (which means that we are just interested in the existence of nontangential limits in $\zeta = 1$). Observe that for $s \in (0, \frac{1}{2})$ we have

$$|1 - r_n e^{i\theta_n}|^2 \simeq (1-r_n)^2 + \theta_n^2 = \frac{1}{2^{2n}} + \frac{1}{2^{2ns}} \simeq \frac{1}{2^{2ns}}, \tag{28}$$

and so when $q > 1$

$$\sum_{n\geq 1} \frac{1-r_n^2}{|1-r_n e^{i\theta_n}|^q} \simeq \sum_{n\geq 1} \frac{1/2^n}{1/2^{nsq}} \simeq \sum_{n\geq 1} 2^{n(sq-1)}. \tag{29}$$

The latter sum is bounded when $q = 2$ which implies in the unweighted situation that every function in the backward shift invariant subspace K_I^2 has a nontangential limit at 1. Note also that since $|g|^2 \in (A_2)$, by Proposition 4 and comments thereafter, $K_I^2(|g|^2)$ imbeds into some K_I^r, $r < 2$. Now taking $q = r' > 2$, where $1/r + 1/r' = 1$, we see that the sum in (29) diverges when $sr' \geq 1$ and converges for $sr' < 1$. So depending on the parameters s and α we can assert continuation or not. It will be clear a posteriori that in our situation r has to be such that $sr' \geq 1$.

Note that $\sigma(I) \cap \mathbb{T} = \{1\}$, which corresponds to the point where g vanishes. Clearly, Λ is an interpolating sequence, and so the sequence $\{k_{\lambda_n}/\|k_{\lambda_n}\|_2\}_n$ is a normalized unconditional basis in K_I^2. This means that we can write $K_I^2 = l^2(\frac{k_{\lambda_n}}{\|k_{\lambda_n}\|_2})$

meaning that $f \in K_I^2$ if and only if

$$f = \sum_{n\geq 1} \alpha_n \frac{k_{\lambda_n}}{\|k_{\lambda_n}\|_2}$$

with $\sum_{n\geq 1}|\alpha_n|^2 < \infty$ (the last sum defines the square of an equivalent norm in K_I^2).

As already mentioned $|g|^2$ is Muckenhoupt (A_2). This implies in particular that we have the local integrability condition $1/g \in L^s(\Gamma)$ for some $s > 2$ and Γ an arc containing the point 1. Moreover, we get from (27)

$$\{A_0(k_{\lambda_n}/\|k_{\lambda_n}\|_2)\}_n = \left\{\frac{k_{\lambda_n}}{\overline{g(\lambda_n)}\|k_{\lambda_n}\|_2}\right\}_n,$$

and $\{k_{\lambda_n}/(\overline{g(\lambda_n)}\|k_{\lambda_n}\|_2)\}_n$ is an unconditional basis in $K_I^2(|g|^2)$ (almost normalized in the sense that $\|A_0(k_{\lambda_n}/\|k_{\lambda_n}\|_2)\|_{|g|^2}$ is comparable to a constant independant of n). Hence for every sequence $\alpha = (\alpha_n)_n$ with $\sum_{n\geq 1}|\alpha_n^2| < \infty$, we have

$$f_\alpha := \sum_{n\geq 1} \frac{\alpha_n}{g(\lambda_n)} \frac{k_{\lambda_n}}{\|k_{\lambda_n}\|_2} \in K_I^2(|g|^2).$$

To fix the ideas we will now pick $\alpha_n = 1/n^{1/2+\varepsilon}$ for some $\varepsilon > 0$ so that $\sum_n \alpha_n k_{\lambda_n}/\|k_{\lambda_n}\|_2$ is in K_I^2, and hence $f_\alpha \in K_I^2(|g|^2)$. Let us show that f_α does not have a nontangential limit in 1. Fix $t \in (0,1)$. Then

$$f_\alpha(t) = \sum_n \frac{\alpha_n}{g(\lambda_n)} \frac{k_{\lambda_n}(t)}{\|k_{\lambda_n}\|_2}.$$

We have $\|k_{\lambda_n}\|_2 = 1/\sqrt{1-|\lambda_n|^2} \simeq 2^{n/2}$. Also as in (28),

$$|g(\lambda_n)| = |1-\lambda_n|^\alpha \simeq \theta_n^\alpha = \frac{1}{2^{ns\alpha}}.$$

Changing the arguments of the α_n's and renormalizing, we can suppose that

$$\frac{\alpha_n}{\overline{g(\lambda_n)}\|k_{\lambda_n}\|_2} = \frac{2^{n(s\alpha-1/2)}}{n^{1/2+\varepsilon}}$$

Let us compute the imaginary part of f_α in t. Observe that the imaginary part of $1/(1-t\overline{\lambda_n})$ is negative. More precisely, assuming $t \in [\frac{1}{2}, 1)$ and $n \geq N_0$,

$$\operatorname{Im}\frac{1}{1-t\overline{\lambda_n}} = \operatorname{Im}\frac{1-t\lambda_n}{|1-t\overline{\lambda_n}|^2} = \frac{-tr_n\sin\theta_n}{|1-t\overline{\lambda_n}|^2} \simeq \frac{-\theta_n}{|1-t\overline{\lambda_n}|^2} = \frac{-1/2^{ns}}{|1-t\overline{\lambda_n}|^2}$$

Also for $n \geq N = \log_2\big(1/(1-t)\big)$, we have $1-t \geq 1/2^n$ and $r_n = 1-1/2^n \geq t$, so that for these n

$$\begin{aligned}|1-t\overline{\lambda_n}|^2 &\simeq (1-tr_n)^2 + \theta_n^2 \leq (1-t^2)^2 + \theta_n^2 \leq 4(1-t)^2 + \theta_N^2 \\ &\leq 4(1-t)^2 + c(1-t)^{2s} \lesssim (1-t)^{2s}\end{aligned}$$

So

$$\begin{aligned}|\operatorname{Im} f_\alpha(t)| &= \left|\operatorname{Im}\sum_n \frac{\alpha_n}{g(\lambda_n)} \frac{k_{\lambda_n}(t)}{\|k_{\lambda_n}\|_2}\right| \gtrsim \sum_{n\geq\log_2(1/(1-t))} \frac{2^{n(s\alpha-1/2)}}{n^{1/2+\varepsilon}} \frac{1/2^{ns}}{(1-t)^{2s}} \\ &\gtrsim \frac{1}{(1-t)^{2s}} \sum_{n\geq\log_2(1/(1-t))} \frac{1}{2^{\gamma n}} \simeq \frac{1}{(1-t)^{2s}} \frac{1}{2^{\gamma\log_2(1/(1-t))}} \\ &\gtrsim (1-t)^{\gamma-2s},\end{aligned}$$

where $\gamma = s + \frac{1}{2} - s\alpha + \delta$ for an arbitrarily small δ (this compensates the term $n^{1/2+\varepsilon}$). So $\gamma - 2s = \frac{1}{2} - s(1+\alpha) + \delta$ which can be made negative by choosing s closely enough to $\frac{1}{2}$.

We conclude that the function f_α is not bounded in 1 and thus cannot have a nontangential limit in $\zeta = 1$

References

1. P. R. Ahern and D. N. Clark, *Invariant subspaces and analytic continuation in several variables*, J. Math. Mech. **19** (1969/1970), 963–969.
2. ———, *Radial limits and invariant subspaces*, Amer. J. Math. **92** (1970), 332–342.
3. ———, *Radial nth derivatives of Blaschke products*, Math. Scand. **28** (1971), 189–201.
4. ———, *On inner functions with H^p-derivative*, Michigan Math. J. **21** (1974), 115–127.
5. A. B. Aleksandrov, *Invariant subspaces of shift operators. An axiomatic approach*, Zap. Nauchn. Sem. Leningrad. Otdel. Mat. Inst. Steklov. (LOMI) **113** (1981), Issled. Linein. Oper. Teorii Funktsii. 11, 7–26 (Russian); English transl., J. Soviet Math. **22** (1983), no. 6, 1695–1708.
6. ———, *Inner functions and related spaces of pseudocontinuable functions*, Zap. Nauchn. Sem. Leningrad. Otdel. Mat. Inst. Steklov. (LOMI) **170** (1989), Issled. Linein. Oper. Teorii Funktsii. 17, 7–33 (Russian, with English summary); English transl., J. Soviet Math. **63** (1993), no. 2, 115–129.
7. ———, *On the existence of angular boundary values of pseudocontinuable functions*, Zap. Nauchn. Sem. S.-Peterburg. Otdel. Mat. Inst. Steklov. (POMI) **222** (1995), Issled. po Linein. Oper. i Teor. Funktsii. 23, 5–17 (Russian, with English and Russian summaries); English transl., J. Math. Sci. (New York) **87** (1997), no. 5, 3781–3787.
8. ———, *Embedding theorems for coinvariant subspaces of the shift operator.* II, Zap. Nauchn. Sem. S.-Peterburg. Otdel. Mat. Inst. Steklov. (POMI) **262** (1999), Issled. po Linein. Oper. i Teor. Funkts. 27, 5–48 (Russian); English transl., J. Math. Sci. (New York) **110** (2002), no. 5, 2907–2929.
9. A. Aleman, S. Richter, and W. T. Ross, *Pseudocontinuations and the backward shift*, Indiana Univ. Math. J. **47** (1998), no. 1, 223–276.
10. J. M. Anderson and J. Rovnyak, *On generalized Schwarz–Pick estimates*, Mathematika **53** (2006), no. 1, 161–168.
11. A. D. Baranov, *Weighted Bernstein inequalities and embedding theorems for model subspaces*, Algebra i Analiz **15** (2003), no. 5, 138–168 (Russian); English transl., St. Petersburg Math. J. **15** (2004), no. 5, 733–752.
12. ———, *Bernstein-type inequalities for shift-coinvariant subspaces and their applications to Carleson embeddings*, J. Funct. Anal. **223** (2005), no. 1, 116–146.
13. ———, *Stability of bases and frames of reproducing kernels in model spaces*, Ann. Inst. Fourier (Grenoble) **55** (2005), no. 7, 2399–2422.
14. ———, *Completeness and Riesz bases of reproducing kernels in model subspaces*, Int. Math. Res. Not. (2006), Art. ID 81530.
15. ———, *Compact and Schatten class c of star-invariant subspaces in the disc*, Izv. Math., to appear, available at `arXiv:0712.0684`.
16. A. D. Baranov, E. Fricain, and J. Mashreghi, *Weighted norm inequalities for de Branges–Rovnyak spaces and their applications*, Amer J. Math., to appear, available at `arXiv:0802.0789`.
17. S. N. Bernstein, *Leçons sur les propriétés extrémales et la meilleure approximation des fonctions analytiques d'une variable réelle*, Gauthier-Villars, Paris, 1926.
18. A. Beurling, *Analytic continuation across a linear boundary*, Acta Math. **128** (1972), no. 3-4, 153–182.
19. R. P. Boas Jr., *Entire functions*, Pure Appl. Math., vol. 5, Academic Press, New York, 1954.
20. V. Bolotnikov and A. Kheifets, *A higher order analogue of the Carathéodory–Julia theorem*, J. Funct. Anal. **237** (2006), no. 1, 350–371.
21. L. Carleson, *An interpolation problem for bounded analytic functions*, Amer. J. Math. **80** (1958), 921–930.

22. ———, *Interpolations by bounded analytic functions and the corona problem*, Ann. of Math. (2) **76** (1962), 547–559.
23. N. Chevrot, E. Fricain, and D. Timotin, *On certain Riesz families in vector-valued de Branges–Rovnyak spaces*, J. Math. Anal. Appl. **355** (2009), no. 1, 110–125.
24. J. A. Cima and A. L. Matheson, *On Carleson embeddings of star-invariant supspaces*, Quaest. Math. **26** (2003), no. 3, 279–288.
25. J. A. Cima and W Ross T., *The backward shift on the Hardy space*, Math, Surveys Monogr., vol. 79, Amer. Math. Soc., Providence, RI, 2000.
26. D. N. Clark, *One dimensional perturbations of restricted shifts*, J. Analyse Math. **25** (1972), 169–191.
27. W. S. Cohn, *Carleson measures for functions orthogonal to invariant subspaces*, Pacific J. Math. **103** (1982), no. 2, 347–364.
28. ———, *Radial limits and star invariant subspaces of bounded mean oscillation*, Amer. J. Math. **108** (1986), no. 3, 719–749.
29. ———, *Carleson measures and operators on star-invariant subspaces*, J. Operator Theory **15** (1986), no. 1, 181–202.
30. L. de Branges and J. Rovnyak, *Canonical models in quantum scattering theory*, Perturbation Theory and its Applications in Quantum Mechanics (Madison, WI, 1965), Wiley, New York, 1966, pp. 295–392.
31. ———, *Square summable power series*, Holt, Rinehart and Winston, New York, 1966.
32. R. G. Douglas, H. S. Shapiro, and A. L. Shields, *Cyclic vectors and invariant subspaces for the backward shift operator*, Ann. Inst. Fourier (Grenoble) **20** (1970), no. 1, 37–76.
33. P. L. Duren, *Theory of H^p spaces*, Pure Appl. Math., vol. 38, Academic Press, New York, 1970.
34. K. M. Dyakonov, *Entire functions of exponential type and model subspaces in H^p*, Zap. Nauchn. Sem. Leningrad. Otdel. Mat. Inst. Steklov. (LOMI) **190** (1991), Issled. po Linein. Oper. i Teor. Funktsii. 19, 81–100 (Russian); English transl., J. Math. Sci. **71** (1994), no. 1, 2222–2233.
35. ———, *Kernels of Toeplitz operators, smooth functions and Bernstein-type inequalities*, Zap. Nauchn. Sem. S.-Peterburg. Otdel. Mat. Inst. Steklov. (POMI) **201** (1992), Issled. po Linein. Oper. Teor. Funktsii. 20, 5–21 (Russian); English transl., J. Math. Sci. **78** (1996), no. 2, 131–141.
36. ———, *Continuous and compact embeddings between star-invariant subspaces*, Complex Analysis, Operators, and Related Topics, Oper. Theory Adv. Appl., vol. 113, Birkhäuser, Basel, 2000, pp. 65–76.
37. ———, *Differentiation in star-invariant subspaces.* I: *Boundedness and compactness*, J. Funct. Anal. **192** (2002), no. 2, 364–386.
38. ———, *Differentiation in star-invariant subspaces.* II: *Schatten class criteria*, J. Funct. Anal. **192** (2002), no. 2, 387–409.
39. ———, *Two theorems on star-invariant subspaces of BMOA*, Indiana Univ. Math. J. **56** (2007), no. 2, 643–658.
40. ———, *Meromorphic functions and their derivatives: equivalence of norms*, Indiana Univ. Math. J. **57** (2008), no. 4, 1557–1571.
41. ———, *Toeplitz operators and arguments of analytic functions*, Math. Ann. **344** (2009), no. 2, 353–380.
42. K. Dyakonov and D. Khavinson, *Smooth functions in star-invariant subspaces*, Recent Advances in Operator-Related Function Theory, Contemp. Math., vol. 393, Amer. Math. Soc., Providence, RI, 2006, pp. 59–66.
43. E. Fricain, *Bases of reproducing kernels in de Branges spaces*, J. Funct. Anal. **226** (2005), no. 2, 373–405.
44. E. Fricain and J. Mashreghi, *Boundary behavior of functions in the de Branges–Rovnyak spaces*, Complex Anal. Oper. Theory **2** (2008), no. 1, 87–97.
45. ———, *Integral representation of the n-th derivative in de Branges–Rovnyak spaces and the norm convergence of its reproducing kernel*, Ann. Inst. Fourier (Grenoble) **58** (2008), no. 6, 2113–2135.
46. A. Hartmann, *Analytic continuations and embeddings*, preprint.
47. A. Hartmann, D. Sarason, and K. Seip, *Surjective Toeplitz operators*, Acta Sci. Math. (Szeged) **70** (2004), no. 3-4, 609–621.

48. A. Hartmann and K. Seip, *Extremal functions as divisors for kernels of Toeplitz operators*, J. Funct. Anal. **202** (2003), no. 2, 342–362.
49. E. Hayashi, *Classification of nearly invariant subspaces of the backward shift*, Proc. Amer. Math. Soc. **110** (1990), no. 2, 441–448.
50. D. Hitt, *Invariant subspaces of $\mathcal{H}^2$ of an annulus*, Pacific J. Math. **134** (1988), no. 1, 101–120.
51. S. V. Hruščëv, N. K. Nikol′skiĭ, and B. S. Pavlov, *Unconditional bases of exponentials and of reproducing kernels*, Complex Analysis and Spectral Theory (Leningrad, 1979/1980), Lecture Notes in Math., vol. 864, Springer, Berlin, 1981, pp. 214–335.
52. M. T. Jury, *Reproducing kernels, de Branges–Rovnyak spaces, and norms of weighted composition operators*, Proc. Amer. Math. Soc. **135** (2007), no. 11, 3669–3675.
53. V. V. Kapustin, *Real functions in weighted Hardy spaces*, Zap. Nauchn. Sem. S.-Peterburg. Otdel. Mat. Inst. Steklov. (POMI) **262** (1999), Issled. po Linein. Oper. i Teor. Funkts. 27, 138–146 (Russian, with English and Russian summaries); English transl., J. Math. Sci. (New York) **110** (2002), no. 5, 2986–2990.
54. B. Ya. Levin, *Lectures on entire functions*, Transl. Math. Monogr., vol. 150, Amer. Math. Soc., Providence, RI, 1996.
55. M. B. Levin, *An estimate of the derivative of a meromorphic function on the boundary of the domain*, Dokl. Akad. Nauk SSSR **216** (1974), 495–497 (Russian).
56. James W. Moeller, *On the spectra of some translation invariant spaces*, J. Math. Anal. Appl. **4** (1962), 276–296.
57. B. Muckenhoupt, *Weighted norm inequalities for the Hardy maximal function*, Trans. Amer. Math. Soc. **165** (1972), 207–226.
58. F. Nazarov and A. Volberg, *The Bellman function, the two-weight Hilbert transform, and embeddings of the model spaces K_θ*, J. Anal. Math. **87** (2002), 385–414.
59. N. K. Nikol′skiĭ, *Operators, functions, and systems: an easy reading.* Vol. 1: *Hardy, Hankel, and Toeplitz*, Math. Surveys Monogr., vol. 92, Amer. Math. Soc., Providence, RI, 2002; Vol. 2: *Model operators and systems*, vol. 93.
60. A. Poltoratski and D. Sarason, *Aleksandrov–Clark measures*, Recent Advances in Operator-Related Function Theory, Contemp. Math., vol. 393, Amer. Math. Soc., Providence, RI, 2006, pp. 1–14.
61. R. Rochberg, *Toeplitz operators on weighted H^p spaces*, Indiana Univ. Math. J. **26** (1977), no. 2, 291–298.
62. W. T. Ross and H. S. Shapiro, *Generalized analytic continuation*, Univ. Lecture Ser., vol. 25, Amer. Math. Soc., Providence, RI, 2002.
63. D. Sarason, *Shift-invariant spaces from the Brangesian point of view*, The Bieberbach Conjecture (West Lafayette, IN, 1985), Math. Surveys Monogr., vol. 21, Amer. Math. Soc., Providence, RI, 1986, pp. 153–166.
64. Donald Sarason, *Nearly invariant subspaces of the backward shift*, Contributions to operator theory and its applications (Mesa, AZ, 1987), Oper. Theory Adv. Appl., vol. 35, Birkhäuser, Basel, 1988, pp. 481–493.
65. D. Sarason, *Exposed points in H^1*. I, The Gohberg Anniversary Collection, Vol. II (Calgary, AB, 1988), Oper. Theory Adv. Appl., vol. 41, Birkhäuser, Basel, 1989, pp. 485–496.
66. ———, *Kernels of Toeplitz operators*, Toeplitz Operators and Related Topics (Santa Cruz, CA, 1992), Oper. Theory Adv. Appl., vol. 71, Birkhäuser, Basel, 1994, pp. 153–164.
67. ———, *Sub-Hardy Hilbert spaces in the unit disk*, Univ. Arkansas Lecture Notes Math. Sci., vol. 10, Wiley, New York, 1994.
68. H. S. Shapiro, *Some observations concerning weighted polynomial approximation of holomorphic functions*, Mat. Sb. (N.S.) **73 (115)** (1967), 320–330 (Russian).
69. J. E. Shapiro, *Relative angular derivatives*, J. Operator Theory **46** (2001), no. 2, 265–280.
70. ———, *More relative angular derivatives*, J. Operator Theory **49** (2003), no. 1, 85–97.
71. B. Sz.-Nagy and C. Foias, *Analyse harmonique des opérateurs de l'espace de Hilbert*, Masson, Paris, 1967.
72. A. L. Volberg, *Thin and thick families of rational fractions*, Complex Analysis and Spectral Theory (Leningrad, 1979/1980), Lecture Notes in Math., vol. 864, Springer, Berlin, 1981, pp. 440–480.
73. A. L. Volberg and S. R. Treil, *Embedding theorems for invariant subspaces of the inverse shift operator*, Zap. Nauchn. Sem. Leningrad. Otdel. Mat. Inst. Steklov. (LOMI) **149** (1986),

Issled. Linein. Teor. Funktsii. 15, 38 – 51 (Russian); English transl., J. Soviet Math. **42** (1988), no. 2, 1562 – 1572.

Institut Camille Jordan, Université Claude Bernard Lyon 1, 43, boulevard du 11 novembre 1918, 69622 Villeurbanne Cedex, France
E-mail address: fricain@math.univ-lyon1.fr

Institut Mathématiques de Bordeaux, Université Bordeaux 1, 351, cours de la Libération, 33405 Talence Cedex, France
E-mail address: hartmann@math.univ-bordeaux1.fr

Centre de Recherches Mathématiques
CRM Proceedings and Lecture Notes
Volume **51**, 2010

The Search for Singularities of Solutions to the Dirichlet Problem: Recent Developments

Dmitry Khavinson and Erik Lundberg

Abstract. This is a survey article based on an invited talk delivered by the first author at the CRM workshop on Hilbert Spaces of Analytic Functions held at CRM, Université de Montréal, December 8 – 12, 2008.

1. The main question

Let Ω be a smoothly bounded domain in $\mathbb{R}^n$. Consider the Dirichlet problem (DP) in Ω of finding the function u, say, $\in C^2(\Omega)\bigcap C(\overline{\Omega})$ and satisfying

$$\begin{cases} \Delta u = 0 \\ u|_\Gamma = v \end{cases}, \tag{1.1}$$

where $\Delta = \sum_{j=1}^n \partial^2/\partial x_j^2$ and $\Gamma := \partial\Omega$, $v \in C(\Gamma)$. It is well known since the early 20th century from works of Poincaré, C. Neumann, Hilbert, and Fredholm that the solution u exists and is unique. Also, since u is harmonic in Ω, hence real-analytic there, no singularities can appear in Ω. Moreover, assuming $\Gamma := \partial\Omega$ to consist of real-analytic hypersurfaces, the more recent and difficult results on "elliptic regularity" assure us that if the data v is real-analytic in a neighborhood of $\overline{\Omega}$ then u extends as a real-analytic function across $\partial\Omega$ into an open neighborhood Ω' of $\overline{\Omega}$. In two dimensions, this can be done using the reflection principle. In higher dimensions, the boundary can be biholomorphically "flattened," but this leads to a general elliptic operator for which the reflection principle does not apply. Instead, analyticity must be shown by directly verifying convergence of the power series representing the solution through difficult estimates on the derivatives (see [14]).

Question. Suppose the data v is a restriction to Γ of a "very good" function, say an entire function of variables $x_1, x_2, \dots, x_n$. In other words, the data presents no reasons whatsoever for the solution u of (1.1) to develop singularities.

(i) Can we then assert that all solutions u of (1.1) with entire data $v(x)$ are also entire?

2000 *Mathematics Subject Classification.* Primary 31B20; Secondary 30H20.
Both authors gratefully acknowledge partial support from the National Science Foundation.
This is the final form of the paper.

(ii) If singularities do occur, they must be caused by geometry of Γ interacting with the differential operator Δ. Can we then find data v_0 that would force the worst possible scenario to occur? More precisely, for any entire data v, the set of possible singularities of the solution u of (1.1) is a subset of the singularity set of u_0, the solution of (1.1) with data v_0.

2. The Cauchy problem

An inspiration to this program launched by H. S. Shapiro and the first author in [22] comes from reasonable success with a similar program in the mid 1980's regarding the analytic Cauchy problem (CP) for elliptic operators, in particular, the Laplace operator. For the latter, we are seeking a function u with $\Delta u = 0$ near Γ and satisfying the initial conditions

$$\begin{cases} (u-v)|_\Gamma = 0 \\ \nabla(u-v)|_\Gamma = 0 \end{cases}, \tag{2.1}$$

where v is assumed to be real-analytic in a neighborhood of Γ. Suppose as before that the data v is a "good" function (e.g., a polynomial or an entire function). In that context, the techniques developed by J. Leray [26] in the 1950s (and jointly with L. Gårding and T. Kotake [15]) together with the works of P. Ebenfelt [11], G. Johnsson [18], and, independently, by B. Sternin and V. Shatalov [33] in Russia and their school produced a more or less satisfactory understanding of the situation. To mention briefly, the answer (for the CP) to question (i) in two dimensions is essentially "never" unless Γ is a line while for (ii) the data mining all possible singularities of solutions to the CP with entire data is $v_0 = |x|^2 = \sum x_j^2$ (see [19–21, 34] and references therein).

3. The Dirichlet problem: When does entire data imply entire solution?

Let us raise question (i) again for the Dirichlet problem: Does (real) entire data v imply entire solution u of (1.1)?

In this section and the next, $\mathbb{P}$ will denote the space of polynomials and $\mathbb{P}_N$ the space of polynomials of degree $\leq N$. The following pretty fact goes back to the 19th century and can be associated with the names of E. Heine, G. Lamé, M. Ferrers, and probably many others (cf. [20]). The proof is from [22] (cf. [2, 3]).

Proposition 3.1. *If* $\Omega := \{x : \sum x_j^2/a_j^2 - 1 < 0, a_1 > \cdots > a_n > 0\}$ *is an ellipsoid, then any DP with a polynomial data of degree* N *has a polynomial solution of degree* $\leq N$.

Proof. Let $q(x) = \sum x_j^2/a_j^2 - 1$ be the defining function for $\Gamma := \partial\Omega$. The (linear) map $T\colon P \to \Delta(qP)$ sends the finite-dimensional space $\mathbb{P}_N$ into itself. T is injective (by the maximum principle) and, therefore, surjective. Hence, for any P, $\deg P \geq 2$ we can find P_0, $\deg P_0 \leq \deg P - 2$. $TP_0 = \Delta(qP_0) = \Delta P$. $u = P - qP_0$ is then the desired solution. □

The following result was proved in [22].

Theorem 3.2. *Any solution to DP* (1.1) *in an ellipsoid* Ω *with entire data is also entire.*

Later on, D. Armitage sharpened the result by showing that the order and the type of the data are carried over, more or less, to the solution [1]. The following conjecture has also been formulated in [22].

Conjecture 3.3. *Ellipsoids are the only bounded domains in* $\mathbb{R}^n$ *for which Theorem* 3.2 *holds, i.e., ellipsoids are the only domains in which entire data implies entire solution for the DP* (1.1).

In 2005 H. Render [30] proved this conjecture for all *algebraically* bounded domains Ω defined as bounded components of $\{\phi(x) < 0, \phi \in \mathbb{P}_N\}$ such that $\{\phi(x) = 0\}$ is a *bounded* set in $\mathbb{R}^n$ or, equivalently, the senior homogeneous part $\phi_N(x)$ of ϕ is elliptic, i.e., $|\phi_N(x)| \geq C|x|^N$ for some constant C. For $n = 2$, an easier version of this result was settled in 2001 by M. Chamberland and D. Siegel [6]. At the beginning of the next section we will outline their argument, which establishes similar results as Render's for the following modified conjecture.

Conjecture 3.4. *Ellipsoids are the only surfaces for which polynomial data implies polynomial solution.*

Remark. We will return to Render's theorem below. For now let us note that, unfortunately, it already tells us nothing even in 2 dimensions for many perturbations of a unit disk, e.g., $\Omega := \{x \in \mathbb{R}^2 : x^2 + y^2 - 1 + \varepsilon h(x,y) < 0\}$ where, say, h is a harmonic polynomial of degree > 2.

4. When does polynomial data imply polynomial solution?

Let $\gamma = \{\phi(x) = 0\}$ be a bounded, irreducible algebraic curve in $\mathbb{R}^2$. If the DP posed on γ has polynomial solution whenever the data is a polynomial, then as Chamberland and Siegel observed, (a) γ is an ellipse or (b) there exists data $f \in \mathbb{P}$ such that the solution $u \in \mathbb{P}$ of DP has $\deg u > \deg f$.

In case (b) $u - f|_\gamma = 0$ implies that ϕ divides $u - f$ by Hilbert's Nullstelensatz, and, since $\deg u = M > \deg f$, $u_M = \phi_k g_l$ where ϕ_k and u_M are the senior homogeneous terms of ϕ and u respectively. The senior term of u must have the form $u_M = az^M + b\bar{z}^M$ since u_M is harmonic. Hence, u_M factors into linear factors and so must ϕ_k. Hence γ is unbounded. This gives the following result [6].

Theorem 4.1. *Suppose* $\deg \phi > 2$ *and* ϕ *is square-free. If the Dirichlet problem posed on* $\{\phi = 0\}$ *has a polynomial solution for each polynomial data, then the senior part of* ϕ*, which we denote by* ϕ_N*, of order* N*, factors into real linear terms, namely,*

$$\phi_N = \prod_{j=0}^{n} (a_j x - b_j y),$$

where a_j, b_j *are some real constants and the angles between the lines* $a_j x - b_j y = 0$*, for all* j*, are rational multiples of* π*.*

This theorem settles Conjecture 3.4 for bounded domains $\Omega \subseteq \{\phi(x) < 0\}$ such that the set $\{\phi(x) = 0\}$ is bounded in $\mathbb{R}^2$. However, the theorem leaves open simple cases such as $x^2 + y^2 - 1 + \varepsilon(x^3 - 3xy^2)$.

Example. The curve $y(y - x)(y + x) - x = 0$ (see Figure 1) satisfies the necessary condition imposed by the theorem. Moreover, any quadratic data can be matched on it by a harmonic polynomial. For instance, $u = xy(y^2 - x^2)$ solves the

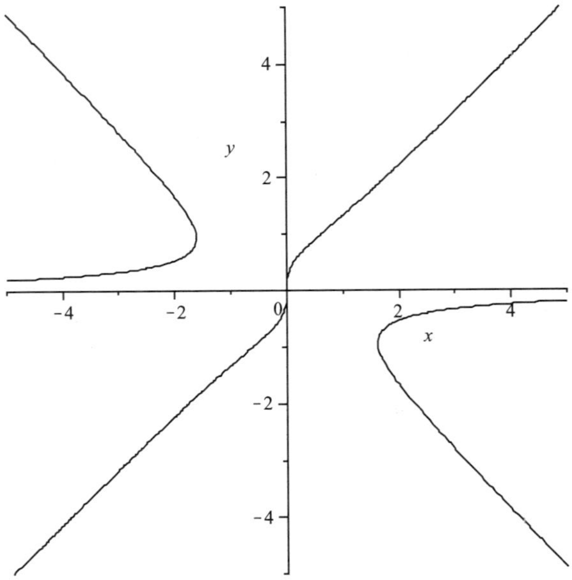

FIGURE 1. A cubic on which any quadratic data can be matched by a harmonic polynomial.

interpolation problem (it is misleading to say "Dirichlet" problem, since there is no bounded component) with data $v(x, y) = x^2$. On the other hand, one can show (nontrivially) that the data x^3 does not have polynomial solution.

5. Dirichlet's problem and orthogonal polynomials

Most recently, N. Stylianopoulos and the first author showed that if for a polynomial data there always exists a polynomial solution of the DP (1.1), with an additional constraint on the degree of the solution in terms of the degree of the data (see below), then Ω is an ellipse [23]. This result draws on the 2007 paper of M. Putinar and N. Stylianopoulos [29] that found a simple but surprising connection between Conjecture 3.4 in $\mathbb{R}^2$ and (Bergman) orthogonal polynomials, i.e. polynomials orthogonal with respect to the inner product $\langle p, q\rangle_\Omega := \int_\Omega p\bar{q}\, dA$, where dA is the area measure. To understand this connection let us consider the following properties:

(1) There exists k such that for a polynomial data of degree n there always exists a polynomial solution of the DP (1.1) posed on Ω of degree $\leq n + k$.

(2) There exists N such that for all m, n, the solution of (1.1) with data $\bar{z}^m z^n$ is a harmonic polynomial of degree $\leq (N-1)m+n$ in z and of degree $\leq (N-1)n+m$ in $\bar{z}$.

(3) There exists N such that orthogonal polynomials $\{p_n\}$ of degree n on Ω satisfy a (finite) $(N + 1)$-recurrence relation, i.e.,

$$zp_n = a_{n+1,n}p_{n+1} + a_{n,n}p_n + \cdots + a_{n-N+1}p_{n-N+1},$$

where $a_{n-j,n}$ are constants depending on n.

(4) The Bergman orthogonal polynomials of Ω satisfy a finite-term recurrence relation, i.e., for every fixed $k > 0$, there exists an $N(k) > 0$, such that $a_{k,n} = \langle zp_n, p_k \rangle = 0$, $n \geq N(k)$.

(5) Conjecture 3.4 holds for Ω.

Putinar and Stylianopoulos noticed that with the additional minor assumption that polynomials are dense in $L^2_a(\Omega)$, properties (4) and (5) are equivalent. Thus, they obtained as a corollary (by way of Theorem 4.1 from the previous section) that the only bounded algebraic sets satisfying property (4) are ellipses. We also have (1) $\Longrightarrow$ (2), (2) $\Longleftrightarrow$ (3), and (3) $\Longrightarrow$ (4). Stylianopoulos and the first author used the equivalence of properties (2) and (3) to prove the following theorem which has an immediate corollary.

Theorem 5.1. *Suppose $\partial\Omega$ is C^2-smooth, and orthogonal polynomials on Ω satisfy a (finite) $(N+1)$-recurrence relation, in other words property (3) is satisfied. Then, $N = 2$ and Ω is an ellipse.*

Corollary 5.2. *Suppose $\partial\Omega$ is a C^2-smooth domain for which there exists N such that for all m, n, the solution of (1.1) with data $\bar{z}^m z^n$ is a harmonic polynomial of degree $\leq (N-1)m + n$ in z and of degree $\leq (N-1)n + m$ in $\bar{z}$. Then $N = 2$ and Ω is an ellipse.*

SKETCH OF PROOF. First, one notes that all the coefficients in the recurrence relation are bounded. Divide both sides of the recurrence relation above by p_n and take the limit of an appropriate subsequence as $n \to \infty$. Known results on asymptotics of orthogonal polynomials (see [35]) give $\lim_{n\to\infty} p_{n+1}/p_n = \Phi(z)$ on compact subsets of $\overline{\mathbb{C}} \setminus \overline{\Omega}$, where $\Phi(z)$ is the conformal map of the exterior of Ω to the exterior of the unit disc. This leads to a *finite* Laurent expansion at ∞ for $\Psi(w) = \Phi^{-1}(w)$. Thus, $\Psi(w)$ is a rational function, so $\tilde{\Omega} := \overline{\mathbb{C}} \setminus \overline{\Omega}$ is an unbounded quadrature domain, and the Schwarz function (cf. [7,37]) of $\partial\Omega$, $S(z)$ $(= \bar{z}$ on $\partial\Omega)$ has a meromorphic extension to $\tilde{\Omega}$. Suppose, for the sake of brevity and to fix the ideas, for example, that $S(z) = cz^d + \sum_{j=1}^{M} c_j/(z - z_j) + f(z)$, where $f \in H^\infty(\tilde{\Omega})$, and $z_j \in \tilde{\Omega}$. Since our hypothesis is equivalent to Ω satisfying property (2) discussed above, the data $\bar{z}P(z) = \bar{z}\prod_{j=1}^{n}(z - z_j)$ has polynomial solution, $g(z) + \overline{h(z)}$ to the DP. On Γ we can replace $\bar{z}$ with $S(z)$. Write $\overline{h(z)} = h^{\#}(\bar{z})$, where $h^{\#}$ is a polynomial whose coefficients are complex conjugates of their counterparts in h. We have on Γ

$$S(z)P(z) = g(z) + h^{\#}(S(z)), \tag{5.1}$$

which is actually true off Γ since both sides of the equation are analytic. Near z_j, the left-hand side of this equation tends to a finite limit (since $S(z)P(z)$ is analytic in $\tilde{\Omega} \setminus \infty$!) while the right-hand side tends to ∞ unless the coefficient c_j is zero. Thus,

$$S(z) = cz^d + f(z). \tag{5.2}$$

Using property (2) again with data $|z|^2 = z\bar{z}$ we can infer that $d = 1$. Hence, $\tilde{\Omega}$ is a null quadrature domain. Sakai's theorem [32] implies now that Ω is an ellipse. □

Remark. It is well-known that families of orthogonal polynomials on the line satisfy a 3-term recurrence relation. P. Duren in 1965 [8] already noted that in $\mathbb{C}$ the only domains with real-analytic boundaries in which polynomials orthogonal with respect to arc-length on the boundary satisfy 3-term recurrence relations are ellipses. L. Lempert [25] constructed peculiar examples of C^∞ nonalgebraic Jordan domains in which no finite recurrence relation for Bergman polynomials holds. Theorem 5.1 shows that actually this is true for *all* C^2-smooth domains except ellipses.

6. Looking for singularities of the solutions to the Dirichlet problem

Once again, inspired by known results in the similar quest for solutions to the Cauchy problem, one could expect, e.g., that the solutions to the DP (1.1) exhibit behavior similar to those of the CP (2.1). In particular, it seemed natural to suggest that the singularities of the solutions to the DP outside Ω are somehow associated with the singularities of the Schwarz potential (function) of $\partial\Omega$ which does indeed completely determine $\partial\Omega$ (cf. [21,37]). It turned out that singularities of solutions of the DP are way more complicated than those of the CP. Already in 1992 in his thesis, P. Ebenfelt showed [9] that the solution of the following "innocent" DP in $\Omega := \{x^4 + y^4 - 1 < 0\}$ (the "TV-screen")

$$\begin{cases} \Delta u = 0 \\ u|_{\partial\Omega} = x^2 + y^2 \end{cases} \tag{6.1}$$

has an infinite discrete set of singularities (of course, symmetric with respect to 90° rotation) sitting on the coordinate axes and running to ∞ (see Figure 2).

To see the difference between analytic continuation of solutions to CP and DP, note that for the former

$$\frac{\partial u}{\partial z}|_{\Gamma:=\partial\Omega} = v_z(z,\bar{z}) = v_z\big(z, S(z)\big), \tag{6.2}$$

and since $\partial u/\partial z$ is analytic, (6.2) allows u_z to be continued everywhere together with v and $S(z)$, the Schwarz function of $\partial\Omega$. For the DP we have on Γ

$$u(z,\bar{z}) = v(z,\bar{z}) \tag{6.3}$$

for $u = f + \bar{g}$ where f and g are analytic in Ω. Hence, (6.3) becomes

$$f(z) + \overline{g(\overline{S(z)})} = v\big(z, S(z)\big). \tag{6.4}$$

Now, $v\big(z, S(z)\big)$ does indeed (for entire v) extend to any domain free of singularities of $S(z)$, but (6.4), even when v is real-valued so that $g = f$, presents a very nontrivial functional equation supported by a rather mysterious piece of information that f is analytic in Ω. (6.4) however gives an insight as to how to capture the DP-solution's singularities by considering the DP as part of a Goursat problem in $\mathbb{C}^2$ (or $\mathbb{C}^n$ in general). The latter Goursat problem can be posed as follows (cf. [36]).

Given a complex-analytic variety $\widehat{\Gamma}$ in $\mathbb{C}^n$, $(\widehat{\Gamma} \cap \mathbb{R}^n = \Gamma := \partial\Omega)$, find u: $\sum_{j=1}^n \partial^2 u/\partial z_j^2 = 0$ near $\widehat{\Gamma}$ (and also in $\Omega \subset \mathbb{R}^n$) so that $u|_{\widehat{\Gamma}} = v$, where v is, say, an entire function of n complex variables. Thus, if $\widehat{\Gamma} := \{\phi(z) = 0\}$, where ϕ is, say, an irreducible polynomial, we can, e.g., ponder the following extension of Conjecture 3.3:

Question. For which polynomials ϕ can every entire function v be split (Fischer decomposition) as $v = u + \phi h$, where $\Delta u = 0$ and u, h are entire functions (cf. [13,36])?

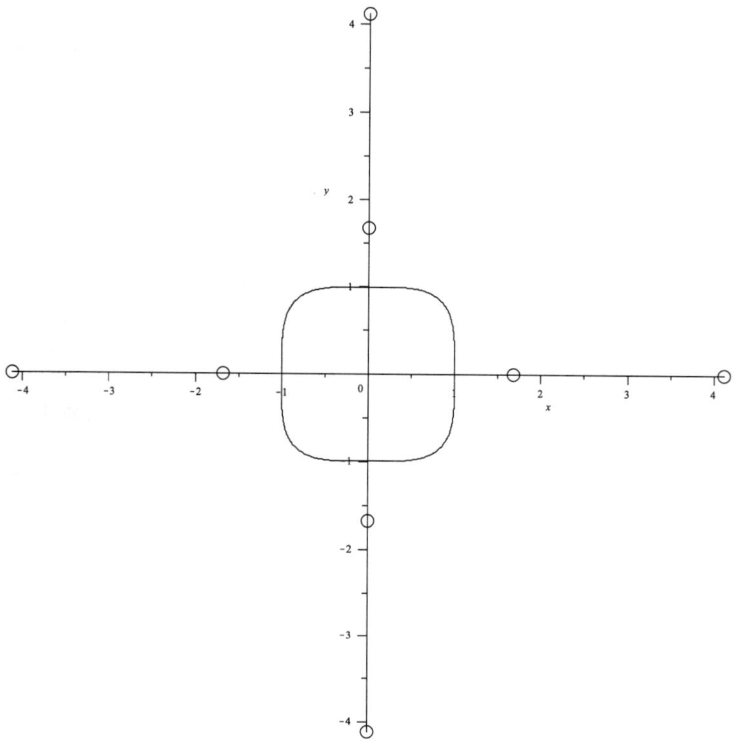

FIGURE 2. A plot of the "TV screen" $\{x^4+y^4=1\}$ along with the first eight singularities (plotted as circles) encountered by analytic continuation of the solution to DP (6.1).

7. Render's breakthrough

Trying to establish Conjecture 3.3 H. Render [30] has made the following ingenious step. He introduced the *real* version of the Fischer space norm

$$\langle f,g\rangle = \int_{\mathbb{R}^n} f\bar{g}e^{-|x|^2}\,dx, \tag{7.1}$$

where f and g are polynomials. Originally, the Fischer norm (introduced by E. Fischer [13]) requires the integration to be carried over all of $\mathbb{C}^n$ and has the property that multiplication by monomials is adjoint to differentiation with the corresponding multi-index (e.g., multiplication by $(\sum_{j=1}^n x_j^2)$ is adjoint to the differential operator Δ). This property is only partially preserved for the real Fischer norm. More precisely [30],

$$\langle \Delta f,g\rangle = \langle f,\Delta g\rangle + 2\big(\deg(f)-\deg(g)\big)\langle f,g\rangle \tag{7.2}$$

for homogeneous f,g.

Suppose u solves the DP with data $|x|^2$ on $\partial\Omega \subseteq \{P=0 : \deg(P)=2k, k>1\}$. Then $u-|x|^2=Pq$ for analytic q, and thus $\Delta^k(Pq)=0$. Using (7.2), this (non-trivially) implies that the real Fischer product $\langle (Pq)_{m+2k}, q_m\rangle$ between all homogeneous parts of degree $m+2k$ and m of Pq and q, respectively, is zero. By a tour de force argument, Render used this along with an added assumption on the senior term of P (see below) to obtain estimates *from below* for the decay of the norms of homogeneous parts of q. This, in turn yields an if-and-only-if criterion for convergence in the real ball of radius R of the series for the solution $u = \sum_{m=0}^{\infty} u_m$, u_m homogeneous of degree m. Let us state Render's main theorem.

Theorem 7.1. *Let P be an irreducible polynomial of degree $2k$, $k > 1$. Suppose P is elliptic, i.e., the senior term P_{2k} of P satisfies $P_{2k}(x) \geq c_P|x|^{2k}$, for some constant c_P. Let ϕ be real analytic in $\{|x| < R\}$, and $\Delta^k(P\phi) = 0$ (at least in a neighborhood of the origin). Then, $R \leq C(P,n) < +\infty$, where C is a constant depending on the polynomial P and the dimension of the ambient space.*

Remark. The assumption in the theorem that P is elliptic is equivalent to the condition that the set $\{P = 0\}$ is bounded in $\mathbb{R}^n$.

Corollary 7.2. *Assume $\partial\Omega$ is contained in the set $\{P = 0\}$, a bounded algebraic set in $\mathbb{R}^n$. Then, if a solution of the DP (1.1) with data $|x|^2$ is entire, Ω must be an ellipsoid.*

PROOF. Suppose not, so $\deg(P) = 2k > 2$, and the following (Fischer decomposition) holds: $|x|^2 = P\phi + u$, $\Delta u = 0$. Hence, $\Delta^k(P\phi) = 0$ and ϕ cannot be analytically continued beyond a *finite* ball of radius $R = C(P) < \infty$, a contradiction. □

Caution. We want to stress again that, unfortunately, the theorem still tells us nothing for say small perturbations of the circle by a nonelliptic term of degree ≥ 3, e.g., $x^2 + y^2 - 1 + \varepsilon(x^3 - 3xy^2)$.

8. Back to $\mathbb{R}^2$: lightning bolts

Return to the $\mathbb{R}^2$ setting and consider as before our boundary $\partial\Omega$ of a domain Ω as (part of) an intersection of an analytic Riemann surface $\widehat{\Gamma}$ in $\mathbb{C}^2$ with $\mathbb{R}^2$. Roughly speaking if say $\partial\Omega$ is a subset of the algebraic curve $\Gamma := \{(x,y) : \phi(x,y) = 0\}$, where ϕ is an irreducible polynomial, then $\widehat{\Gamma} = \{(X,Y) \in \mathbb{C}^2 : \phi(X,Y) = 0\}$. Now look at the Dirichlet problem again in the context of the Goursat problem: Given, say, a polynomial data P, find f, g holomorphic functions of one variable near $\widehat{\Gamma}$ (a piece of $\widehat{\Gamma}$ containing $\partial\Omega \subseteq \widehat{\Gamma} \cap \mathbb{R}^2$) such that

$$u = f(z) + g(w)|_{\widehat{\Gamma}} = P(z,w), \tag{8.1}$$

where we have made the linear change of variables $z = X + iY$, $w = X - iY$ (so $\overline{w} = z$ on $\mathbb{R}^2 = \{(X,Y) : X, Y \text{ are both real}\}$). Obviously, $\Delta u = 4\partial^2/\partial z\partial w = 0$ and u matches P on $\partial\Omega$. Thus, the DP in $\mathbb{R}^2$ has become an interpolation problem in $\mathbb{C}^2$ of matching a polynomial on an algebraic variety by a sum of holomorphic functions in each variable separately. Suppose that for all polynomials P the solutions u of (8.1) extend as analytic functions to a ball $B_\Omega = \{|z|^2 + |w|^2 < R_\Omega\}$ in $\mathbb{C}^2$. Then, if $\widehat{\Gamma} \cap B_\Omega$ is path connected, we can interpolate every polynomial $P(z,w)$ on $\widehat{\Gamma} \cap B_\Omega$ by a holomorphic function of the form $f(z) + g(w)$. Now suppose we can produce a compactly supported measure μ on $\widehat{\Gamma} \cap B_\Omega$ which annihilates *all* functions of the form $f(z) + g(w)$, f, g holomorphic in B_Ω and at the same time does not annihilate all polynomials $P(z,w)$. This would force the solution u of (8.1) to have a singularity in the ball B_Ω in $\mathbb{C}^2$. Then, invoking a theorem of Hayman [17] (see also [20]), we would be able to assert that u cannot be extended as a real-analytic function to the *real* disk B_R in $\mathbb{R}^2$ containing Ω and of radius $\geq \sqrt{2}R$. An example of such annihilating measure supported by the vertices of a "quadrilateral" was independently observed by E. Study [38], H. Lewy [27], and L. Hansen and H. S. Shapiro [16]. Indeed, assign alternating values ± 1 for the measure supported at the four points $p_0 := (z_1, w_1)$, $q_0 := (z_1, w_2)$, $p_1 := (z_2, w_2)$, and $q_1 := (z_2, w_1)$. Then $\int (f+g)\,\mathrm{d}\mu = f(z_1)+g(w_1)-f(z_1)-g(w_2)+f(z_2)+g(w_2)-f(z_2)-g(w_1) = 0$ for all holomorphic functions f and g of one variable. This is an example of a closed lightning bolt (LB) with four vertices. Clearly, the idea can be extended to any even number of vertices.

Definition. A complex closed lightning bolt (LB) of length $2(n+1)$ is a finite set of points (vertices) $p_0, q_0, p_1, q_1, \ldots, p_n, q_n, p_{n+1}, q_{n+1}$ such that $p_0 = p_{n+1}$, and each complex line connecting p_j to q_j or q_j to p_{j+1} has either z or w coordinate fixed and they alternate, i.e., if we arrived at p_j with w coordinate fixed then we follow to q_j with z fixed etc.

For "real" domains lightning bolts were introduced by Arnold and Kolmogorov in the 1950s to study Hilbert's 13th problem (see [24] and the references therein).

The following theorem has been proved in [4] (see also [5]).

Theorem 8.1. *Let Ω be a bounded simply connected domain in $\mathbb{C} \cong \mathbb{R}^2$ such that the Riemann map $\phi\colon \Omega \to \mathbb{D} = \{|z| < 1\}$ is algebraic. Then all solutions of the DP with polynomial data have only algebraic singularities only at branch points of ϕ with the branching order of the former dividing the branching order of the latter iff ϕ^{-1} is a rational function. This in turn is known to be equivalent to Ω being a quadrature domain.*

Idea of proof. The hypotheses imply that the solution $u = f + \bar{g}$ extends as a single-valued meromorphic function into a $\mathbb{C}^2$-neighborhood of $\widehat{\Gamma}$. By another theorem of [4], one can find (unless ϕ^{-1} is rational) a continual family of closed LBs on $\widehat{\Gamma}$ of bounded length avoiding the poles of u. Hence, the measure with alternating values ± 1 on the vertices of any of these LBs annihilates all solutions $u = f(z) + g(w)$ holomorphic on $\widehat{\Gamma}$, but does not, of course, annihilate all polynomials of z, w. Therefore, ϕ^{-1} must be rational, i.e., Ω is a quadrature domain [36]. □

The second author [28] has recently constructed some other examples of LBs on complexified boundaries of planar domains which do not satisfy the hypothesis of Render's theorem. The LBs validate Conjecture 3.3 and produce an estimate regarding how far into the complement $\mathbb{C} \setminus \overline{\Omega}$ the singularities may develop. For instance, the complexification of the cubic, $8x(x^2 - y^2) + 57x^2 + 77y^2 - 49 = 0$ has a lightning bolt with six vertices in the (nonphysical) plane where z and w are real, i.e., x is real and y is imaginary (see Figure 3 for a plot of the cubic in the plane where x and y are real and see Figure 4 for the "nonphysical" slice including the lightning bolt). If the solution with appropriate cubic data is analytically continued in the direction of the closest unbounded component of the curve defining $\partial\Omega$, it will have to develop a singularity before it can be forced to match the data on that component.

9. Concluding remarks, further questions

In two dimensions one of the main results in [4] yields that disks are the only domains for which all solutions of the DP with rational (in x, y) data v are rational. The fact that in a disk every DP with rational data has a rational solution was

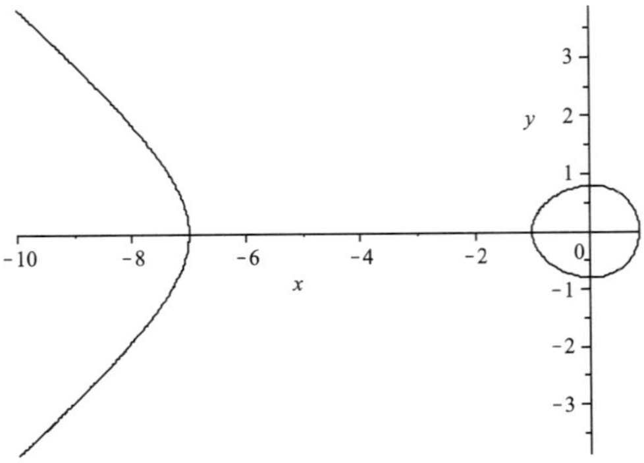

FIGURE 3. A Maple plot of the cubic $8x(x^2 - y^2) + 57x^2 + 77y^2 - 49 = 0$, showing the bounded component and one unbounded component (there are two other unbounded components further away).

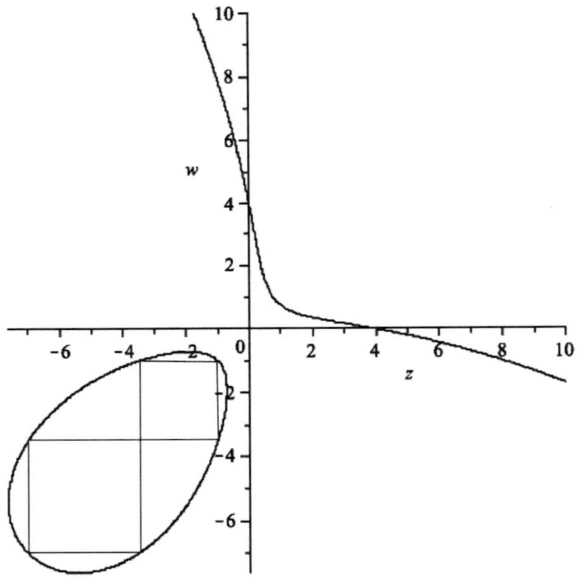

FIGURE 4. A lightning bolt with six vertices on the cubic $2(z + w)(z^2 + w^2) + 67zw - 5(z^2 + w^2) = 49$ in the nonphysical plane with z and w real, i.e. x real and y imaginary.

observed in a senior thesis of T. Fergusson at U. of Richmond [31]. On the other hand, algebraic data may lead to a transcendental solution even in disks (see [10], also cf. [12]). In dimensions 3 and higher, rational data on the sphere (e.g., $v = 1/(x_1 - a)$, $|a| > 1$) yields transcendental solutions of (1.1), although we have not been able to estimate the location of singularities precisely (cf. [10]).

It is still not clear on an intuitive level why ellipsoids play such a distinguished role in providing "excellent" solutions to DP with "excellent" data. A very similar question, important for applications, (which actually inspired the program launched in [22] on singularities of the solutions to the DP) goes back to Raleigh and concerns singularities of solutions of the Helmholtz equation ($[\Delta - \lambda^2]u = 0$, $\lambda \in \mathbb{R}$) instead. (The minus sign will guarantee that the maximum principle holds and, consequently, ensures uniqueness of solutions of the DP.) To the best of our knowledge, this topic remains virtually unexplored.

References

1. D. H. Armitage, *The Dirichlet problem when the boundary function is entire*, J. Math. Anal. Appl. **291** (2004), no. 2, 565–577.
2. S. Axler, P. Gorkin, and K. Voss, *The Dirichlet problem on quadratic surfaces*, Math. Comp. **73** (2004), no. 246, 637–651.
3. J. A. Baker, *The Dirichlet problem for ellipsoids*, Amer. Math. Monthly **106** (1999), no. 9, 829–834.
4. S. R. Bell, P. Ebenfelt, D. Khavinson, and H. S. Shapiro, *On the classical Dirichlet problem in the plane with rational data*, J. Anal. Math. **100** (2006), 157–190.
5. ———, *Algebraicity in the Dirichlet problem in the plane with rational data*, Complex Var. Elliptic Equ. **52** (2007), no. 2-3, 235–244.
6. M. Chamberland and D. Siegel, *Polynomial solutions to Dirichlet problems*, Proc. Amer. Math. Soc. **129** (2001), no. 1, 211–217.
7. P. J. Davis, *The Schwarz function and its applications*, Vol. 17, Math. Assoc. America, Buffalo, NY, 1974. Carus Math. Monogr.
8. P. L. Duren, *Polynomials orthogonal over a curve*, Michigan Math. J. **12** (1965), 313–316.
9. P. Ebenfelt, *Singularities encountered by the analytic continuation of solutions to Dirichlet's problem*, Complex Variables Theory Appl. **20** (1992), no. 1-4, 75–91.
10. P. Ebenfelt, D. Khavinson, and H. S. Shapiro, *Algebraic aspects of the Dirichlet problem*, Quadrature Domains and their Applications, Oper. Theory Adv. Appl., vol. 156, Birkhäuser, Basel, 2005, pp. 151–172.
11. P. Ebenfelt and H. S. Shapiro, *The Cauchy–Kowalevskaya theorem and generalizations*, Comm. Partial Differential Equations **20** (1995), no. 5-6, 939–960.
12. P. Ebenfelt and M. Viscardi, *On the solution of the Dirichlet problem with rational holomorphic boundary data*, Comput. Methods Funct. Theory **5** (2005), no. 2, 445–457.
13. E. Fischer, *Über die Differentiationsprozesse der Algebra*, J. Reine Angew. Math. **148** (1917), 1–78.
14. A. Friedman, *Partial differential equations*, Holt, Rinehart and Winston, New York, 1969.
15. L. Gårding, T. Kotake, and J. Leray, *Uniformisation et développement asymptotique de la solution du problème de Cauchy linéaire, à données holomorphes; analogie avec la théorie des ondes asymptotiques et approchées (Problème de Cauchy,* I *bis et* VI), Bull. Soc. Math. France **92** (1964), 263–361.
16. L. J. Hansen and H. S. Shapiro, *Functional equations and harmonic extensions*, Complex Variables Theory Appl. **24** (1994), no. 1-2, 121–129.
17. W. K. Hayman, *Power series expansions for harmonic functions*, Bull. London Math. Soc. **2** (1970), 152–158.
18. G. Johnsson, *The Cauchy problem in* $\mathbf{C}^N$ *for linear second order partial differential equations with data on a quadric surface*, Trans. Amer. Math. Soc. **344** (1994), no. 1, 1–48.
19. D. Khavinson, *Singularities of harmonic functions in* $\mathbf{C}^n$, Several Complex Variables and Complex Geometry, Part 3 (Santa Cruz, CA, 1989), Proc. Sympos. Pure Math., vol. 52, Amer. Math. Soc., Providence, RI, 1991, pp. 207–217.
20. ———, *Holomorphic partial differential equations and classical potential theory*, Universidad de La Laguna, Departamento de Análisis Matemático, La Laguna, 1996.
21. D. Khavinson and H. S. Shapiro, *The Schwarz potential in* $\mathbb{R}^n$ *and Cauchy's problem for the Laplace equation*, Technical Report TRITA-MAT-1989-36, Royal Institute of Technology, Stockholm.
22. ———, *Dirichlet's problem when the data is an entire function*, Bull. London Math. Soc. **24** (1992), no. 5, 456–468.
23. D. Khavinson and N. Stylianopoulos, *Recurrence relations for orthogonal polynomials and the Khavinson–Shapiro conjecture*, in preparation.
24. S. Ya. Khavinson, *Best approximation by linear superpositions (approximate nomography)*, Transl. Math. Monogr., vol. 159, Amer. Math. Soc., Providence, RI, 1997.

25. L. Lempert, *Recursion for orthogonal polynomials on complex domains*, Fourier Analysis and Approximation Theory, Vol. II (Budapest, 1976), Colloq. Math. Soc. János Bolyai, vol. 19, North-Holland, Amsterdam, 1978, pp. 481–494.
26. J. Leray, *Uniformisation de la solution du problème linéaire analytique de Cauchy près de la variété qui porte les données de Cauchy*, C. R. Acad. Sci. Paris **245** (1957), 1483–1488.
27. H. Lewy, *On the reflection laws of second order differential equations in two independent variables*, Bull. Amer. Math. Soc. **65** (1959), 37–58.
28. E. Lundberg, *Dirichlet's problem and complex lightning bolts*, Comput. Methods Funct. Theory **9** (2009), no. 1, 111–125.
29. M. Putinar and N. S. Stylianopoulos, *Finite-term relations for planar orthogonal polynomials*, Complex Anal. Oper. Theory **1** (2007), no. 3, 447–456.
30. H. Render, *Real Bargmann spaces, Fischer decompositions, and sets of uniqueness for polyharmonic functions*, Duke Math. J. **142** (2008), no. 2, 313–352.
31. W. T. Ross, private communication.
32. M. Sakai, *Null quadrature domains*, J. Analyse Math. **40** (1981), 144–154 (1982).
33. B. Yu. Sternin and V. E. Shatalov, *Legendre uniformization of multivalued analytic functions*, Mat. Sb. (N.S.) **113(155)** (1980), no. 2(10), 263–284 (Russian).
34. B. Yu. Sternin, H. S. Shapiro, and V. E. Shatalov, *Schwarz potential and singularities of the solutions of the branching Cauchy problem*, Dokl. Akad. Nauk **341** (1995), no. 3, 307–309 (Russian).
35. P. K. Suetin, *Polynomials orthogonal over a region and Bieberbach polynomials*, Proc. Steklov Inst. Math., vol. 100, Amer. Math. Soc., Providence, R.I., 1974.
36. H. S. Shapiro, *An algebraic theorem of E. Fischer, and the holomorphic Goursat problem*, Bull. London Math. Soc. **21** (1989), no. 6, 513–537.
37. ———, *The Schwarz function and its generalization to higher dimensions*, Univ. Arkansas Lecture Notes Math. Sci., vol. 9, Wiley, New York, 1992.
38. E. Study, *Einige Elementare Bemerkungen uber den Prozess der analytischen Fortzetsung*, Math. Ann. **63** (1907), 239–245.

Department of Mathematics, University of South Florida, 4202 E. Fowler Ave., Tampa, FL 33617, USA

E-mail address, D. Khavinson: `dkhavins@cas.usf.edu`

E-mail address, E. Lundberg: `elundber@mail.usf.edu`

Centre de Recherches Mathématiques
CRM Proceedings and Lecture Notes
Volume **51**, 2010

Invariant Subspaces of the Dirichlet Space

Omar El-Fallah, Karim Kellay, and Thomas Ransford

ABSTRACT. We present an overview of the problem of describing the invariant subspaces of the Dirichlet space. We also discuss some recent progress in the problem of characterizing the cyclic functions.

1. Introduction

Let X be a Banach space of functions holomorphic in the open unit disk $\mathbb{D}$, such that the shift operator $S\colon f(z) \mapsto zf(z)$ is a continuous map of X into itself. An *invariant subspace* of X is a closed subspace $\mathcal{M}$ of X such that $S\mathcal{M} \subset \mathcal{M}$. Given $f \in X$, we denote by $[f]_X$ the smallest invariant subspace of X containing f, namely

$$[f]_X = \overline{\{pf : p \text{ a polynomial}\}}.$$

We say that f is *cyclic* for X if $[f]_X = X$.

1.1. The Hardy space. This is the case $X = H^2$, where

$$H^2 := \Big\{ f(z) = \sum_{k\geq 0} a_k z^k : \|f\|_{H^2}^2 := \sum_{k\geq 0} |a_k|^2 < \infty \Big\}.$$

Recall that, for every function $f \in H^2 \setminus (0)$, the radial limit $f^*(\zeta) := \lim_{r\to 1^-} f(r\zeta)$ exists a.e. on the unit circle $\mathbb{T}$. The function f has a unique factorization $f = \theta h$, where θ, h are H^2-functions, θ is *inner* (this means that $|\theta^*| = 1$ a.e.), and h is *outer* (which means that $\log|h(0)| = (2\pi)^{-1} \int_0^{2\pi} \log|h^*(\zeta)|\,|d\zeta|$). The inner factor can be expressed as product of a Blaschke product and singular inner factor. More precisely, we have $\theta = cBS_\sigma$, where c is a unimodular constant,

$$B(z) = \prod_n \frac{|a_n|}{a_n} \frac{a_n - z}{1 - \bar{a}_n z}, \qquad \Big(a_n \in \mathbb{D}, \sum_n (1 - |a_n|) < \infty, |a_n|/a_n := 1 \text{ if } a_n = 0 \Big),$$

2000 *Mathematics Subject Classification.* Primary 47A15; Secondary 30C85, 46E20, 47B37.

Key words and phrases. Dirichlet space, invariant subspace, cyclic function, logarithmic capacity.

The research of the third author was partially supported by grants from NSERC (Canada), FQRNT (Québec) and the Canada research chairs program.

This is the final form of the paper.

and

$$S_\sigma(z) = \exp\Big(-\int_{\mathbb{T}} \frac{\zeta+z}{\zeta-z}\,\mathrm{d}\sigma(\zeta)\Big) \qquad (\sigma \geq 0, \sigma \perp \mathrm{d}\theta).$$

For more details, see for example [8, 11, 13].

The invariant subspaces of H^2 are completely described by Beurling's theorem [2]:

Theorem 1. *Let $\mathcal{M} \neq (0)$ be an invariant subspace of H^2. Then $\mathcal{M} = \theta H^2$, where θ is an inner function.*

This result leads immediately to a characterization of cyclic functions for H^2.

Corollary 2. *A function f is cyclic for H^2 if and only if it is an outer function.*

1.2. The Dirichlet space. The *Dirichlet space* is defined by

$$\mathcal{D} := \Big\{ f(z) = \sum_{k\geq 0} a_k z^k : \|f\|_{\mathcal{D}}^2 := \sum_{k\geq 0}(k+1)|a_k|^2 < \infty \Big\}.$$

Clearly $\mathcal{D}$ is a Hilbert space and $\mathcal{D} \subset H^2$. It is called the Dirichlet space because of the close connection with Dirichlet integral:

$$\mathcal{D}(f) := \frac{1}{\pi}\int_{\mathbb{D}} |f'(z)|^2\,\mathrm{d}A(z) = \sum_{k\geq 0} k|a_k|^2.$$

Thus $\|f\|_{\mathcal{D}}^2 = \|f\|_{H^2}^2 + \mathcal{D}(f)$. (The H^2-norm is added to ensure that we get a genuine norm.)

Here are two other formulas for the Dirichlet integral. The first, due to Douglas [7], expresses the integral purely in terms of f^*, and leads to the notion of Besov spaces:

$$\mathcal{D}(f) = \Big(\frac{1}{2\pi}\Big)^2 \int_{\mathbb{T}}\int_{\mathbb{T}} \frac{|f^*(\zeta_1) - f^*(\zeta_2)|^2}{|\zeta_1 - \zeta_2|^2}\,|\mathrm{d}\zeta_1|\,|\mathrm{d}\zeta_2|.$$

The second formula is due to Carleson [5]. Using the factorization above ($f = cBS_\sigma h$), Carleson's formula expresses $\mathcal{D}(f)$ in terms of the data $|h^*|, (a_n)$ and σ. More precisely:

$$\begin{aligned}(1)\quad \mathcal{D}(f) = {}& \frac{1}{2\pi}\int_{\mathbb{T}}\sum_n \frac{1-|a_n|^2}{|\zeta - a_n|^2}|h^*(\zeta)|^2\,|\mathrm{d}\zeta| \\ &+ \frac{1}{2\pi}\int_{\mathbb{T}}\int_{\mathbb{T}} \frac{2}{|\zeta_1-\zeta_2|^2}|h^*(\zeta_1)|^2\,|\mathrm{d}\zeta_1|\,\mathrm{d}\sigma(\zeta_2) \\ &+ \frac{1}{4\pi^2}\int_{\mathbb{T}}\int_{\mathbb{T}} \frac{(|h^*(\zeta_1)|^2 - |h^*(\zeta_2)|^2)(\log|h^*(\zeta_1)| - \log|h^*(\zeta_2)|)}{|\zeta_1-\zeta_2|^2}\,|\mathrm{d}\zeta_1|\,|\mathrm{d}\zeta_2|.\end{aligned}$$

It follows directly from Carleson's formula that, if $f \in \mathcal{D}$, then its outer factor h also belongs to $\mathcal{D}$ and satisfies $\mathcal{D}(h) \leq \mathcal{D}(f)$. A further consequence is that the only inner functions which belong to $\mathcal{D}$ are finite Blaschke products.

In [15, 16], Richter established an analogue of Beurling's theorem for the Dirichlet space. To state his result, we need to introduce a family of Dirichlet-type spaces $\mathcal{D}(\mu)$. Given a finite positive Borel measure μ on the unit circle, we define

$$\mathcal{D}(\mu) := \Big\{ f \in H^2 : \mathcal{D}_\mu(f) := \frac{1}{\pi}\int_{\mathbb{D}} |f'(z)|^2 \varphi_\mu(z)\,\mathrm{d}A(z) < \infty \Big\},$$

where φ_μ is the Poisson transform of μ, namely

$$\varphi_\mu(z) := \frac{1}{2\pi}\int_0^{2\pi} \frac{1-|z|^2}{|\zeta - z|^2}\, d\mu(\zeta).$$

We equip $\mathcal{D}(\mu)$ with the norm $\|\cdot\|_\mu$ defined by $\|f\|_\mu^2 := \|f\|_{H^2}^2 + \mathcal{D}_\mu(f)$. Note that the classical Dirichlet space corresponds to taking μ to be normalized Lebesgue measure m on $\mathbb{T}$. The following theorem was proved by Richter [15, 16].

Theorem 3. *Let $\mathcal{M} \neq (0)$ be an invariant subspace of $\mathcal{D}$. Then there exists $f \in \mathcal{D}$ such that $\mathcal{M} \ominus S\mathcal{M} = \mathbb{C}f$ and $\mathcal{M} = [f]_\mathcal{D} = f\mathcal{D}(|f|^2\, dm)$.*

In particular, the invariant subspaces of $\mathcal{D}$ are all cyclic (of the form $[f]_\mathcal{D}$ for some $f \in \mathcal{D}$). The next theorem, due to Richter and Sundberg [17], goes one step further, expressing such subspaces in terms of invariant subspaces generated by an outer function.

Theorem 4. *Let $f \in \mathcal{D} \setminus \{0\}$, say $f = \theta h$, where θ is inner and h is outer. Then*

$$[f]_\mathcal{D} = \theta[h]_\mathcal{D} \cap \mathcal{D} = [h]_\mathcal{D} \cap \theta H^2.$$

This still leaves us with the problem of describing $[h]_\mathcal{D}$ when h is outer. In particular, it leaves open the problem of characterizing the cyclic functions for $\mathcal{D}$.

2. Dirichlet space and logarithmic capacity

Given a probability measure μ on $\mathbb{T}$, we define its *energy* by

$$I(\mu) := \iint \log\frac{1}{|z-w|}\, d\mu(z)\, d\mu(w).$$

Note that $I(\mu) \in (-\infty, \infty]$. A simple calculation (see [3, p. 294]) shows that

$$I(\mu) = \sum_{n \geq 1} \frac{|\widehat{\mu}(n)|^2}{n}.$$

Hence in fact $I(\mu) \geq 0$, with equality if and only if μ is normalized Lebesgue measure on $\mathbb{T}$.

Given a Borel subset E of $\mathbb{T}$, we define its *capacity* by

$$c(E) := 1/\inf\{I(\mu) : \mu \text{ is a probability measure supported on a compact subset of } E\}.$$

Note that $c(E) = 0$ if and only if E supports no probability measure of finite energy. It is easy to see that

countable $\Longrightarrow$ capacity zero $\Longrightarrow$ Hausdorff dimension zero
$\Longrightarrow$ Lebesgue measure zero.

None of these implications is reversible. A property is said to hold *quasi-everywhere* (q.e.) on $\mathbb{T}$ if it holds everywhere outside a Borel set of capacity zero.

The following result, due to Beurling [1], reveals an important connection between capacity and the Dirichlet space.

Theorem 5. *Let $f \in \mathcal{D}$. Then $f^*(\zeta) := \lim_{r\to 1} f(r\zeta)$ exists q.e. on $\mathbb{T}$, and*

$$c(|f^*| \geq t) \leq \frac{16\|f\|_\mathcal{D}^2}{t^2} \qquad (t \geq 4\|f\|_\mathcal{D}). \tag{2}$$

The inequality (2) is called a weak-type inequality for capacity. The strong-type inequality for capacity is

$$\int_0^\infty c(|f^*| \geq t)\, \mathrm{d}t^2 \leq C\|f\|_{\mathcal{D}}^2, \tag{3}$$

where C is a constant. For a proof, see for example [19]. In Theorem 15 below, we shall exhibit a 'converse' to the strong-type inequality.

Given a Borel subset E of $\mathbb{T}$, we define

$$\mathcal{D}_E := \{f \in \mathcal{D} : f^* = 0 \text{ q.e. on } E\}.$$

The following result is essentially due to Carleson [4]. The simple proof given below is taken from [3], where it is attributed to Joel Shapiro.

Theorem 6. *$\mathcal{D}_E$ is an invariant subspace of $\mathcal{D}$.*

PROOF. We just need to show that $\mathcal{D}_E$ is closed in $\mathcal{D}$, the rest is clear. Let (f_n) be a sequence in $\mathcal{D}_E$ and suppose that $f_n \to f$ in $\mathcal{D}$. By (2), if $t > 0$, then, for all n sufficiently large,

$$c(E \cap \{|f^*| \geq t\}) \leq c(|f^* - f_n^*| \geq t) \leq \frac{16\|f - f_n\|_{\mathcal{D}}^2}{t^2}.$$

Letting $n \to \infty$ and then $t \to 0$, we deduce that $f^* = 0$ q.e. on E. □

Corollary 7. *If $f \in \mathcal{D}$, then $[f]_{\mathcal{D}} \subset \mathcal{D}_{Z(f^*)}$, where $Z(f^*) := \{\zeta \in \mathbb{T} : f^*(\zeta) = 0\}$.*

3. Cyclic functions for the Dirichlet space

Recall that a function $f \in \mathcal{D}$ is cyclic for $\mathcal{D}$ if $[f]_{\mathcal{D}} = \mathcal{D}$. So, from Corollary 7, if f is cyclic for $\mathcal{D}$, then f is outer and $c(Z(f^*)) = 0$. In [3], Brown and Shields conjectured that the converse is also true. In this section we will give some sufficient conditions to ensure cyclicity in the Dirichlet space. For simplicity, we restrict our attention mostly to functions in the disk algebra $A(\mathbb{D})$. To help state these results, we introduce a class $\mathcal{C}$.

Definition 8. The class $\mathcal{C}$ consists of closed subsets E of the unit circle satisfying the following property: every outer function $f \in \mathcal{D} \cap A(\mathbb{D})$ such that $Z(f) \subset E$ is cyclic for $\mathcal{D}$.

For functions in $\mathcal{D} \cap A(\mathbb{D})$, the Brown–Shields conjecture thus becomes:

Conjecture 9. *If E is a closed subset of $\mathbb{T}$ with $c(E) = 0$, then $E \in \mathcal{C}$.*

It was proved by Hedenmalm and Shields [12] that every countable closed subset of $\mathbb{T}$ belongs to $\mathcal{C}$. Richter and Sundberg [18] subsequently obtained a more general version of this result which also covered the case of functions not necessarily in $A(\mathbb{D})$. The first examples of uncountable sets in $\mathcal{C}$ were given by the present authors in [9, 10], and our aim is now to give an overview of this work.

Let $\Gamma \subset \mathbb{T}$ be a union of disjoint arcs. We denote by $\partial\Gamma$ the boundary of Γ in the unit circle. Given an outer function f, we associate to it the outer function f_Γ defined by

$$f_\Gamma(z) := \exp\left(\frac{1}{2\pi}\int_\Gamma \frac{\zeta + z}{\zeta - z} \log|f^*(\zeta)|\, |\mathrm{d}\zeta|\right).$$

The following lemma can be obtained from Carleson's formula (1) (for the details, see [10]).

Lemma 10. *Let $f \in \mathcal{D}$ be a bounded outer function. There is a constant C, which depends only on f, such that, for every union of disjoint arcs Γ of $\mathbb{T}$, and every function $g \in \mathcal{D}$ satisfying $|g^*(\zeta)| \le d(\zeta, \partial\Gamma)$ a.e., we have $f_\Gamma g \in \mathcal{D}$ and*

$$\|f_\Gamma g\|_{\mathcal{D}} \le C\|g\|_{\mathcal{D}}.$$

Theorem 11. *Let $f \in \mathcal{D}$ be an outer function, and set $E := \{\zeta \in \mathbb{T} : \liminf_{z\to\zeta}|f(z)| = 0\}$. If $g \in \mathcal{D}$ and $|g^*(\zeta)| \le d(\zeta, E)$ a.e. on $\mathbb{T}$, then $g \in [f]_{\mathcal{D}}$.*

Idea of proof. We employ a method developed by Korenblum [14]. Let $(\gamma_k)_{k\ge1}$ be the connected components of $\mathbb{T}\backslash E$, and set $\Gamma_n := \cup_{k\ge n}\gamma_k$. First we prove that $gf_{\Gamma_n} \in [f]_{\mathcal{D}}$ for all n. Then, using Lemma 10, we show that $\sup_n\|gf_{\Gamma_n}\|_{\mathcal{D}} < \infty$. Therefore we can extract a subsequence of (gf_{Γ_n}) that converges weakly to g. It follows that $g \in [f]_{\mathcal{D}}$. □

Note that the existence of a function $g \in H^2 \setminus (0)$ such that $|g^*(\zeta)| \le d(\zeta, E)$ implies that

$$\int_{\mathbb{T}} \log(1/d(\zeta, E))\, |d\zeta| < \infty. \tag{4}$$

A closed subset E of $\mathbb{T}$ satisfying (4) is known as a *Carleson set.*

Let $E \subset \mathbb{T}$ be a Carleson set. We denote by f_E the outer function associated to the distance function $d(\zeta, E)$:

$$f_E(z) := \exp\left(\frac{1}{2\pi}\int_{\mathbb{T}} \frac{\zeta+z}{\zeta-z}\log(d(\zeta, E))\, |d\zeta|\right). \tag{5}$$

Using Carleson's formula (1), it is not difficult to prove that $f_E \in \mathcal{D}$. As consequences of Theorem 11, we obtain the following results.

Corollary 12. *Let E be a Carleson set. Then $E \in \mathcal{C}$ if and only if f_E is cyclic for $\mathcal{D}$.*

Corollary 13. *If $E, F \in \mathcal{C}$ and if E is a Carleson set, then $E \cup F \in \mathcal{C}$.*

Idea of proof. Let f be an outer function in $\mathcal{D} \cap A(\mathbb{D})$ such that $Z(f) \subset E \cup F$. We have to prove that f is cyclic for $\mathcal{D}$. Let $\Gamma_n = \cup_{k\ge n}\gamma_k$, where the (γ_k) are the connected components of $\mathbb{T} \setminus E$. We can write

$$ff_E^2 = f_{\Gamma_n} f_E f_{\mathbb{T}\setminus\Gamma_n} f_E.$$

Note that $\partial(\mathbb{T} \setminus \Gamma_n)$ is a finite set. Since $F \in \mathcal{C}$, we have $F \cup \partial(\mathbb{T} \setminus \Gamma_n) \in \mathcal{C}$. From the fact that $Z(f_{\mathbb{T}\setminus\Gamma_n} f_E) \subset F \cup \partial(\mathbb{T} \setminus \Gamma_n)$, we deduce that $f_{\mathbb{T}\setminus\Gamma_n} f_E$ is cyclic in $\mathcal{D}$. Thus $f_{\Gamma_n} f_E \in [f]_{\mathcal{D}}$. Using Lemma 10, we obtain that $\|f_{\Gamma_n} f_E\|_{\mathcal{D}}$ is uniformly bounded, so $f_E \in [f]_{\mathcal{D}}$. Since $E \in \mathcal{C}$, it follows that f_E is cyclic, and therefore so is f. □

We now turn our attention to Conjecture 9. We shall establish two partial results in this direction, both of them sufficient conditions for a closed set E to belong to the class $\mathcal{C}$.

3.1. A capacitary sufficient condition. Given a closed set $E \subset \mathbb{T}$ and $t > 0$, we write $E_t := \{\zeta \in \mathbb{T} : d(\zeta, E) \leq t\}$, and denote by $|E_t|$ the Lebesgue measure of E_t. It is clear that $c(E) = \lim_{t\to 0^+} c(E_t)$. In the following theorem we prove that, if $c(E_t)$ goes to zero sufficiently rapidly as $t \to 0^+$, then $E \in \mathcal{C}$.

Theorem 14. *Let E be a closed subset of $\mathbb{T}$. If*

$$\int_0 c(E_t)\frac{\log\log(1/t)}{t\log(1/t)}\,\mathrm{d}t < \infty, \tag{6}$$

then $E \in \mathcal{C}$.

The proof of Theorem 14 is based on the following converse (proved in [9]) to the strong-type inequality for capacity (3).

Theorem 15. *Let E be a proper closed subset of $\mathbb{T}$, and let $\eta\colon (0,\pi] \to \mathbb{R}^+$ be a continuous, decreasing function. Then the following are equivalent:*

(i) *there exists $h \in \mathcal{D}$ such that*

$$|h^*(\zeta)| \geq \eta\big(d(\zeta, E)\big) \qquad \text{q.e. on } \mathbb{T};$$

(ii) *there exists $h \in \mathcal{D}$ such that*

$$\operatorname{Re} h^*(\zeta) \geq \eta\big(d(\zeta,E)\big) \quad \text{and} \quad |\operatorname{Im} h^*(\zeta)| \leq \pi/4 \qquad \text{q.e. on } \mathbb{T};$$

(iii) *E and η satisfy*

$$\int_0 c(E_t)\,\mathrm{d}\eta^2(t) > -\infty. \tag{7}$$

Idea of the proof of Theorem 14. Suppose that condition (6) holds. By Theorem 15, there exists $h \in \mathcal{D}$ such that

$$\operatorname{Re} h^*(\zeta) \geq \log\log\big(d(\zeta,E)\big) \quad \text{and} \quad |\operatorname{Im} h^*(\zeta)| \leq \pi/4 \qquad \text{q.e. on } \mathbb{T}.$$

We consider the analytic semigroup $\varphi_\lambda := \exp(-\lambda e^h)$ for $|\arg(\lambda)| < \pi/4$. It has the following properties:

(a) the map $\lambda \mapsto \varphi_\lambda$ is holomorphic from $\{\lambda : |\arg(\lambda)| < \pi/4\}$ to $\mathcal{D}$;
(b) $\lim_{t\to 0^+} \|\varphi_t - 1\|_{\mathcal{D}} = 0$;
(c) $|\varphi_\lambda(z)| = O(d(z,E))$ for large λ.

Let $f \in \mathcal{D} \cap A(\mathbb{D})$ be an outer function such that $Z(f) \subset E$, and let ψ be an element of the dual space of $\mathcal{D}$ which is orthogonal to $[f]_{\mathcal{D}}$. It follows from property (c) and Theorem 11 that $\varphi_t \in [f]_{\mathcal{D}}$ for large t. Thus $\langle p\varphi_t, \psi\rangle = 0$ for every polynomial p. Using property (a), we can extend this equality to all $t > 0$. Property (b) then implies that $\langle p, \psi\rangle = 0$. Hence $\psi = 0$ and f is cyclic. □

3.2. A geometric sufficient condition. By geometric condition, we mean a condition expressed in terms of $|E_t|$. It is well known (see for example [6]) that there exists a constant $C > 0$ such that, for every closed set $E \subset \mathbb{T}$, we have

$$\int_0^\pi \frac{\mathrm{d}s}{|E_s|} \leq \frac{C}{c(E)}. \tag{8}$$

In particular, if $\int_0^\pi \mathrm{d}t/|E_t| = \infty$, then $c(E) = 0$. In the case of Cantor-type sets the converse is also true [6]. Using (8) with E_t in place of E, we obtain

$$\int_t^\pi \frac{\mathrm{d}s}{|E_s|} \leq \frac{C}{c(E_t)}.$$

So, from Theorem 14, if E satisfies

$$\int_0 \frac{|E_t|}{(t\log(1/t))^2}\,\mathrm{d}t < \infty,$$

then $E \in \mathcal{C}$. The following theorem gives a more precise condition.

Theorem 16. *Let E be a closed subset of $\mathbb{T}$ such that $|E_t| = O(t^\epsilon)$ for some $\epsilon > 0$. If*

$$\int_0 \frac{\mathrm{d}t}{|E_t|} = \infty,$$

then $E \in \mathcal{C}$.

To give an idea of the proof of this theorem, we need to digress slightly and introduce a generalization of the functions f_E defined in (5). Let E be a closed subset of $\mathbb{T}$ of Lebesgue measure zero, and let $w\colon (0,\pi] \to \mathbb{R}^+$ be a continuous function such that

(9) $$\int_{\mathbb{T}} |\log w(d(\zeta,E))|\,|\mathrm{d}\zeta| < \infty.$$

We shall denote by f_w the outer function satisfying

(10) $$|f_w^*(\zeta)| = w(d(\zeta,E)) \quad \text{a.e.}$$

Functions of this kind were already studied, for example, by Carleson in [4], in the course of his construction of outer functions in $A^k(\overline{\mathbb{D}})$ with prescribed zero sets. The following result gives a two-sided estimate for the Dirichlet integral of certain of these functions (for the details of the proof, see [10]).

Theorem 17. *Let E be a closed subset of $\mathbb{T}$ of measure zero, let $w\colon (0,\pi] \to \mathbb{R}^+$ be an increasing function such that (9) holds, and let f_w be the outer function given by (10). Suppose further that there exists $\gamma > 2$ such that $t \mapsto w(t^\gamma)$ is concave. Then*

(11) $$\mathcal{D}(f_w) \asymp \int_{\mathbb{T}} w'(d(\zeta,E))^2 d(\zeta,E)\,|\mathrm{d}\zeta|,$$

where the implied constants depend only on γ. In particular, $f_w \in \mathcal{D}$ if and only if the integral in (11) is finite.

Idea of the proof of Theorem 16. We first suppose that E is regular, in the sense that $|E_t| \asymp \psi(t)$, where ψ is a function such that $\psi(t)/t^\alpha$ is increasing for some $\alpha \in (\frac{1}{2}, 1)$.

For $\delta \in (0,1)$, define $w_\delta\colon (0,\pi] \to \mathbb{R}^+$ by

$$w_\delta(t) := \begin{cases} \frac{\delta^\alpha}{\psi(\delta)} t^{1-\alpha}, & 0 < t \le \delta, \\ A_\delta - \log \int_t^\pi \mathrm{d}s/\psi(s), & \delta < t \le \eta_\delta, \\ 1, & \eta_\delta < t \le \pi. \end{cases}$$

Here, the constants A_δ, η_δ are chosen to make w_δ continuous. Note that w_δ is increasing, and one can show that $t \mapsto w_\delta(t^\gamma)$ is concave if $\gamma > 1/(1-\alpha)$ and δ is sufficiently small. Using Theorem 17, we obtain that $\limsup_{\delta\to 0} \mathcal{D}(f_{w_\delta}) < \infty$. From the condition $\int_0 \mathrm{d}t/|E_t| = \infty$, it is easy to see that $\lim_{\delta\to 0}\eta_\delta = 0$, which implies that $\lim_{\delta\to 0}|f_{w_\delta}^*| = 1$ a.e. and that $\lim_{\delta\to 0}|f_{w_\delta}(0)| = 1$. Putting these facts together, we deduce that $f_{w_\delta} \to 1$ weakly in $\mathcal{D}$ as $\delta \to 0$.

Now, for each $\delta \in (0,1)$, the quotient $w_\delta(t)/t^{1-\alpha}$ is bounded, so $f_{w_\delta}/f_E^{1-\alpha}$ is bounded on $\mathbb{D}$. By [17, Lemma 2.4], it follows that $f_{w_\delta} \in [f_E^{1-\alpha}]_{\mathcal{D}}$. Letting $\delta \to 0$, we obtain $1 \in [f_E^{1-\alpha}]_{\mathcal{D}}$. Also, by [17, Theorem 4.3], we have $[f_E^{1-\alpha}]_{\mathcal{D}} = [f_E]_{\mathcal{D}}$. Hence f_E is cyclic and $E \in \mathcal{C}$.

When E is not regular, the result is still true, but now there is an extra step in the proof. To obtain the function ψ, we first need to regularize $|E_t|$, using a quantitative form of M. Riesz's rising-sun lemma. For the details, we refer to [10]. □

References

1. A. Beurling, *Ensembles exceptionnels*, Acta Math. **72** (1940), 1 – 13.
2. ———, *On two problems concerning linear transformations in Hilbert space*, Acta Math. **81** (1949), 239 – 255.
3. L. Brown and A. L. Shields, *Cyclic vectors in the Dirichlet space*, Trans. Amer. Math. Soc. **285** (1984), no. 1, 269 – 303.
4. L. Carleson, *Sets of uniqueness for functions regular in the unit circle*, Acta Math. **87** (1952), 325 – 345.
5. ———, *A representation formula for the Dirichlet integral*, Math. Z. **73** (1960), 190 – 196.
6. ———, *Selected problems on exceptional sets*, Van Nostrand Mathematical Studies, vol. 13, Van Nostrand, Princeton, NJ – Toronto, ON – London, 1967.
7. J. Douglas, *Solution of the problem of Plateau*, Trans. Amer. Math. Soc. **33** (1931), no. 1, 263 – 321.
8. P. L. Duren, *Theory of H^p spaces*, Pure Appl. Math., vol. 38, Academic Press, New York, 1970.
9. O. El-Fallah, K. Kellay, and T. Ransford, *Cyclicity in the Dirichlet space*, Ark. Mat. **44** (2006), no. 1, 61 – 86.
10. ———, *On the Brown – Shields conjecture for cyclicity in the Dirichlet space*, Adv. Math. **222** (2009), no. 6, 2196 – 2214.
11. J. B. Garnett, *Bounded analytic functions*, 1st ed., Grad. Texts in Math., vol. 236, Springer, New York, 2007.
12. H. Hedenmalm and A. Shields, *Invariant subspaces in Banach spaces of analytic functions*, Michigan Math. J. **37** (1990), no. 1, 91 – 104.
13. P. Koosis, *Introduction to H_p spaces*, London Math. Soc. Lecture Note Ser., vol. 40, Cambridge Univ. Press, Cambridge, 1980.
14. B. I. Korenblum, *Invariant subspaces of the shift operator in a weighted Hilbert space*, Mat. Sb. (N.S.) **89(131)** (1972), 110 – 137 (Russian); English transl., Math. USSR-Sb. **18** (1972), 111 – 138.
15. S. Richter, *Invariant subspaces of the Dirichlet shift*, J. Reine Angew. Math. **386** (1988), 205 – 220.
16. ———, *A representation theorem for cyclic analytic two-isometries*, Trans. Amer. Math. Soc. **328** (1991), no. 1, 325 – 349.
17. S. Richter and C. Sundberg, *Multipliers and invariant subspaces in the Dirichlet space*, J. Operator Theory **28** (1992), no. 1, 167 – 186.
18. ———, *Invariant subspaces of the Dirichlet shift and pseudocontinuations*, Trans. Amer. Math. Soc. **341** (1994), no. 2, 863 – 879.
19. Z. Wu, *Carleson measures and multipliers for Dirichlet spaces*, J. Funct. Anal. **169** (1999), no. 1, 148 – 163.

Département de Mathématiques, Université Mohamed V, B.P. 1014, Rabat, Morocco
E-mail address: elfallah@fsr.ac.ma

Laboratoire d'Analyse, Topologie, Probabilités, Centre de Mathématiques et Informatique, Université de Provence, 39, rue F. Joliot-Curie, 13453 Marseille Cedex 13, France
E-mail address: kellay@cmi.univ-mrs.fr

Département de mathématiques et de statistique, Université Laval, Québec, QC G1V 0A6, Canada
E-mail address: ransford@mat.ulaval.ca

Centre de Recherches Mathématiques
CRM Proceedings and Lecture Notes
Volume **51**, 2010

Arguments of Zero Sets in the Dirichlet Space

Javad Mashreghi, Thomas Ransford, and Mahmood Shabankhah

Abstract. We characterize the unimodular sequences $(e^{i\theta_n})_{n\geq 1}$ such that $(r_n e^{i\theta_n})_{n\geq 1}$ is a zero set for the Dirichlet space for every positive Blaschke sequence $(r_n)_{n\geq 1}$. The principal tool is a characterization of Carleson sets in terms of their convergent subsequences.

1. Introduction

The *Dirichlet space* $\mathcal{D}$ is the set of functions f, holomorphic on the open unit disk $\mathbb{D}$, for which

$$\mathcal{D}(f) := \frac{1}{\pi}\int_{\mathbb{D}} |f'(z)|^2 \,dx\,dy < \infty.$$

If $f(z) = \sum_{n=0}^{\infty} a_n z^n$, then $\mathcal{D}(f) = \sum_{n=1}^{\infty} n|a_n|^2$. Hence $\mathcal{D}$ is properly contained in the Hardy space

$$H^2 := \Big\{ f(z) = \sum_{n=0}^{\infty} a_n z^n : \|f\|_{H^2}^2 := \sum_{n=0}^{\infty} |a_n|^2 < \infty \Big\}.$$

$\mathcal{D}$ is a Hilbert space with respect to the norm $\|\cdot\|_{\mathcal{D}}$ defined by $\|f\|_{\mathcal{D}}^2 := \mathcal{D}(f) + \|f\|_{H^2}^2$.

Let X be a space of holomorphic functions of $\mathbb{D}$. A sequence $(z_n)_{n\geq 1} \subset \mathbb{D}$ is said to be a *zero sequence* for X if there exists function $f \in X$, not identically zero, which vanishes on z_n, $n \geq 1$. We do not require that the (z_n) be the only zeros of f. If $(z_n)_{n\geq 1}$ is not a zero sequence for X, then we call it a *uniqueness sequence* for X.

Zero sequences of the Hardy space H^2 are completely characterized: a sequence $(z_n)_{n\geq 1} \subset \mathbb{D}$ is a zero sequence for H^2 if and only if it satisfies the Blaschke condition

$$\sum_{n=1}^{\infty} (1 - |z_n|) < \infty. \tag{1}$$

2000 *Mathematics Subject Classification.* Primary 30C15; Secondary 30D50, 30D55, 31C25.
Key words and phrases. Zero sets, Dirichlet space, Carleson sets, Blaschke sets.
First author partially supported by NSERC (Canada), FQRNT (Québec).
Second author partially supported by NSERC, FQRNT and the Canada research chairs program.
Third author supported by an NSERC Canada graduate scholarship.
This is the final form of the paper.

Since $\mathcal{D} \subset H^2$, this is also a necessary condition for a zero sequence for $\mathcal{D}$. However, it is far from being sufficient. Indeed, the complete characterization of the zero sequences of $\mathcal{D}$ is still an open problem.

The first breakthrough in this direction was the pioneering work of Carleson [4]. He showed that if a sequence $(z_n)_{n\geq 1}$ in $\mathbb{D}$ satisfies

$$\sum_{n=1}^{\infty}\left(\frac{1}{-\log(1-|z_n|)}\right)^{1-\varepsilon} < \infty$$

for some $\varepsilon > 0$, then $(z_n)_{n\geq 1}$ is a zero sequence for $\mathcal{D}$. Using a completely different approach, Shapiro and Shields [10] obtained showed that this result remains true even with $\varepsilon = 0$.

Thus, if a sequence $(r_n)_{n\geq 1}$ in $[0,1)$ satisfies

$$\sum_{n=1}^{\infty}\frac{1}{-\log(1-r_n)} < \infty, \tag{2}$$

then $\left(r_n e^{i\theta_n}\right)_{n\geq 1}$ is a zero sequence for $\mathcal{D}$ for *every* choice of $\left(e^{i\theta_n}\right)_{n\geq 1}$. Later on, Nagel, Rudin and Shapiro [8] showed that, if (2) is not satisfied, then there exists a sequence $\left(e^{i\theta_n}\right)_{n\geq 1}$ for which $\left(r_n e^{i\theta_n}\right)_{n\geq 1}$ is a uniqueness sequence for $\mathcal{D}$. Putting these results together, we conclude that $\left(r_n e^{i\theta_n}\right)_{n\geq 1}$ is a zero sequence for $\mathcal{D}$ for *every choice* of $\left(e^{i\theta_n}\right)_{n\geq 1}$ if and only if (2) holds.

The main purpose of this article is to prove the following theorem, which can be considered as the dual of the last statement. It was already stated as a remark in [6, p. 704, line 12], but without detailed proof.

Theorem 1. *Let $\left(e^{i\theta_n}\right)_{n\geq 1}$ be a sequence in $\mathbb{T}$. Then $\left(r_n e^{i\theta_n}\right)_{n\geq 1}$ is a zero sequence for $\mathcal{D}$ for every positive Blaschke sequence $(r_n)_{n\geq 1}$ if and only if $\overline{\{e^{i\theta_n} : n\geq 1\}}$ is a Carleson set.*

We recall that a closed subset F of $\mathbb{T}$ is a *Carleson set* if

$$\int_{\mathbb{T}} \log \operatorname{dist}(\zeta, F)\,|d\zeta| > -\infty, \tag{3}$$

where "dist" is measured with respect to arclength distance. These sets were first discovered by Beurling [1] and then studied thoroughly by Carleson. Carleson [3] showed that the condition (3) characterizes the zero sets of $f|_{\mathbb{T}}$ for $f \in \mathcal{A}^1$, where $\mathcal{A}^1 := \{f \in C^1(\overline{\mathbb{D}}) : f \text{ is holomorphic on } \mathbb{D}\}$. If F is a closed subset of $\mathbb{T}$ of Lebesgue measure zero whose complementary arcs are denoted by I_n, $n \geq 1$, then (3) is equivalent to

$$\sum_n |I_n| \log|I_n| > -\infty. \tag{4}$$

Our proof of Theorem 1 is based on the following theorem, which we believe is of interest in its own right. In particular, it characterizes Carleson sets in terms of their convergent subsequences.

Theorem 2. *Let E be a subset of $\mathbb{T}$. Then $\overline{E}$ is a Carleson set if and only if the closure of every convergent sequence in E is a Carleson set.*

Theorem 2 is proved (in a more general form) in Section 2, and Theorem 1 is deduced from it in §3. Finally, in §4 we relate these results to the notion of Blaschke sets.

2. Proof of Theorem 2

We shall in fact prove Theorem 2 in a more general form. Let $\omega : [0,\pi] \to [0,\infty]$ be a continuous, decreasing function such that $\omega(0) = \infty$ and $\int_0^\pi \omega(t)\,dt < \infty$. A closed subset F of $\mathbb{T}$ is an *ω-Carleson set* if

$$\int_{\mathbb{T}} \omega\big(\operatorname{dist}(\zeta, F)\big)\,|d\zeta| < \infty. \tag{5}$$

If F is a closed subset of $\mathbb{T}$ of measure zero, then condition (5) is equivalent to

$$\sum_{n=1}^{\infty} \int_0^{|I_n|/2} \omega(t)\,dt < \infty,$$

where $(I_n)_{n\ge1}$ are the components of $\mathbb{T}\setminus F$ (see [7, Proposition A.1]). The classical Carleson sets correspond to the case $\omega(t) = \log^+(1/t)$, and so Theorem 2 is a special case of the following result.

Theorem 3. *Let ω be as above and let E be a subset of $\mathbb{T}$. Then $\overline{E}$ is an ω-Carleson set if and only if the closure of every convergent sequence in E is an ω-Carleson set.*

Proof. Set $F = \overline{E}$. We need to show that F is an ω-Carleson set.

We first show that the closure of every convergent sequence in F is an ω-Carleson set. As the union of two ω-Carleson sets is again one, it suffices to consider sequences converging to a limit from one side. Suppose that $\big(e^{i\theta_n}\big)_{n\ge1} \subset F$, where $\theta_1 < \theta_2 < \dots$ and $\theta_n \to \theta_0$ as $n \to \infty$. Since $\lim_{x\to0^+} \int_0^x \omega(t)\,dt = 0$, we may choose a positive sequence $(\eta_n)_{n\ge1}$, such that

$$\sum_{n=1}^{\infty} \int_0^{\eta_n} \omega(t)\,dt < \infty.$$

Set $\delta_n := \min\{\eta_n, \eta_{n-1}, (\theta_{n+1}-\theta_n)/2, (\theta_n-\theta_{n-1})/2\}$, $n \ge 2$, and choose $e^{i\phi_n} \in E$ with $\phi_n \in (\theta_n - \delta_n, \theta_n + \delta_n)$, $n \ge 2$. Then $\phi_2 < \phi_3 < \cdots$ and $\phi_n \to \theta_0$ as $n \to \infty$. Since $\big(e^{i\phi_n}\big)_{n\ge2}$ is a convergent sequence in E, it follows from the assumption that $\overline{\big\{e^{i\phi_n} : n \ge 2\big\}}$ is an ω-Carleson set, and therefore

$$\sum_{n=2}^{\infty} \int_0^{(\phi_{n+1}-\phi_n)/2} \omega(t)\,dt < \infty.$$

As $\theta_{n+1} - \theta_n \le \phi_{n+1} - \phi_n + 2\eta_n$ and $\omega(t)$ is a decreasing function,

$$\int_0^{(\theta_{n+1}-\theta_n)/2} \omega(t)\,dt \le \int_0^{(\phi_{n+1}-\phi_n)/2} \omega(t)\,dt + \int_0^{\eta_n} \omega(t)\,dt.$$

Therefore

$$\sum_{n\ge1} \int_0^{(\theta_{n+1}-\theta_n)/2} \omega(t)\,dt < \infty,$$

and $\overline{\big\{e^{i\theta_n} : n \ge 1\big\}}$ is an ω-Carleson set.

We next show that F has zero Lebesgue measure. Suppose, on the contrary, that $|F| > 0$. We claim that there exists a positive sequence $(\varepsilon_n)_{n\ge1}$ such that

$$\sum_{n=1}^{\infty} \varepsilon_n \le |F| \qquad \text{and} \qquad \sum_{n=1}^{\infty} \int_0^{\varepsilon_n/2} \omega(t)\,dt = \infty. \tag{6}$$

Indeed, since $(1/x)\int_0^x \omega(t)\,dt \to \infty$ as $x \to 0^+$, we may choose integers N_i, $i \geq 1$, such that

$$\int_0^{|F|/(2^{i+1}N_i)} \omega(t)\,dt \geq 1/N_i \qquad (i \geq 1).$$

Then, the sequence

$$(\varepsilon_n)_{n\geq 1} = \Big(\dots, \underbrace{\frac{|F|}{2^i N_i}, \dots, \frac{|F|}{2^i N_i}}_{N_i \text{ times}}, \dots\Big)$$

satisfies (6). Now choose $(\theta_n)_{n\geq 1}$ inductively as follows. Pick θ_1 such that $e^{i\theta_1} \in F$. If $\theta_1, \dots, \theta_n$ have already been selected, choose θ_{n+1} as small as possible such that $\theta_{n+1} \geq \theta_n + \varepsilon_n$ and $e^{i\theta_{n+1}} \in F$. Note that $\theta_n < \theta_1 + 2\pi$ for all n, for otherwise F would be covered by the finite set of closed arcs $\big[e^{i\theta_j}, e^{i(\theta_j+\varepsilon_j)}\big]_{j=1}^n$, contradicting the fact that $\sum_{j=1}^n \varepsilon_j < |F|$. Thus $\big(e^{i\theta_n}\big)_{n\geq 1}$ is a convergent sequence in F. Also

$$\sum_n \int_0^{(\theta_{n+1}-\theta_n)/2} \omega(t)\,dt \geq \sum_n \int_0^{\varepsilon_n/2} \omega(t)\,dt = \infty,$$

so $\overline{\{e^{i\theta_n} : n \geq 1\}}$ is not an ω-Carleson set. This contradicts what we proved in the previous paragraph. So we conclude that $|F| = 0$, as claimed.

Finally, we prove that F is an ω-Carleson set. Once again, we proceed by contradiction. If F is not an ω-Carleson set, then, as it has measure zero, it follows that

$$\sum_{n=1}^{\infty} \int_0^{|I_n|/2} \omega(t)\,dt = \infty, \tag{7}$$

where $(I_n)_{n\geq 1}$ are the components of $\mathbb{T}\backslash F$. Denote by $e^{i\theta_n}$ the midpoint of I_n, where $\theta_n \in [0, 2\pi]$. A simple compactness argument shows that there exists $\theta \in [0, 2\pi]$ such that, for all $\delta > 0$,

$$\sum_{\theta_n \in (\theta-\delta, \theta+\delta)} \int_0^{|I_n|/2} \omega(t)\,dt = \infty.$$

We can therefore extract a subsequence (I_{n_j}) such that $\theta_{n_j} \to \theta$ and

$$\sum_j \int_0^{|I_{n_j}|/2} \omega(t)\,dt = \infty.$$

The endpoints of the I_{n_j} then form a convergent sequence in F whose closure is not an ω-Carleson set, contradicting what we proved earlier. We conclude that F is indeed an ω-Carleson set. □

3. Proof of Theorem 1

If $\overline{\{e^{i\theta_n} : n \geq 1\}}$ is a Carleson set then, by [9, Theorem 1.2], for every Blaschke sequence $(r_n)_{n\geq 1}$ in $[0, 1)$, there exists $f \in \mathcal{A}^\infty$, $f \not\equiv 0$, which vanishes on $\big(r_n e^{i\theta_n}\big)_{n\geq 1}$. Here $\mathcal{A}^\infty := \{f \in C^\infty(\overline{\mathbb{D}}) : f \text{ is holomorphic on } \mathbb{D}\}$. In particular, $\big(r_n e^{i\theta_n}\big)_{n\geq 1}$ is a zero sequence for $\mathcal{D}$.

For the converse, we use a technique inspired by an argument in [5]. Suppose that $\big(r_n e^{i\theta_n}\big)_{n\geq 1}$ is a zero sequence for $\mathcal{D}$ for every Blaschke sequence $(r_n)_{n\geq 1} \subset$

$[0,1)$. Let $\left(e^{i\theta_{n_j}}\right)_{j\geq 1}$ be a convergent subsequence of $\left(e^{i\theta_n}\right)_{n\geq 1}$. We shall show that $\overline{\left\{e^{i\theta_{n_j}} : j \geq 1\right\}}$ is a Carleson set. As the union of two Carleson sets is again a Carleson set, it suffices to consider the case when $\theta_{n_1} < \theta_{n_2} < \dots$ and $\theta_{n_j} \to \theta_0$, as $j \to \infty$. Let $\delta_j := \theta_{n_{j+1}} - \theta_{n_j}$, $j \geq 1$. Consider the Blaschke sequence

$$z_j := (1-\delta_j)e^{i\theta_{n_j}}, \qquad (j \geq 1).$$

By hypothesis, this is a zero sequence for $\mathcal{D}$, and as such it therefore satisfies

$$\int_0^{2\pi} \log\left(\sum_{j=1}^{\infty} \frac{1-|z_j|^2}{\left|e^{i\theta}-z_j\right|^2}\right) d\theta < \infty,$$

(see for example [5, p. 313, equation (2)]). On the other hand, for $\theta \in (\theta_{n_k}, \theta_{n_{k+1}})$, we have $\left|e^{i\theta} - z_k\right| \leq 2\delta_k$ which gives

$$\frac{1-|z_k|^2}{\left|e^{i\theta}-z_k\right|^2} \geq \frac{1}{4\delta_k},$$

and consequently

$$\int_0^{2\pi} \log\left(\sum_{j=1}^{\infty} \frac{1-|z_j|^2}{\left|e^{i\theta}-z_j\right|^2}\right) d\theta \geq \sum_{k=1}^{\infty} \int_{\theta_{n_k}}^{\theta_{n_{k+1}}} \log\frac{1}{4\delta_k}\, d\theta = \sum_{k=1}^{\infty} \delta_k \log\frac{1}{4\delta_k}.$$

We conclude that $\sum_k \delta_j \log\delta_k > -\infty$. Thus $\overline{\left\{e^{i\theta_{n_k}} : k \geq 1\right\}}$ is indeed a Carleson set. Now apply Theorem 2 to obtain the desired result. □

4. Blaschke sets

Let X be a space of holomorphic functions on $\mathbb{D}$ and let A be a subset of $\mathbb{D}$. We say that A is a *Blaschke set* for X if every Blaschke sequence in A (perhaps with repetitions) is a zero sequence for X. Blaschke sets for $\mathcal{A}^\infty$ were characterized by Taylor and Williams [11], and for $\mathcal{D}$ by Bogdan [2]. The following theorem summarizes their results and takes them a little further.

Theorem 4. *Let A be a subset of $\mathbb{D}$. The following statements are equivalent.*

(a) *A is a Blaschke set for $\mathcal{D}$.*
(b) *A is a Blaschke set for $\mathcal{A}^\infty$.*
(c) *Every convergent Blaschke sequence in A is a zero sequence for $\mathcal{D}$.*
(d) *Every convergent Blaschke sequence in A is a zero sequence for $\mathcal{A}^\infty$.*
(e) *For every Blaschke sequence $\left(r_n e^{i\theta_n}\right)_{n\geq 1}$ in A, $\overline{\left\{e^{i\theta_n} : n \geq 1\right\}}$ is a Carleson set.*
(f) *The Euclidean distance* $\operatorname{dist}(\zeta, A)$ *satisfies*

$$\int_{\mathbb{T}} \log \operatorname{dist}(\zeta, A)\, |d\zeta| > -\infty.$$

PROOF. The equivalence of (a), (b), (e) and (f) was already known. Indeed, (a) $\iff$ (f) is Bogdan's theorem [2, Theorem 1], and (b) $\iff$ (f) is due to Taylor and Williams [11, Theorem 1]. Also (b) $\iff$ (e) follows from a result of Nelson [9, Theorem 1.2].

The new element is the equivalence of these conditions with (c) and (d). It is obvious that (b) $\implies$ (d) and that (d) $\implies$ (c), so it suffices to prove that (c) $\implies$ (e).

Assume that (c) holds. Let $(r_n e^{i\theta_n})$ be a Blaschke sequence in A, and let $(e^{i\theta_{n_j}})$ be a convergent subsequence of $(e^{i\theta_n})$. Set $B := \{r_{n_j} e^{i\theta_{n_j}} : j \geq 1\}$. Then every Blaschke sequence in B is a convergent Blaschke sequence in A, so by hypothesis (c) it is a zero sequence for $\mathcal{D}$. In other words B is a Blaschke set for $\mathcal{D}$. By the equivalence of (a) and (e), but applied to B in place of A, it follows that $\overline{\{e^{i\theta_{n_j}} : j \geq 1\}}$ is a Carleson set. By virtue of Theorem 2, we deduce that $\overline{\{e^{i\theta_n} : n \geq 1\}}$ is Carleson set. Thus (e) holds, and the proof is complete. □

Remark. Theorem 2 can be deduced, in turn, from Theorem 4. Let E be a subset of $\mathbb{T}$ such that the closure of every convergent sequence in E is a Carleson set. Define

$$A := \{re^{i\theta} : r \in [0,1), e^{i\theta} \in E\}.$$

Let $(r_n e^{i\theta_n})_{n\geq 1}$ be a convergent Blaschke sequence in A. Then $(e^{i\theta_n})$ is a convergent sequence in E, so $\overline{\{e^{i\theta_n} : n \geq 1\}}$ is a Carleson set. By [9, Theorem 1.2], the sequence $(r_n e^{i\theta_n})$ is a zero sequence for $\mathcal{A}^\infty$. To summarize, we have shown that A satisfies condition (d) in Theorem 4. Therefore A also satisfies condition (e). From this it follows easily that $\overline{E}$ is a Carleson set.

References

1. A. Beurling, *Ensembles exceptionnels*, Acta Math. **72** (1940), 1–13.
2. K. Bogdan, *On the zeros of functions with finite Dirichlet integral*, Kodai Math. J. **19** (1996), no. 1, 7–16.
3. L. Carleson, *Sets of uniqueness for functions regular in the unit circle*, Acta Math. **87** (1952), 325–345.
4. ———, *On the zeros of functions with bounded Dirichlet integrals*, Math. Z. **56** (1952), 289–295.
5. J. G. Caughran, *Two results concerning the zeros of functions with finite Dirichlet integral*, Canad. J. Math. **21** (1969), 312–316.
6. ———, *Zeros of analytic functions with infinitely differentiable boundary values*, Proc. Amer. Math. Soc. **24** (1970), 700–704.
7. O. El-Fallah, K. Kellay, and T. Ransford, *Cyclicity in the Dirichlet space*, Ark. Mat. **44** (2006), no. 1, 61–86.
8. A. Nagel, W. Rudin, and J. H. Shapiro, *Tangential boundary behavior of functions in Dirichlet-type spaces*, Ann. of Math. (2) **116** (1982), no. 2, 331–360.
9. J. D. Nelson, *A characterization of zero sets for A^∞*, Michigan Math. J. **18** (1971), 141–147.
10. H. S. Shapiro and A. L. Shields, *On the zeros of functions with finite Dirichlet integral and some related function spaces*, Math. Z. **80** (1962), 217–229.
11. B. A. Taylor and D. L. Williams, *Zeros of Lipschitz functions analytic in the unit disc*, Michigan Math. J. **18** (1971), 129–139.

Département de mathématiques et de statistique, Université Laval, 1045, avenue de la Médecine, Québec, QC G1V 0A6, Canada

E-mail address, J. Mashreghi: javad.mashreghi@mat.ulaval.ca

E-mail address, T. Ransford: ransford@mat.ulaval.ca

E-mail address, M. Shabankhah: mahmood.shabankhah.1@ulaval.ca

Centre de Recherches Mathématiques
CRM Proceedings and Lecture Notes
Volume **51**, 2010

Questions on Volterra Operators

Jaroslav Zemánek

Consider the classical Volterra operator

$$(Vf)(t) = \int_0^t f(s)\,\mathrm{d}s$$

on the Lebesgue spaces $L^p(0,1)$, $1 \le p \le \infty$, and its complex analogue

$$(Wf)(z) = \int_0^z f(\lambda)\,\mathrm{d}\lambda$$

on the Hardy spaces H^p on the unit disc, $1 \le p \le \infty$.

The important Allan – Pedersen relation

$$S^{-1}(I - V)S = (I + V)^{-1},$$

where $(Sf)(t) = \mathrm{e}^t f(t)$, $f \in L^p(0,1)$, $1 \le p \le \infty$ was noticed in [1] and extended by an elegant induction to

$$S^{-1}(I - mV)S = (I - (m-1)V)(I + V)^{-1}$$

in [10], for $m = 1, 2, \dots$. Analogously, we have the corresponding complex formulas. In fact, the formula

$$S^{-1}(I - zV)S = (I - (z-1)V)(I + V)^{-1}$$

is true for any complex number z. Indeed, from the Allan – Pedersen relation we have

$$S^{-1}VS = I - (I + V)^{-1} = V(I + V)^{-1},$$

and then

$$S^{-1}(I - zV)S = I - zV(I + V)^{-1} = (I - (z-1)V)(I + V)^{-1}.$$

Since $\|(I + V)^{-1}\|_2 = 1$ on $L^2(0,1)$ by [5, Problem 150], it follows that every operator of the form $I - tV$, with $t \ge 0$, is power-bounded on $L^2(0,1)$. This in turn implies, as observed in [10] by using [3, Lemma 2.1] and [9, Theorem 4.5.3], that

$$\|(I - V)^n - (I - V)^{n+1}\|_2 = \mathcal{O}(n^{-1/2}) \qquad \text{as } n \to \infty. \tag{1}$$

2000 *Mathematics Subject Classification.* Primary 47G10; Secondary 47A10, 47A12, 47A35.

This paper was supported in part by the European Community project TODEQ (MKTD-CT-2005-030042).

This is the final form of the paper.

The exact order of the norms of powers $(I - V)^n$ on $L^p(0,1)$ and of their consecutive differences was obtained in [8], connecting the power boundedness on $L^2(0,1)$ with the result $\|(I-V)^n\|_1 = \mathcal{O}(n^{1/4})$ on $L^1(0,1)$ obtained in [6].

In particular, the operator $I - V$ is power-bounded on $L^p(0,1)$, $1 \le p \le \infty$, if and only if $p = 2$. Similarly, the order (1) holds on $L^2(0,1)$ only.

More generally, the following characterization has recently been obtained by Yu. Lyubich [7]. Let $\phi(z)$ be an analytic function on a disc around zero such that $\phi(0) = 1$, $\phi(z) \not\equiv 1$. Then $\phi(V)$ is power-bounded on $L^p(0,1)$ if and only if $\phi'(0) < 0$ and $p = 2$.

If $p \neq 2$, a sequence of functions was found in [8] on which the powers of the operator $I - V$ increase correspondingly. However, by the Banach–Steinhaus theorem, plenty of single functions should exist in $L^p(0,1)$, $p \neq 2$, on which the powers of $I - V$ are not bounded.

Question 1. Find a function f in $L^p(0,1)$, $p \neq 2$, such that

$$\sup_{n=1,2,\dots} \|(I-V)^n f\|_p = \infty.$$

We were not able to do that! Only indirectly, J. Sánchez-Álvarez showed that on the function

$$f(x) = x^{\beta - 1}$$

with $(p-1)/p < \beta < \frac{1}{2}$ and $1 \le p < 2$ we have

$$\limsup_{n\to\infty} n^{\frac{1}{2}} \|[(I-V)^n - (I-V)^{n+1}]f\|_p = \infty,$$

by considering the subsequence $n = 4m^2$. So (1) does not hold, hence the operator $I - V$ cannot be power-bounded (the same reasoning as above that led to (1)). If $p = 2$, then such a β does not exist, thus no contradiction.

The complex operator $I - W$ is not power-bounded on H^2. It was observed by V. I. Vasyunin and S. Torba that the norms of the polynomials $(I - W)^n 1$ (i.e., the Euclidean norms of their coefficients) increase very fast.

Question 2. Is it possible to characterize the space H^2 among the spaces H^p, $1 \le p \le \infty$, in terms of the growth of the operator norms of the powers of $I - W$?

The numerical range of the operator V on the Hilbert space $L^2(0,1)$ is described in [5, Problem 166]. Since the operator $I + V$ preserves positivity of functions, the numerical ranges of all the powers are symmetric with respect to the real axis. Moreover, they are not bounded (since the operator $I + V$ is not power-bounded, see [12]) and not contained in the right half-plane $\mathrm{Re} z \ge 0$ (since the operator V is not self-adjoint, see [4] and some references therein).

Question 3. What is the union of the numerical ranges of the powers of the operator $I + V$ on $L^2(0,1)$? Is it all the complex plane? What about the operators $I \pm W$ on H^2? Or, even $I - V$ on $L^2(0,1)$?

The numerical ranges of $(I + V)^n$ on $L^2(0,1)$ were approximately determined on computer by I. Domanov, for $n = 1, 2, \dots, 7$. It turns out that these seven sets are increasing, for $n = 2$ catching 1 as an interior point, and for $n = 5$ already reaching the negative half-plane $\mathrm{Re} z < 0$.

Question 4. Fix a half-line l starting at the spectral point 1 of the operator $I+V$ on $L^2(0,1)$, and denote by l_n the length of the intersection of l with the numerical range of $(I+V)^n$. What is the behaviour of l_n with respect to n? Does the limit l_n/n exist? Does it depend on the direction of l?

It is interesting to observe, as pointed out by M. Lin, that the power boundedness of operators is very unstable even on segments: the operator

$$(1-\alpha)(I-V)+\alpha(I+V^2)=\alpha V^2-(1-\alpha)V+I$$

on $L^2(0,1)$ is power-bounded for $0\le\alpha<1$ by [11, Theorem 5], but not for $\alpha=1$ by [10, Theorem 3].

Question 5. Does there exist a quasi-nilpotent operator Q such that the operators $I-tQ$ are power-bounded for $0\le t<1$, but not for $t=1$?

Let

$$(Mf)(t)=tf(t)$$

be the multiplication operator on $L^2(0,1)$. It was shown in [2, Example 3.3] that the Volterra operator V belongs to the radical of the Banach algebra generated by M and V. Consequently, all the products in M and V, involving at least one factor V, as well as their linear combinations, are quasi-nilpotent operators (which does not seem to be obvious from the spectral radius formula!). Hence all these candidates can be tested in place of Q above.

Moreover, a number of variants of Question 5 can be considered for various versions of the well-known Kreiss resolvent condition (studied, e.g., in [8]). The answer is particularly elegant for the Ritt condition [10, Proposition 2].

I wish to thank the referee for his interesting comments.

References

1. G. R. Allan, *Power-bounded elements and radical Banach algebras*, Linear Operators (Warsaw, 1994), Banach Center Publ., vol. 38, Polish Acad. Sci., Warsaw, 1997, pp. 9–16.
2. J. Bračič, R. Drnovšek, Yu. B. Farforovskaya, E. L. Rabkin, and J. Zemánek, *On positive commutators*, Positivity, to appear.
3. S. R. Foguel and B. Weiss, *On convex power series of a conservative Markov operator*, Proc. Amer. Math. Soc. **38** (1973), 325–330.
4. A. Gomilko, I. Wróbel, and J. Zemánek, *Numerical ranges in a strip*, Operator Theory 20, Theta Ser. Adv. Math., vol. 6, Theta, Bucharest, 2006, pp. 111–121.
5. P. R. Halmos, *A Hilbert space problem book*, Van Nostrand, Princeton, NJ – Toronto, ON – London, 1967.
6. E. Hille, *Remarks on ergodic theorems*, Trans. Amer. Math. Soc. **57** (1945), 246–269.
7. Yu. Lyubich, *The power boundedness and resolvent conditions for functions of the classical Volterra operator*, Studia Math. **196** (2010), 41–63.
8. A. Montes-Rodríguez, J. Sánchez-Álvarez, and J. Zemánek, *Uniform Abel–Kreiss boundedness and the extremal behaviour of the Volterra operator*, Proc. London Math. Soc. (3) **91** (2005), no. 3, 761–788.
9. O. Nevanlinna, *Convergence of iterations for linear equations*, Lectures Math. ETH Zurich, Birkhäuser, Basel, 1993.
10. D. Tsedenbayar, *On the power boundedness of certain Volterra operator pencils*, Studia Math. **156** (2003), no. 1, 59–66.
11. D. Tsedenbayar and J. Zemánek, *Polynomials in the Volterra and Ritt operators*, Topological Algebras, Their Applications, and Related Topics, Banach Center Publ., vol. 67, Polish Acad. Sci., Warsaw, 2005, pp. 385–390.

12. J. Zemánek, *On the Gel′fand – Hille theorems*, Functional Analysis and Operator Theory (Warsaw, 1992), Banach Center Publ., vol. 30, Polish Acad. Sci., Warsaw, 1994, pp. 369–385.

INSTITUTE OF MATHEMATICS, POLISH ACADEMY OF SCIENCES, P.O. BOX 21, 00-956 WARSAW 10, POLAND

E-mail address: zemanek@impan.pl

Centre de Recherches Mathématiques
CRM Proceedings and Lecture Notes
Volume **51**, 2010

Nonhomogeneous Div-Curl Decompositions for Local Hardy Spaces on a Domain

Der-Chen Chang, Galia Dafni, and Hong Yue

ABSTRACT. Let $\Omega \subset \mathbb{R}^n$ be a Lipschitz domain. We prove div-curl type lemmas for the local spaces of functions of bounded mean oscillation on Ω, $\mathrm{bmo}_r(\Omega)$ and $\mathrm{bmo}_z(\Omega)$, resulting in decompositions for the corresponding local Hardy spaces $h^1_z(\Omega)$ and $h^1_r(\Omega)$ into nonhomogeneous div-curl quantities.

1. Div-curl lemmas for Hardy spaces and BMO on $\mathbb{R}^n$

This article is an outgrowth, among many others, of the results of Coifman, Lions, Meyer and Semmes ([7]) which connected the div-curl lemma, part of the theory of compensated compactness developed by Tartar and Murat, to the theory of real Hardy spaces in $\mathbb{R}^n$ (see [10]). In particular, denote by $H^1(\mathbb{R}^n)$ the space of distributions (in fact L^1 functions) f on $\mathbb{R}^n$ satisfying

$$\mathcal{M}_\phi(f) \in L^1(\mathbb{R}^n) \tag{1}$$

for some fixed choice of Schwartz function ϕ with $\int \phi = 1$, with the maximal function $\mathcal{M}_\phi$ defined by

$$\mathcal{M}_\phi(f)(x) = \sup_{0<t<\infty} |f * \phi_t(x)|, \qquad \phi_t(\cdot) = t^{-n}\phi(t^{-1}\cdot).$$

One version of the results in [7] states that for exponents p, q with $1 < p < \infty$, $1/p + 1/q = 1$, and vector fields $\vec{V}$ in $L^p(\mathbb{R}^n, \mathbb{R}^n)$, $\vec{W}$ in $L^q(\mathbb{R}^n, \mathbb{R}^n)$ with

$$\operatorname{div} \vec{V} = 0, \qquad \operatorname{curl} \vec{W} = 0$$

2000 *Mathematics Subject Classification.* Primary 42B30; Secondary 46E99, 30D50.

Key words and phrases. Bounded mean oscillation, Hardy spaces, Lipschitz domain, div-curl lemma, div curl decomposition.

The first author is partially supported by a Hong Kong RGC competitive earmarked research grant #600607 and a competitive research grant at Georgetown University.

The second author is partially supported by the Natural Sciences and Engineering Research Council of Canada.

The third author is partially supported by the Natural Sciences and Engineering Research Council, Canada, and the Centre de recherches mathématiques, Montréal.

This is the final form of the paper.

in the sense of distributions, the scalar (dot) product $f = \vec{V}\cdot\vec{W}$ belongs to $H^1(\mathbb{R}^n)$. Moreover, one can bound the H^1 norm (defined, say, as the L^1 norm of $\mathcal{M}_\phi(f)$) by $\|\vec{V}\|_{L^p}\|\vec{W}\|_{L^q}$.

While a local version of this result, in terms of H^1_{loc}, is given in [7], in order to obtain norm estimates we use instead the local Hardy space $h^1(\mathbb{R}^n)$. This was defined by Goldberg (see [11]) by replacing the maximal function in (1) by its "local" version

$$m_\phi(f)(x) = \sup_{0<t<1} |f * \phi_t(x)|. \tag{2}$$

Again the norm can be given by $\|m_\phi(f)\|_{L^1(\mathbb{R}^n)}$ and different choices of ϕ give equivalent norms. In addition, we can replace the number 1 in (2) by any finite constant without changing the space.

For this space the following nonhomogeneous versions of the div-curl lemma can be shown (these are special cases of Theorems 3 and 4 in [8]):

Theorem 1 ([8]). *Suppose $\vec{v}$ and $\vec{w}$ are vector fields on $\mathbb{R}^n$ satisfying*

$$\vec{V} \in L^p(\mathbb{R}^n)^n, \quad \vec{W} \in L^q(\mathbb{R}^n)^n, \qquad 1 < p < \infty, \quad \frac{1}{p} + \frac{1}{q} = 1.$$

(a) *Assume*

$$\operatorname{div}\vec{V} = f \in L^p(\mathbb{R}^n), \qquad \operatorname{curl}\vec{W} = 0 \tag{3}$$

in the sense of distributions. Then $\vec{V}\cdot\vec{W}$ belongs to the local Hardy space $h^1(\mathbb{R}^n)$ with

$$\|\vec{V}\cdot\vec{W}\|_{h^1(\mathbb{R}^n)} \le C(\|\vec{V}\|_{L^p(\mathbb{R}^n)} + \|f\|_{L^p(\mathbb{R}^n)})\|\vec{W}\|_{L^q(\mathbb{R}^n)}. \tag{4}$$

(b) *If $M^{n\times n}$ denotes the space of n-by-n matrices over $\mathbb{R}$ and*

$$\operatorname{div}\vec{V} = 0, \quad \operatorname{curl}\vec{W} = B \in L^q(\mathbb{R}^n, M^{n\times n}) \tag{5}$$

in the sense of distributions, then $\vec{V}\cdot\vec{W}$ belongs to the local Hardy space $h^1(\mathbb{R}^n)$ with

$$\|\vec{V}\cdot\vec{W}\|_{h^1(\mathbb{R}^n)} \le C\|\vec{V}\|_{L^p(\mathbb{R}^n)}\left[\|\vec{W}\|_{L^q(\mathbb{R}^n)} + \sum_{i,j}\|B_{ij}\|_{L^q(\mathbb{R}^n)}\right]. \tag{6}$$

Before continuing further, let us make clear what we mean by the divergence and curl of a vector field in the sense of distributions. Let Ω be an open subset of $\mathbb{R}^n$, and suppose $\vec{v} = (v_1, \ldots, v_n)$ with v_i locally integrable on Ω. For a locally integrable function f on Ω, one says that $\operatorname{div}\vec{v} = f$ in the sense of distributions on Ω if

$$\int_\Omega \vec{v}\cdot\vec{\nabla}\varphi = -\int_\Omega f\varphi \tag{7}$$

for all $\varphi \in C_0^\infty(\Omega)$ (i.e., smooth functions with compact support in Ω).

Similarly, if $\vec{w} = (w_1, \ldots, w_n)$ with w_i locally integrable on Ω , and B is an $n \times n$ matrix of locally integrable functions B_{ij} on Ω, we say $\operatorname{curl}\vec{w} = B$ in the sense of distributions on Ω if

$$\int_\Omega w_i\frac{\partial\varphi}{\partial x_j} - w_j\frac{\partial\varphi}{\partial x_i} = -\int_\Omega B_{ij}\varphi \tag{8}$$

for all $i, j \in \{1, \ldots, n\}$ and all $\varphi \in C_0^\infty(\Omega)$. If the components of $\vec{v}$ or $\vec{w}$ are sufficiently smooth, these definitions are equivalent to the classical notions of divergence and curl via integration by parts.

Recall that C. Fefferman [9] identified the dual of the real Hardy space H^1 with the space BMO of functions of bounded mean oscillation, introduced by John and Nirenberg [13]. In the local case, Goldberg [11] showed that the dual of $h^1(\mathbb{R}^n)$ can be identified with the space $\mathrm{bmo}(\mathbb{R}^n)$, the Banach space of locally integrable functions f which satisfy

$$\|f\|_{\mathrm{bmo}} := \sup_{|I|\leq 1} \frac{1}{|I|} \int_I |f - f_I| + \sup_{|I|>1} \frac{1}{|I|} \int_I |f| < \infty. \tag{9}$$

Here I can be used to denote either balls or cubes with sides parallel to the axes, $|I|$ denotes Lebesgue measure (volume) and f_I is the mean of f over I, i.e., $(1/|I|) \int_I f$. As in the case of h^1, the upper-bound 1 on the size of the cubes in the definition can be replaced by any other finite nonzero constant, resulting in an equivalent norm.

In [5], the authors prove (in Theorem 2.2) a kind of dual version to the div-curl lemmas in Theorem 1, which is a local analogue of a result proved in [7] for BMO: for $G \in \mathrm{bmo}(\mathbb{R}^n)$,

$$\|G\|_{\mathrm{bmo}} \approx \sup_{\vec{V},\vec{W}} \int_{\mathbb{R}^n} G\vec{V} \cdot \vec{W}, \tag{10}$$

where the supremum is taken over all vector fields $\vec{V}, \vec{W}$ as above, satisfying (3), with $\|\vec{V}\|_{L^p}$, $\|f\|_{L^p}$ and $\|\vec{W}\|_{L^q}$ all bounded by 1. Here, and below, one must obviously consider only real-valued functions g in bmo.

Moreover, the same equation (10) holds if the vector fields, instead of (3), satisfy (5) with $\|B_{ij}\|_{L^q} \leq 1$ for all $i, j \in \{1, \ldots, n\}$, as well as $\|\vec{V}\|_{L^p}, \|\vec{W}\|_{L^q} \leq 1$.

As a consequence of these results, one is able to show (see [5, Theorem 3.1]) a decomposition of functions in $h^1(\mathbb{R}^n)$ into nonhomogeneous div-curl quantities $\vec{V} \cdot \vec{W}$ of the type found in Theorem 1, part (a) or part (b).

The goal of this paper is to prove analogues of (10) for functions in local bmo spaces on a domain Ω, and obtain decomposition results for the local Hardy spaces. This was done in the case of BMO and with homogeneous, L^2 div-curl quantities in [3], and independently by Lou [16]. In [1] homogeneous div-curl results on domains were stated under the assumption that one of the vector fields is a gradient, and extended to Hardy–Sobolev spaces. Related work may be found in [12, 17].

In the next section we introduce some definitions of Hardy spaces and BMO on domains, as well as explain the boundary conditions for equations (7) and (8). The statements and proofs of our results are contained in Section 3.

2. Preliminary definitions for a domain Ω

For the moment we will just assume Ω is an open subset of $\mathbb{R}^n$, but often we will restrict ourselves to a Lipschitz domain, i.e., one whose boundary is made up, piecewise, of Lipschitz graphs.

Miyachi [19] defined Hardy spaces on Ω by letting $\delta(x) = \mathrm{dist}(x, \partial\Omega)$, replacing the maximal function $\mathcal{M}$ in (1) by

$$\mathcal{M}_{\phi,\Omega}(f)(x) = \mathrm{M}_{\phi,\delta(x)}(f)(x) = \sup_{0<t<\delta(x)} |f * \phi_t(x)|,$$

for $f \in L^1_{\text{loc}}(\Omega)$, and requiring $\mathrm{M}_{\phi,\Omega}(f) \in L^1(\Omega)$. The space of such functions was later denoted by $H^1_r(\Omega)$ in [6], since when the boundary is sufficiently nice (say Lipschitz), $H^1_r(\Omega)$ can be identified with the quotient space of restrictions to Ω of functions in $H^1(\mathbb{R}^n)$ (see [6, 19]). Moreover,

$$\|f\|_{H^1_r(\Omega)} := \|\mathrm{M}_{\phi,\Omega}(f)\|_{L^1(\Omega)} \approx \inf\{\|F\|_{H^1(\mathbb{R}^n)} : F|_\Omega = f\}.$$

The space $h^1_r(\Omega)$, corresponding to restrictions to Ω of functions in $h^1(\mathbb{R}^n)$, can be defined by replacing $\delta(x) = \operatorname{dist}(x, \partial\Omega)$ in Miyachi's definition by $\delta(x) = \min\big(\delta, \operatorname{dist}(x, \partial\Omega)\big)$, for some fixed finite $\delta > 0$. Since different choices of δ give equivalent norms, when Ω is bounded one can choose $\delta > \operatorname{diam}(\Omega)$, so $h^1_r(\Omega)$ is the same as $H^1_r(\Omega)$ (with norm equivalence involving constants depending on Ω).

For Ω a Lipschitz domain, the dual of $h^1_r(\Omega)$ (see [19] for the case of H^1 and BMO, and [2]) can be identified with the subspace

$$\mathrm{bmo}_z(\Omega) = \big\{g \in \mathrm{bmo}(\mathbb{R}^n) : \operatorname{supp}(g) \subset \overline{\Omega}\big\}.$$

Analogously, one can consider the subspace

$$h^1_z(\Omega) = \big\{g \in h^1(\mathbb{R}^n) : \operatorname{supp}(g) \subset \overline{\Omega}\big\}.$$

This was originally done in [15] in the case of H^1 functions supported on a closed subset with certain geometric properties, and later in [6] for a Lipschitz domain and in [4] for a domain with smooth boundary, in connection with boundary value problems. For a bounded domain Ω, $H^1_z(\Omega)$ and $h^1_z(\Omega)$ do not coincide since functions in H^1_z must satisfy a vanishing moment condition over the whole domain Ω, while those in h^1_z do not.

The dual of $h^1_z(\Omega)$ can be identified with $\mathrm{bmo}_r(\Omega)$, defined by requiring the supremum in (9) to be taken only over cubes I contained in Ω. In fact, one can actually require the cubes to satisfy $2I \subset \Omega$. This space was originally studied, in the case of BMO, by Jones [14], who showed that when the boundary of Ω is sufficiently nice, $\mathrm{BMO}_r(\Omega)$ is the same as the quotient space of restrictions to Ω of functions in $\mathrm{BMO}(\mathbb{R}^n)$. This holds in particular when Ω is a Lipschitz domain, and is also true in the case of bmo_r, with

$$\|g\|_{\mathrm{bmo}_r(\Omega)} \approx \inf\{\|G\|_{\mathrm{bmo}(\mathbb{R}^n)} : G|_\Omega = g\}.$$

Note that when Ω is a bounded domain, every element of $\mathrm{BMO}_r(\Omega)$ is in $\mathrm{bmo}_r(\Omega)$, but the bmo_r norm depends also on the norm of the function in $L^1(\Omega)$.

Since elements of $h^1_z(\Omega)$ are controlled in norm up to the boundary, when discussing div-curl quantities in this space one needs to consider the "boundary values" of the vector fields $\vec{v}$ and $\vec{w}$. As these vector fields are only defined in $L^p(\Omega)$ and do not have traces on the boundary, the appropriate boundary conditions are expressed, as in the case of Dirichlet and Neumann boundary value problems, by specifying the class of test functions. In particular, for the equations

$$\operatorname{div} \vec{v} = f \qquad \text{and} \qquad \operatorname{curl} \vec{w} = B,$$

we now require (7) and (8) to hold in the case when the test functions do not have compact support in Ω. This is equivalent to saying that the vector fields

$$\vec{V} = \begin{cases} \vec{v} & \text{in } \Omega \\ \vec{0} & \text{in } \mathbb{R}^n \setminus \Omega \end{cases} \tag{11}$$

and

$$\vec{W} = \begin{cases} \vec{w} & \text{in } \Omega \\ \vec{0} & \text{in } \mathbb{R}^n \setminus \Omega \end{cases} \tag{12}$$

satisfy $\operatorname{div} \vec{V} = f$ and $\operatorname{curl} \vec{W} = B$ in the sense of distributions on $\mathbb{R}^n$, with f and B vanishing outside of Ω.

When the boundary $\partial\Omega$ of Ω is sufficiently smooth, let $\vec{\mathbf{n}} = (\eta_1, \dots, \eta_n)$ denote the outward unit normal vector. If the vector fields are sufficiently smooth (so as to have a trace on the $\partial\Omega$), we can integrate by parts in (7) and (8). If φ does not have compact support in Ω, the boundary values of $\vec{v}$ must satisfy $\vec{\mathbf{n}} \cdot \vec{v} = 0$, and in the case of a bounded domain, this also entails $\int_\Omega f = 0$, while for $\vec{w}$ it must hold that on $\partial\Omega$

$$w_j \eta_i = w_i \eta_j,$$

meaning that $\vec{w}$ is colinear with $\vec{\mathbf{n}}$.

We will denote these conditions as follows. Let Ω be a Lipschitz domain and suppose f and the components of the vector fields $\vec{v}$ and $\vec{w}$ are locally integrable on Ω. As in the statement of the Neumann problem on Ω, write

$$\begin{cases} \operatorname{div} \vec{v} = f & \text{in } \Omega, \\ \int_\Omega f = 0 & \text{if } \Omega \text{ is bounded}, \\ \vec{\mathbf{n}} \cdot \vec{v} = 0 & \text{on } \partial\Omega \end{cases} \tag{13}$$

to indicate that (7) holds for all $\varphi \in C_0^\infty(\mathbb{R}^n)$, and

$$\begin{cases} \operatorname{curl} \vec{w} = B & \text{in } \Omega, \\ \vec{\mathbf{n}} \times \vec{w} = 0 & \text{on } \partial\Omega \end{cases} \tag{14}$$

to indicate that (8) holds for all $i, j \in \{1, \dots, n\}$ and all $\varphi \in C_0^\infty(\mathbb{R}^n)$.

3. Div-curl lemmas for local Hardy spaces and BMO on a domain

In order to prove an analogue of (10) for $\mathrm{bmo}_z(\Omega)$, one needs the following versions of the nonhomogeneous div-curl lemma for $h^1_r(\Omega)$. The first is a special case of Theorem 7 in [8]:

Theorem 2 ([8]). *Suppose $\vec{v}$ and $\vec{w}$ are vector fields on an open set $\Omega \subset \mathbb{R}^n$, satisfying*

$$\vec{v} \in L^p(\Omega, \mathbb{R}^n), \quad \vec{w} \in L^q(\Omega, \mathbb{R}^n), \qquad 1 < p < \infty, \quad \frac{1}{p} + \frac{1}{q} = 1,$$

and

$$\operatorname{div} \vec{v} = f \in L^p(\Omega), \quad \operatorname{curl} \vec{w} = 0$$

in the sense of distributions on Ω. Then $\vec{v} \cdot \vec{w}$ belongs to the local Hardy space $h^1_r(\Omega)$ with

$$\|\vec{v} \cdot \vec{w}\|_{h^1_r(\Omega)} \leq C(\|\vec{v}\|_{L^p(\Omega)} + \|f\|_{L^p(\Omega)})\|\vec{w}\|_{L^q(\Omega)}. \tag{15}$$

The second is a domain version of Theorem 4 in [8], whose proof we shall adapt below:

Theorem 3. *Suppose $\vec{v}$ and $\vec{w}$ are vector fields on an open set $\Omega \subset \mathbb{R}^n$, satisfying*

$$\vec{v} \in L^p(\Omega, \mathbb{R}^n), \quad \vec{w} \in L^q(\Omega, \mathbb{R}^n), \qquad 1 < p < \infty, \quad \frac{1}{p} + \frac{1}{q} = 1,$$

and

$$\operatorname{div} \vec{v} = 0, \qquad \operatorname{curl} \vec{w} = B \in L^q(\Omega; M^{n\times n})$$

in the sense of distributions on Ω. Then $\vec{v}\cdot\vec{w}$ belongs to the local Hardy space $h^1_r(\Omega)$ with

$$\|\vec{v} \cdot \vec{w}\|_{h^1_r(\Omega)} \leq C\|\vec{v}\|_{L^p(\Omega)}\big(\|\vec{w}\|_{L^q(\Omega)} + \sum_{i,j}\|B_{ij}\|_{L^q(\Omega)}\big). \tag{16}$$

Proof. Consider a point $x \in \Omega$ and a cube Q^x_l, centered at x and of sidelength $l > 0$, depending on x. We choose $l = \min\big(1, \operatorname{dist}(x, \partial\Omega)\big)/\sqrt{n}$, which guarantees Q^x_l lies inside Ω. Without loss of generality assume $Q^x_l = [0, l]^n$. Writing $\vec{v} = (v_1, \ldots, v_n)$, and fixing i, we solve $-\Delta u_i = v_i$ with mixed boundary conditions: on the two faces $x_i = 0$ and $x_i = l$ we impose Neumann boundary values

$$\frac{\partial u_i}{\partial x_i} = 0,$$

and on the other faces (corresponding to $x_j = 0$ and $x_j = l$, $j \neq i$) Dirichlet boundary values $u_i = 0$. This can be done by expanding in multiple Fourier series (with even coefficients in x_i and odd coefficients in x_j, $j \neq i$). By the Marcinkiewicz multiplier theorem (see [18, Theorem 4]) the second derivatives of the solution u_i will be bounded in $L^\alpha(Q^x_l)$ by $\|v_i\|_{L^\alpha(Q^x_l)}$, for every $\alpha \leq p$, $i = 1, \ldots, n$. Note that by the homogeneity of the multipliers, the constants will be independent of l. Since we have taken $l \leq 1$, we also get that $\|u_i\|_{W^{2,p}(Q^x_l)} \leq C\|v_i\|_{L^p(Q^x_l)}$ with a constant independent of l.

Set $\vec{U} = (u_1, \ldots, u_n)$ and consider the function $\operatorname{div} \vec{U} \in W^{1,p}(Q^x_l)$. This function satisfies

$$\Delta(\operatorname{div} \vec{U}) = -\operatorname{div} \vec{v} = 0$$

in the sense of distributions on Q^x_l, since $Q^x_l \subset \Omega$, and moreover

$$\operatorname{div} \vec{U} = \sum \frac{\partial u_i}{\partial x_i} = 0$$

on the boundary, by the choice of boundary conditions above. By the uniqueness of the solution of the Dirichlet problem in $W^{1,p}_0(Q^x_l)$, we must have $\operatorname{div} \vec{U} = 0$ on Q^x_l. Let A be the matrix $\operatorname{curl} \vec{U}$, i.e.,

$$A_{ij} = \frac{\partial u_i}{\partial x_j} - \frac{\partial u_j}{\partial x_i}.$$

These are functions in $W^{1,p}(Q^x_l)$ with first derivatives bounded in the $L^\alpha(Q^x_l)$-norm by $\|v_i\|_{L^\alpha(Q^x_l)}$, for every $\alpha \leq p$.

Now writing $\vec{A}_j$ for the jth column of the matrix A, we have, in the sense of distributions on Q^x_l,

$$\operatorname{div} \vec{A}_j = \sum_{i=1}^n \left(\frac{\partial^2 u_i}{\partial x_i \partial x_j} - \frac{\partial^2 u_j}{\partial x_i^2}\right) = \frac{\partial}{\partial x_j} \operatorname{div} \vec{U} - \Delta u_j = v_j, \tag{17}$$

for each $j = 1, \ldots, n$. Taking the dot product with $\vec{w}$ and recalling that we identify $\operatorname{curl}\vec{w}$, in the sense of distributions on Ω, with a matrix B whose components are in $L^q(\Omega)$, we have

$$\begin{aligned}\vec{v}\cdot\vec{w} &= \sum_{j=1}^{n}(\operatorname{div}\vec{A}_j)w_j = \sum_{j=1}^{n}\operatorname{div}(\vec{A}_j w_j) - \sum_{i,j} A_{ij}\frac{\partial w_j}{\partial x_i}\\ &= \sum_{j=1}^{n}\operatorname{div}(\vec{A}_j w_j) + \sum_{i<j} A_{ij}B_{ij},\end{aligned}$$

again in the sense of distributions on Q_l^x.

Take $\phi \in C^\infty$ with support in $B\big(0, 1/(2\sqrt{n})\big)$ and $\int \phi = 1$, and define, for $0 < t \le \min\big(1, \operatorname{dist}(x, \partial\Omega)\big)$, ϕ_t^x by $\phi_t^x(y) = t^{-n}\phi\big(t^{-1}(x-y)\big)$. Since $l = \min\big(1, \operatorname{dist}(x,\partial\Omega)\big)/\sqrt{n}$ we have

$$\operatorname{supp}(\phi_t^x) \subset B(x, t/2\sqrt{n}) \subset Q_l^x \subset \Omega.$$

Denote $B(x, t/2\sqrt{n})$ by $\widetilde{B_t^x}$.

We integrate $\vec{v}\cdot\vec{w}$ against ϕ_t^x, noting that equation (17) holds even if we change $\vec{A}_j$ by adding a vector field which is constant on Q_l^x. In particular we modify each A_{ij} by subtracting its average $(A_{ij})_{\widetilde{B_t^x}}$ over $\widetilde{B_t^x}$. Integration by parts yields:

$$\begin{aligned}\int \phi_t^x(\vec{v}\cdot\vec{w}) = -\sum_{i,j}\int t^{-(n+1)}\frac{\partial\phi}{\partial y_i}\big(t^{-1}(x-y)\big)\big(A_{ij}(y) - (A_{ij})_{\widetilde{B_t^x}}\big)w_j(y)\,dy\\ +\sum_{i<j}\int t^{-n}\phi\big(t^{-1}(x-y)\big)\big(A_{ij} - (A_{ij})_{\widetilde{B_t^x}}\big)B_{ij}.\end{aligned}$$

Take α, β with $1 < \alpha < p$, $1 < \beta < q$ and $1/\alpha + 1/\beta = 1 + 1/n$. The Sobolev–Poincaré inequality in $\widetilde{B_t^x}$, together with the fact that $t \le 1$, gives (see the proof of Lemma II.1 in [7]):

$$\begin{aligned}|\phi_t * (\vec{v}\cdot\vec{w})(x)| &\le C\|\vec{\nabla}\phi\|_\infty \sum_{i,j}\Big(t^{-n}\int_{\widetilde{B_t^x}}|\vec{\nabla}A_{ij}|^\alpha\Big)^{1/\alpha}\Big(t^{-n}\int_{\widetilde{B_t^x}}|\vec{w}|^\beta\Big)^{1/\beta}\\ &\quad + C\|\phi\|_\infty\sum_{i,j}\Big(t^{-n}\int_{\widetilde{B_t^x}}|\vec{\nabla}A_{ij}|^\alpha\Big)^{1/\alpha}\Big(t^{-n}\int_{\widetilde{B_t^x}}|B_{ij}|^\beta\Big)^{1/\beta}\\ &\le C_\phi \mathrm{M}(|\vec{v}|^\alpha)(x)^{1/\alpha}\Big[\mathrm{M}(|\vec{w}|^\beta)(x)^{1/\beta} + \sum_{i,j}\mathrm{M}(|B_{ij}|^\beta)(x)^{1/\beta}\Big].\end{aligned}$$

Here the Hardy–Littlewood maximal function on $\mathbb{R}^n$, denoted by M, is applied to the functions $|\vec{v}|^\alpha, |\vec{w}|^\beta$ and $|B_{ij}|^\beta$ by extending them by zero outside Ω. The constant depends on the choice of ϕ but not on t or x.

Recalling that the maximal function is bounded on $L^r(\mathbb{R}^n)$, $r > 1$, we conclude that:

$$\begin{aligned}\int_\Omega \sup_{0<t<\operatorname{dist}(x,\partial\Omega)} &|\phi_t * (\vec{v}\cdot\vec{w})(x)|\,\mathrm{d}x\\ &\le C_\phi\Big(\int_\Omega (\mathrm{M}(|\vec{v}|^\alpha)(x))^{p/\alpha}\,\mathrm{d}x\Big)^{1/p}\\ &\quad\times\Big[\Big(\int_\Omega (\mathrm{M}(|\vec{w}|^\beta)(x))^{q/\beta}\,\mathrm{d}x\Big)^{1/q}+\sum_{i,j}\Big(\int_\Omega (\mathrm{M}(|B_{ij}|^\beta)(x))^{q/\beta}\,\mathrm{d}x\Big)^{1/q}\Big]\\ &\le C_\phi\|\vec{v}\|_{L^p(\Omega)}\Big[\|\vec{w}\|_{L^q(\Omega)}+\sum_{i,j}\|B_{ij}\|_{L^q(\Omega)}\Big].\end{aligned}$$

This shows $\vec{v}\cdot\vec{w}\in h^1_r(\Omega)$, and (16) holds. □

Lemma 4. *Suppose $\vec{v}$ and $\vec{w}$ are vector fields on a Lipschitz domain $\Omega\subset\mathbb{R}^n$, satisfying the hypotheses of either Theorem* 2 *or Theorem* 3, *but with the conditions on the divergence and the curl satisfied in the stronger sense of* (13) *and* (14). *Then $\vec{v}\cdot\vec{w}\in h^1_z(\Omega)$ with norm bounded as in* (15) *or* (16).

Proof. Given such vector fields $\vec{v}$ and $\vec{w}$ on Ω, define the zero extensions $\vec{V}$ and $\vec{W}$ as in (11) and (12). The L^p and L^q norms of $\vec{V}$ and $\vec{W}$ are the same as those of $\vec{v}$ and $\vec{w}$ on Ω. Moreover, conditions (13) and (14) guarantee that $\vec{V}$ and $\vec{W}$ satisfy (3) (respectively (5)) in the sense of distributions on $\mathbb{R}^n$. Therefore, by using Theorem 1, part (a) (respectively part (b)), we can conclude that $\vec{V}\cdot\vec{W}\in h^1(\mathbb{R}^n)$ with the appropriate bound on its norm. But $\vec{V}\cdot\vec{W}$ is equal to zero outside Ω and is $\vec{v}\cdot\vec{w}$ on Ω, hence this is a function in $h^1_z(\Omega)$. The h^1_z norm is the same as the h^1 norm and the bounds can be given in terms of the L^p and L^q norms of the relevant quantities on Ω. □

Now we can proceed to state and prove the local bmo versions of the div-curl lemma on a Lipschitz domain:

Theorem 5. *Let $\Omega\subset\mathbb{R}^n$ be a Lipschitz domain.*

(a) *If $g\in\mathrm{bmo}_z(\Omega)$, then*

$$\|g\|_{\mathrm{bmo}_z}\approx\sup_{\vec{v},\vec{w}}\int_\Omega g\,\vec{v}\cdot\vec{w}, \tag{18}$$

where the supremum is taken over all vector fields $\vec{v}\in L^p(\Omega,\mathbb{R}^n)$, $\vec{w}\in L^q(\Omega,\mathbb{R}^n)$, $\|\vec{v}\|_{L^p(\Omega)}\le 1$, $\|\vec{w}\|_{L^q(\Omega)}\le 1$, satisfying (3) *in the sense of distributions on Ω, with $\|f\|_{L^p(\Omega)}\le 1$.*

(b) *If $g\in\mathrm{bmo}_z(\Omega)$, then equation* (18) *holds with the supremum now taken over all vector fields $\vec{v}\in L^p(\Omega,\mathbb{R}^n)$, $\vec{w}\in L^q(\Omega,\mathbb{R}^n)$, $\|\vec{v}\|_{L^p(\Omega)}\le 1$, $\|\vec{w}\|_{L^q(\Omega)}\le 1$, satisfying* (5) *in the sense of distributions on Ω, with $\|B_{ij}\|_{L^q}(\Omega)\le 1$ for all $i,j\in 1,\dots,n$.*

(c) *If $g\in\mathrm{bmo}_r(\Omega)$ then*

$$\|g\|_{\mathrm{bmo}_r}\approx\sup_{\vec{v},\vec{w}}\int_\Omega g\,\vec{v}\cdot\vec{w},$$

the supremum being taken over all vector fields $\vec{v}$ and $\vec{w}$ as in part (a) *or in part* (b), *but satisfying the stronger conditions* (13) *and* (14).

PROOF. Let $g \in \mathrm{bmo}_z(\Omega)$ (real-valued) and take $\vec{v}, \vec{w}$ as in the hypotheses of part (a) (respectively part (b)). By Theorem 2 (resp. Theorem 3), the dot product $\vec{v} \cdot \vec{w}$ belongs to $h^1_r(\Omega)$ with norm bounded by a constant. The duality of $\mathrm{bmo}_z(\Omega)$ with $h^1_r(\Omega)$ then gives

$$\int_\Omega g\,\vec{v} \cdot \vec{w} \leq C\|g\|_{\mathrm{bmo}_z}.$$

Conversely, by the nature of $\mathrm{bmo}_z(\Omega)$, the extension G of g to $\mathbb{R}^n$ by setting it zero outside Ω is in $\mathrm{BMO}(\mathbb{R}^n)$ with $\|G\|_{\mathrm{bmo}} \approx \|g\|_{\mathrm{bmo}_z}$. Hence, by (10), one has

$$\|g\|_{\mathrm{bmo}_z} \approx \sup_{\vec{V},\vec{W}} \int_{\mathbb{R}^n} G\vec{V} \cdot \vec{W} = \sup_{\vec{V},\vec{W}} \int_\Omega g\vec{V}\big|_\Omega \cdot \vec{W}\big|_\Omega,$$

where the supremum is taken over all vector fields $\vec{V} \in L^p(\mathbb{R}^n, \mathbb{R}^n)$, $\vec{W} \in L^q(\mathbb{R}^n, \mathbb{R}^n)$, $\|\vec{V}\|_{L^p} \leq 1$, $\|\vec{W}\|_{L^q} \leq 1$, satisfying (3) (resp. (5)) in the sense of distributions on $\mathbb{R}^n$. The restrictions $\vec{v} = \vec{V}\big|_\Omega$, $\vec{w} = \vec{W}\big|_\Omega$ satisfy the same conditions in Ω, proving the inequality $\lesssim$ in (18).

If $g \in \mathrm{bmo}_r(\Omega)$ and $\vec{v}, \vec{w}$ are as in part (c), by Lemma 4 $\vec{v} \cdot \vec{w} \in h^1_z(\Omega)$ with norm bounded by a constant, so the duality of bmo_r and h^1_z implies

$$\int_\Omega g\,\vec{v} \cdot \vec{w} \leq C\|g\|_{\mathrm{bmo}_r}\|\vec{v} \cdot \vec{w}\|_{h^1_z} \leq C\|g\|_{\mathrm{bmo}_r}.$$

This shows that

$$\sup_{\vec{v},\vec{w}} \int_\Omega g\,\vec{v} \cdot \vec{w} \leq C\|g\|_{\mathrm{bmo}_r}.$$

It remains to prove the other direction, i.e.,

$$\|g\|_{\mathrm{bmo}_r} \leq C' \sup_{\vec{v},\vec{w}} \int_\Omega g\,\vec{v} \cdot \vec{w}.$$

The left-hand side is given by

$$\sup_{\substack{Q \subset \Omega \\ |Q| \leq 1}} \frac{1}{|Q|} \int_Q |g(x) - g_Q|\, dx + \sup_{\substack{Q \subset \Omega \\ |Q| > 1}} \frac{1}{|Q|} \int_Q |g(x)|\, dx.$$

As explained in the proof of Theorem 2.1 in [3] (for the case of $\mathrm{BMO}_r(\Omega)$ but the same arguments apply to $\mathrm{bmo}_r(\Omega)$), it suffices to take the supremum over cubes Q satisfying $\widetilde{Q} = 2Q \subset \Omega$ (or with some constant C_Ω replacing 2). In that case it just remains to point out that in the proof of estimate (10) in [5] (see the proof of Theorem 2.2., Case I), it was shown that for a ball $B \subset \mathbb{R}^n$ with radius bounded by 1,

$$\left(\frac{1}{|B|} \int_B |g(x) - g_B|^2\, dx\right)^{1/2} \leq C_n \sup \int g\,\vec{v} \cdot \vec{w},$$

where the supremum is taken over all vector fields $\vec{v}, \vec{w} \in C_0^\infty(\widetilde{B})$ with $\|\vec{v}\|_{L^p} \leq 1$, $\|\vec{w}\|_{L^q} \leq 1$ and $\operatorname{div} \vec{v} = 0, \operatorname{curl} \vec{w} = 0$. There we took $\widetilde{B} = 2B$ but the argument immediately applies to $\widetilde{B} = C_\Omega B$ for some $C_\Omega > 1$. Note that if $\widetilde{B} \subset \Omega$, such vector fields will vanish on the boundary $\partial\Omega$ and thus satisfy the boundary conditions (13) and (14).

Similarly, for a ball $B \subset \mathbb{R}^n$ with radius larger than 1, we showed in [5] (see the proof of Theorem 2.2., Case I) that

$$\left(\frac{1}{|B|}\int_B |g(x)|^2\,\mathrm{d}x\right)^{1/2} \leq C_n \sup \int g\,\vec{v}\cdot\vec{w},$$

where this time the supremum can be taken over all vector fields $\vec{v}, \vec{w} \in C_0^\infty(\widetilde{B})$ with $\|\vec{v}\|_{L^p} \leq 1$, $\|\vec{w}\|_{L^q} \leq 1$ satisfying the relaxed div-curl conditions (3), or alternatively the supremum can be taken over such vector fields satisfying (5). Again such vector fields will automatically satisfy (13) and (14) — the boundary conditions follow from the vanishing on the boundary and the condition $\int_\Omega \operatorname{div}\vec{v} = 0$, in the case of bounded Ω, follows from the divergence theorem since we are now dealing with smooth functions. □

Finally we arrive at the desired nonhomogeneous div-curl decompositions for the local Hardy spaces on Ω. These are corollaries of Theorem 5 and follow from the duality between bmo_z and h^1_r (respectively bmo_r and h^1_z) by using Lemmas III.1 and III.2 in [7]:

Theorem 6. *Let $\Omega \subset \mathbb{R}^n$ be a Lipschitz domain and $1 < p < \infty$, $1/p + 1/q = 1$.*

(a) *For a function f in $h^1_r(\Omega)$, there exists a sequence of scalars $\{\lambda_k\}$ with $\sum_{k=1}^\infty |\lambda_k| < \infty$, and sequences of vector fields $\{\vec{v}_k\}$ in $L^p(\Omega, \mathbb{R}^n)$, $\{\vec{w}_k\}$ in $L^q(\Omega, \mathbb{R}^n)$ with $\|\vec{v}_k\|_{L^p}$, $\|\vec{w}_k\|_{L^q} \leq 1$ for all k, satisfying, for each k, condition* (3) *in the sense of distributions on Ω, so that*

$$f = \sum_{k=1}^\infty \lambda_k \vec{v}_k \cdot \vec{w}_k.$$

(b) *The same result holds as in part* (a) *but with $\vec{v}_k$ and $\vec{w}_k$ satisfying* (5) *instead of* (3), *for each k.*

(c) *For a function $f \in h^1_z(\Omega)$, there exists a sequence of scalars $\{\lambda_k\}$ with $\sum_{k=1}^\infty |\lambda_k| < \infty$, and sequences of vector fields $\{\vec{v}_k\}$ and $\{\vec{w}_k\}$, as in part* (a) *or part* (b), *but satisfying the div-curl conditions in the stronger sense of* (13) *for each $\vec{v}_k$ and* (14) *for each $\vec{w}_k$, so that*

$$f = \sum_{k=1}^\infty \lambda_k \vec{v}_k \cdot \vec{w}_k.$$

Remark. As pointed out in Section 2, when the domain Ω is bounded the "local" Hardy space $h^1_r(\Omega)$ coincides with $H^1_r(\Omega)$ and similarly for $\mathrm{BMO}_z(\Omega)$ and $\mathrm{bmo}_z(\Omega)$. By taking the constants in the definitions and proofs sufficiently large (depending on the size of Ω), we do not need to deal with the case of "large" balls or cubes, so everything reverts to the homogeneous case. As previously mentioned, this case was dealt with in [3] and [16], but only for $p = q = 2$, so the current results are a generalization of the older ones.

References

1. P. Auscher, E. Russ, and P. Tchamitchian, *Hardy Sobolev spaces on strongly Lipschitz domains of $\mathbb{R}^n$*, J. Funct. Anal. **218** (2005), no. 1, 54–109.
2. D.-C. Chang, *The dual of Hardy spaces on a bounded domain in $\mathbf{R}^n$*, Forum Math. **6** (1994), no. 1, 65–81.

3. D.-C. Chang, G. Dafni, and C. Sadosky, *A div-curl lemma in BMO on a domain*, Harmonic analysis, signal processing, and complexity, Progr. Math., vol. 238, Birkhäuser, Boston, MA, 2005, pp. 55–65.
4. D.-C. Chang, G. Dafni, and E. M. Stein, *Hardy spaces, BMO, and boundary value problems for the Laplacian on a smooth domain in* $\mathbf{R}^n$, Trans. Amer. Math. Soc. **351** (1999), no. 4, 1605–1661.
5. D.-C. Chang, G. Dafni, and Hong Yue, *A div-curl decomposition for the local Hardy space*, Proc. Amer. Math. Soc. **137** (2009), no. 10, 3369–3377.
6. D.-C. Chang, S. G. Krantz, and E. M. Stein, H^p *theory on a smooth domain in* $\mathbf{R}^N$ *and elliptic boundary value problems*, J. Funct. Anal. **114** (1993), no. 2, 286–347.
7. R. Coifman, P.-L. Lions, Y. Meyer, and S. Semmes, *Compensated compactness and Hardy spaces*, J. Math. Pures Appl. (9) **72** (1993), no. 3, 247–286.
8. G. Dafni, *Nonhomogeneous div-curl lemmas and local Hardy spaces*, Adv. Differential Equations **10** (2005), no. 5, 505–526.
9. C. Fefferman, *Characterizations of bounded mean oscillation*, Bull. Amer. Math. Soc. **77** (1971), 587–588.
10. C. Fefferman and E. M. Stein, H^p *spaces of several variables*, Acta Math. **129** (1972), no. 3-4, 137–193.
11. D. Goldberg, *A local version of real Hardy spaces*, Duke Math. J. **46** (1979), no. 1, 27–42.
12. J. Hogan, C. Li, A. McIntosh, and K. Zhang, *Global higher integrability of Jacobians on bounded domains*, Ann. Inst. H. Poincaré Anal. Non Linéaire **17** (2000), no. 2, 193–217.
13. F. John and L. Nirenberg, *On functions of bounded mean oscillation*, Comm. Pure Appl. Math. **14** (1961), 415–426.
14. P. W. Jones, *Extension theorems for BMO*, Indiana Univ. Math. J. **29** (1980), no. 1, 41–66.
15. A. Jonsson, P. Sjögren, and H. Wallin, *Hardy and Lipschitz spaces on subsets of* $\mathbf{R}^n$, Studia Math. **80** (1984), no. 2, 141–166.
16. Z. Lou, *Div-curl type theorems on Lipschitz domains*, Bull. Austral. Math. Soc. **72** (2005), no. 1, 31–38.
17. Z. Lou and A. McIntosh, *Hardy spaces of exact forms on Lipschitz domains in* $\mathbb{R}^N$, Indiana Univ. Math. J. **53** (2004), no. 2, 583–611.
18. J. Marcinkiewicz, *Sur les multiplicateurs des séries de Fourier*, Studia Math. **8** (1939), 78–91.
19. A. Miyachi, H^p *spaces over open subsets of* $\mathbf{R}^n$, Studia Math. **95** (1990), no. 3, 205–228.

Department of Mathematics, Georgetown University, Washington DC, 20057, USA
E-mail address: chang@math.georgetown.edu

Department of Mathematics and Statistics, Concordia University, 1455 de Maisonneuve Blvd. W, Montréal, QC H3G 1M8, Canada
E-mail address: gdafni@mathstat.concordia.ca

Department of Mathematics and Statistics, Concordia University, 1455 de Maisonneuve Blvd. W, Montréal, QC H3G 1M8, Canada
Current address: Department of Mathematics and Informatics, Trine University, Angola, IN 46703, USA
E-mail address: yueh@trine.edu

Centre de Recherches Mathématiques
CRM Proceedings and Lecture Notes
Volume **51**, 2010

On the Bohr Radius for Simply Connected Plane Domains

Richard Fournier and Stephan Ruscheweyh

ABSTRACT. We give various estimates and discuss sharpness questions for a generalized Bohr radius applicable to simply connected domains of the complex plane.

1. Introduction and statement of the results

Let $\mathbb{D} = \{z \mid |z| < 1\}$ be the open unit disc in the complex plane and $H(\mathbb{D})$ the class of analytic maps on $\mathbb{D}$. Let $f \in H(\mathbb{D})$ with $f(z) = \sum_{k=0}^{\infty} a_k z^k$ and $f(\mathbb{D}) \subseteq \overline{\mathbb{D}}$. Then,

$$\sum_{k=0}^{\infty} |a_k| r^k \leq 1 \qquad \text{if } r \leq \tfrac{1}{3}.$$

This result was first obtained by Harald Bohr [2] in 1914 with the constant $\frac{1}{3}$ replaced by $\frac{1}{6}$ and later improved by M. Riesz and others who established that in this context the constant $\frac{1}{3}$ is best possible; different proofs were later published while similar problems were considered for H_p spaces or more abstract spaces or else in the context of several complex variables by a number of authors. We refer to a paper [6] and a book [7] by Kresin and Maz′ya for a rather complete survey of recent and less recent related results.

Our point of view is the following: we consider a simply connected domain $\mathcal{D}$ with $\mathbb{D} \subseteq \mathcal{D}$ and define the Bohr constant $B = B_{\mathcal{D}}$ as

$$\sup\left\{ r \in (0,1) : \sum_{k=0}^{\infty} |a_k| r^k \leq 1 \text{ for all } f(z) := \sum_{k=0}^{\infty} a_k z^k \in \mathcal{B}(\mathcal{D}),\ z \in \mathbb{D} \right\}$$

while $\mathcal{B}(\mathcal{D})$ is the class of functions $f \in H(\mathcal{D})$ such that $f(\mathcal{D}) \subseteq \overline{\mathbb{D}}$. Clearly $B_{\mathbb{D}} = \frac{1}{3}$ coincides with the classical Bohr radius and we wish to estimate $B_{\mathcal{D}}$ for more general domains $\mathcal{D}$. Our main results are the following:

2000 *Mathematics Subject Classification.* Primary 30B10; Secondary 30A10.
Key words and phrases. Bohr radius, power series, plane simply connected domains.
This is the final form of the paper.

Theorem 1. *Let* $0 \le \gamma < 1$ *and* $\mathcal{D}_\gamma$ *the disc* $\{w : |w + \gamma/(1-\gamma)| < 1/(1-\gamma)\}$. *Then* $B_{\mathcal{D}_\gamma} = \rho_\gamma := (1+\gamma)/(3+\gamma)$ *and* $\sum_{k=0}^\infty |a_k|\rho_\gamma^k = 1$ *holds for a function* $f(z) = \sum_{k=0}^\infty a_k z^k$ *in* $\mathcal{B}(\mathcal{D}_\gamma)$ *if and only if* $f \equiv c$ *with* $|c| = 1$.

Theorem 2. *Let* $\mathcal{D}$ *be a simply connected domain with* $\mathcal{D} \supseteq \mathbb{D}$ *and let*

$$\lambda := \lambda(\mathcal{D}) := \sup_{\substack{f \in \mathcal{B}(\mathcal{D}) \\ k \ge 1}} \left\{ \frac{|a_k|}{1 - |a_0|^2} : a_0 \not\equiv f(z) = \sum_{k=0}^\infty a_k z^k, z \in \mathbb{D} \right\}.$$

Then $1/(1+2\lambda) \le B_{\mathcal{D}}$ *and the equality* $\sum_{k=0}^\infty |a_k|(1/(1+2\lambda))^k = 1$ *holds for a function* $f(z) = \sum_{k=0}^\infty a_k z^k$ *in* $\mathcal{B}(\mathcal{D})$ *if and only if* $f \equiv c$ *with* $|c| = 1$.

In the case of a convex domain $\mathcal{D}$, it is possible to estimate $B_{\mathcal{D}}$ in terms of the conformal radius:

Theorem 3. *Let* $\mathcal{D}$ *be a convex domain with* $\mathcal{D} \supseteq \mathbb{D}$ *and* $F(z) := A_1 z + o(z)$ *a conformal map of* $\mathcal{D}$ *onto* $\mathbb{D}$ *with* $A_1 > 0$. *Then*

$$\beta := \min\left(1, \frac{1}{4A_1}\right) \le B_{\mathcal{D}}.$$

Our next result is a limiting case of Theorem 1, but should also be compared to Theorem 3.

Corollary 4. *Let* P *denote the half-plane* $\{z \mid \mathrm{Re}(z) < 1\}$ *and* $f(z) = \sum_{n=0}^\infty a_n z^n$ *for* $|z| < 1$ *where* $f \in \mathcal{B}(P)$. *Then*

$$\sum_{n=0}^\infty |a_n| r^n \le 1 \qquad \text{if } 0 \le r \le \tfrac{1}{2}. \tag{1}$$

2. Proofs

The proof of Theorem 1 is based on the following result which may be of independent interest:

Lemma 5. *Let* $a \in \mathbb{D}$ *and* $f \in \mathcal{B}(\mathbb{D})$ *with*

$$f(z) = \sum_{k=0}^\infty \alpha_k (z-a)^k, \qquad |z-a| \le 1 - |a|.$$

Then

$$\sum_{k=0}^\infty |\alpha_k| r^k \le 1 \qquad \text{if } r \le r_0 := \frac{1-|a|^2}{3+|a|}. \tag{2}$$

Furthermore, r_0 *is the largest number with this property, and*

$$\sum_{k=0}^\infty |\alpha_k| r_0^k = 1$$

if and only if $f \equiv c$ *with* $|c| = 1$.

PROOF. We make use of the following estimate (see [9] for a proof; this estimate is of course an extension of the Schwarz-Pick inequality and has shown to be useful in a number of situations):

$$|\alpha_k| \le (1+|a|)^{k-1} \frac{1-|\alpha_0|^2}{(1-|a|^2)^k}, \qquad k \ge 1, \tag{3}$$

and note that $|\alpha_0| \le 1$. We get

$$\sum_{k=0}^{\infty} |\alpha_k| r^k \le |\alpha_0| + \frac{(1-|\alpha_0|^2)r}{((1+|a|)(1-|a|-r))}$$

and the last quantity is seen to be less or equal to 1 for all admissible α_0 if and only if $2r/(1+|a|)(1-|a|-r) \le 1$. This leads to (2).

Again using (3) it is easily established that

$$\sum_{k=0}^{\infty} |\alpha_k| r_0^k = 1 \iff (|\alpha_0| = 1 \text{ and } \alpha_k = 0,\ k \ge 1). \tag{4}$$

Further, the functions

$$f_b(z) = \frac{z-b}{1-\bar{b}z} := \sum_{k=0}^{\infty} \alpha_k(a,b)(z-a)^k, \qquad b \in \mathbb{D},$$

belong to $\mathcal{B}(\mathbb{D})$ and one has

$$\sum_{k=0}^{\infty} |\alpha_k(a,b)| r^k \le 1 \qquad \text{for all } b \in \mathbb{D} \tag{5}$$

if and only if $r \le r_0$ which implies that the constant r_0 is indeed optimal with respect to the statement of the lemma. □

Proof of Theorem 1. For some $0 \le \gamma \le 1$ let $f \in \mathcal{B}(\mathcal{D}_\gamma)$ such that $f(z) = \sum_{k=0}^{\infty} a_k z^k$, $z \in \mathbb{D}$. Then $g(z) := f\big((z-\gamma)/(1-\gamma)\big)$ belongs to $\mathcal{B}(\mathbb{D})$ and if $|z-\gamma| \le |1-\gamma|$, we have

$$g(z) = f\left(\frac{z-\gamma}{1-\gamma}\right) = \sum_{k=0}^{\infty} \frac{a_k}{(1-\gamma)^k}(z-\gamma)^k.$$

The lemma now gives,

$$\sum_{k=0}^{\infty} |a_k| \left(\frac{\rho}{1-\gamma}\right)^k \le 1 \qquad \text{if } \rho \le \frac{1-\gamma^2}{3+\gamma},$$

and $(1-\gamma^2)/(3+\gamma) \le B_{\mathcal{D}_\gamma}$ follows together with the statement of equality which is a consequence of (4). The functions $f \in \mathcal{B}(\mathcal{D}_\gamma)$ defined by $f\big((z-\gamma)/(1-\gamma)\big) = f_b(z)$, $b, z \in \mathbb{D}$, may be used as in (5) to show that the constant $(1+\gamma)/(3+\gamma)$ is best possible with respect to Theorem 1, i.e. $B_{\mathcal{D}_\gamma} = (1+\gamma)/(3+\gamma)$. □

Proof of Corollary 4. Under the hypothesis, $f \in \mathcal{B}(\mathcal{D}_\gamma)$ for all γ in $[0,1)$ and the result follows from Theorem 1 by letting $\gamma \to 1$. The constant $\frac{1}{2}$ is best possible as can be seen from the Taylor expansion at the origin of the functions

$$g_a(z) = \frac{z/(z-2) - a}{1 - \bar{a}z/(z-2)}, \qquad \mathrm{Re}(z) < 1,\ a \in \mathbb{D}.$$

We omit the details.

Due to the limiting process used in the above argument, the case of equality in (1) is no more a simple consequence of (4) and a separate proof must be given. We first apply inequality (3) to obtain

$$|a_k| \le \frac{1-|\alpha_0|^2}{1+\gamma}, \qquad k \ge 1, \tag{6}$$

for any $g \in \mathcal{B}(\mathcal{D}_\gamma)$, $g(z) = \sum_{k=0}^{\infty} \alpha_k z^k$, $z \in \mathbb{D}$.

This leads to

$$|a_k| \le \tfrac{1}{2}(1 - |a_0|^2), \quad k \ge 1$$

for any $f \in \mathcal{B}(P) = \bigcap_{0 \le \gamma < 1} \mathcal{B}(\mathcal{D}_\gamma)$ such that $f(z) = \sum_{k=0}^{\infty} a_k z^k$, $z \in \mathbb{D}$. If, in particular, $\sum_{k=0}^{\infty} |a_k| (\frac{1}{2})^k = 1$ for f as above, we obtain

$$1 \le |a_0| + \frac{1}{2}(1 - |a_0|^2) \sum_{k=1}^{\infty} \left(\frac{1}{2}\right)^k = |a_0| + \frac{1 - |a_0|^2}{2}$$

and $|a_0| = 1$, i.e., f is a constant function of modulus 1. □

Proof of Theorem 2. The proof of the main statement is rather straightforward: if f belongs to $\mathcal{B}(\mathcal{D})$ with $f(z) = \sum_{k=0}^{\infty} a_k z^k$ in $\mathbb{D}$, then for any $r \in (0, 1)$

$$\sum_{k=0}^{\infty} |a_k| r^k \le |a_0| + (1 - |a_0|^2) \frac{\lambda r}{1 - r}$$

so that $\sum_{k=0}^{\infty} |a_k| r^k \le 1$ when $2\lambda r/(1-r) \le 1$, i.e., when $r \le 1/(1+2\lambda)$. This clearly shows that $B_{\mathcal{D}} \ge 1/(1 + 2\lambda)$. As in the last step of the proof of Corollary 4, the equality $\sum_{k=0}^{\infty} |a_k| (1/1 + 2\lambda)^k = 1$ holds only for constant functions of modulus 1. □

Remark. It does not seem entirely trivial to characterize domains $\mathcal{D}$ for which $B_{\mathcal{D}} = 1/(1+2\lambda)$. It is not hard to see that this indeed is the case (see (6) for a hint in this direction) when $\mathcal{D} = \mathcal{D}_\gamma$, $0 \le \gamma < 1$. Next we exhibit a class of domains for which this is definitely not true. Consider a sufficiently regular simply connected domain $\mathcal{D} \supseteq \mathbb{D}$ such that $\partial\mathcal{D} \cap \partial\mathbb{D} \ne \varnothing$, together with the associated conformal map F of $\mathcal{D}$ onto $\mathbb{D}$ with $F(0) = 0$, $F'(0) =: A_1 > 0$. The inverse map $F^{-1} \colon \mathbb{D} \to \mathcal{D}$ satisfies $F^{-1}(u) = (1/A_1)u + o(u)$ and by the classical growth theorem,

$$A_1 |F^{-1}(u)| \ge \frac{|u|}{(1 + |u|)^2}, \qquad u \in \mathbb{D}. \tag{7}$$

Our hypothesis on $\partial\mathbb{D}$ and $\partial\mathbb{D}$ implies that for some sequence of elements $u_j \in \mathbb{D}$ we have $|u_j| \to 1$ and $|F^{-1}(u_j)| \to 1$. It then follows from (7) that $A_1 \ge \frac{1}{4}$.

Let $f(z) = \sum_{k=0}^{\infty} a_k z^k$, $z \in \mathbb{D}$, for some f in $\mathcal{B}(\mathcal{D})$ where $\mathcal{D}$ is to be determined later. We have, thanks to estimates due to Avkhadiev andWirths [1, pp.60 – 75],

$$|a_n| \le \frac{4^{n-1/2}}{\sqrt{n+1}} A_1^n (1 - |a_0|^2), \qquad n \ge 2.$$

This estimate is actually also valid for $n = 1$: For if $w(u) := f\big(F^{-1}(u)\big)$, $u \in \mathbb{D}$, then $w \in \mathcal{B}(\mathbb{D})$ and $a_0 = w(0)$, $a_1 = w'(0) A_1$ and

$$\frac{|a_1|}{1 - |a_0|^2} = \frac{|w'(0)|}{1 - |w(0)|^2} A_1 \le A_1 < \sqrt{2} A_1.$$

Therefore

$$\sum_{n=0}^{\infty} |a_n| r^n \le |a_0| + (1 - |a_0|^2) \sum_{n=1}^{\infty} \frac{4^{n-1/2} A_1^n}{\sqrt{n+1}} r^n \le 1$$

if $\sum_{n=1}^{\infty} (4A_1 r)^n / \sqrt{n+1} \le 1$. Let X be the unique root in $(0, 1)$ of the equation

$$\sum_{n=1}^{\infty} \frac{X^n}{\sqrt{n+1}} = 1.$$

We now produce domains $\mathcal{D}$ for which $4A_1/(1+2\lambda) < X$: this will clearly imply that for such domains $1/(1+2\lambda) < B_{\mathcal{D}}$.

Since $A_1 = F'(0) = |F'(0)|/(1-|F(0)|^2) \leq \lambda$, it shall be sufficient to identify domains $\mathcal{D} \supseteq \mathbb{D}$ with $4A_1/(1+2A_1) < X$, i.e., $A_1 < 1/(4-2X)$. Since $\frac{1}{4} < 1/(4-2X) < \frac{1}{2}$, it follows that any simply connected domain $\mathcal{D}$ containing the plane $P_{1,0}$ (for which $A_1 = \frac{1}{2}$) and close enough, in the sense of kernel convergence, to the slit plane $\mathbb{C} \setminus [1,\infty)$ (for which $A_1 = \frac{1}{4}$) will serve as an example.

PROOF OF THEOREM 3. Let $\mathcal{D}$ be a convex domain, $\mathcal{D} \supseteq \mathbb{D}$ and F the conformal map as above with $F(0) = 0$, $F'(0) =: A_1 > 0$.

If $f \in \mathcal{B}(\mathcal{D})$ and $f(z) = \sum_{k=0}^{\infty} a_k z^k$, $z \in \mathbb{D}$, another estimate due to Avkhadiev and Wirths [1, pp. 60–76] yields

$$|a_k| \leq 2^{k-1} A_1^k (1-|a_0|^2), \qquad k \geq 2, \tag{8}$$

and as in the proof of Theorem 2, this extends to $k = 1$.

First assume that $A_1 \leq \frac{1}{4}$; then by (8),

$$\sum_{k=0}^{\infty} |a_k| r^k \leq |a_0| + \frac{1-|a_0|^2}{2} \sum_{k=1}^{\infty} (2A_1)^k r^k \leq |a_0| + (1-|a_0|^2)\frac{A_1}{1-2A_1}$$

and this last quantity is easily seen to be less or equal to 1 for all admissible a_0.

When $A_1 > \frac{1}{4}$ and $r \leq 1/(4A_1)$, we also obtain from (8)

$$\begin{aligned}\sum_{k=0}^{\infty} |a_k| r^k &\leq |a_0| + \frac{1-|a_0|^2}{2} \sum_{k=1}^{\infty} (2A_1 r)^k \\ &\leq |a_0| + \frac{1-|a_0|^2}{2} \sum_{k=1}^{\infty} \left(\frac{1}{2}\right)^k = |a_0| + \frac{1-|a_0|^2}{2} \leq 1\end{aligned}$$

and the result follows. It should also be clear from our arguments that the equality case $\sum_{k=0}^{\infty} |a_k| \beta^k = 1$ can occur only when f is a constant function of modulus one. □

Remarks. (1) When $A_1 > \frac{1}{4}$, it again does not seem easy to characterize the convex domains $\mathcal{D}$ for which $\beta = 1/(4A_1) = B_{\mathcal{D}}$. Indeed, $1/(4A_1) < B_{\mathcal{D}} = \frac{1}{3}$ if $\mathcal{D}$ is the unit disc $\mathbb{D}$ and $1/(4A_1) = B_{\mathcal{D}} = 1/(1+2\lambda) = \frac{1}{2}$ if $\mathcal{D}$ is the half-plane P. Further, we sometimes have $1/(4A_1) < 1/(1+2\lambda)$ (as in the case of the unit disc) and $1/(4A_1) > 1/(1+2\lambda)$ (as in the case where $\mathcal{D}$ is a disc centered at the origin with radius > 4).

(2) Theorem 3 does not necessarily hold for non-convex domains. Let $\mathcal{D}$ be the slit plane $\mathbb{C} \setminus (-\infty, -1]$; then $F(z)$ is the inverse of $4k(z)$ where $k(z) := z/(1-z)^2$ is the Koebe map and $A_1 = \frac{1}{4}$ with $F(z) = \frac{1}{4}z - \frac{1}{8}z^2 + \frac{5}{32}z^3 + \cdots$. For $0 < a < 1$ define $f \in \mathcal{B}(\mathcal{D})$ by

$$\begin{aligned} f(z) &= \frac{F(z)+a}{1+aF(z)} \\ &= a - \frac{(1-a^2)}{4} z + \frac{(1-a^2)(a+2)}{16} z^2 + \frac{(1-a^2)(a^2+4a+10)}{64} z^3 + \cdots \end{aligned}$$

where $z \in \mathbb{D}$. Then, for $a = .9$, we have

$$a + \frac{1-a^2}{4} + \frac{(1-a^2)(a+2)}{16} + \frac{(1-a^2)(a^2+4a+10)}{64} = 1.0247\ldots > 1,$$

which shows that $B_{\mathcal{D}} < 1 = \beta$.

3. Conclusion

Given a domain $\mathcal{D} \supseteq \mathbb{D}$, the determination of the function

$$M(r,\mathcal{D}) := \sup \sum_{n=0}^{\infty} |a_n| r^n, \qquad 0 \leq r < 1, \tag{9}$$

(here the sup is taken over all functions $f(z) = \sum_{n=0}^{\infty} a_n z^n$, $z \in \mathbb{D}$, in $\mathcal{B}(\mathcal{D})$) may be seen as a generalized Bohr problem. Because the coefficients a_n of functions f in $\mathcal{B}(\mathcal{D})$ are uniformly bounded, it should be clear that $M(r) = 1$ for r sufficiently small and indeed $M(r) = 1$ for $0 \leq r \leq B_{\mathcal{D}}$ where of course $\frac{1}{3} \leq B_{\mathcal{D}}$.

Very little seems to be known about the function $M(r,\mathcal{D})$ in general; a result of Bombieri [3] (see also [8] for related matters) says that $M(r,\mathbb{D}) = (3 - \sqrt{8(1-r^2)})/r$ when $\frac{1}{3} \leq r \leq 1/\sqrt{2}$ (Bombieri studies in fact the inverse function of $M(r,\mathbb{D})$). It also follows from the results of Bombieri that $M(r,\mathbb{D}) \leq 1/\sqrt{1-r^2}$ if $0 < r < 1$ and a recent result due to Bombieri and Bourgain [4] says that $M(r,\mathbb{D}) < 1/\sqrt{1-r^2}$ if $1/\sqrt{2} < r < 1$; in [4], a deeper result implies that $1/\sqrt{1-r^2}$ is in some sense the sharp order of growth of $M(r,\mathbb{D})$ as $r \to 1$.

Think of $\mathcal{B}(\mathcal{D})$ as a topological vector space endowed with the topology of uniform convergence on compact subsets of $\mathcal{D}$. It follows from a simple compacity argument that the sup in (9) is indeed a maximum and there exists for each $r \in (0,1)$ a function $f_r(z) := \sum_{n=0}^{\infty} a_n(f_r) z^n$ such that $M(r,\mathbb{D}) = \sum_{n=0}^{\infty} |a_n(f_r)| r^n$. Bombieri has proved in [3] that f_r is a disc automorphism (i.e., a Blaschke product of order 1) when $\frac{1}{3} < r \leq 1/\sqrt{2}$.

This last result can be extended to general domains $\mathcal{D}$ in the following fashion; let, given $r \in (0,1)$, $f_r(z) = \sum_{n=0}^{\infty} a_n(f_r) z^n$ with $a_n(f_r) = |a_n(f_r)| e^{i\theta_n}$ where $\theta_n = \theta_n(r)$ is an angle in $[0, 2\pi)$. We define a linear functional L over $\mathcal{B}(\mathcal{D})$ by

$$L(f) = \sum_{n=0}^{\infty} a_n(f) e^{-i\theta_n} r^n \qquad \text{if } f(z) = \sum_{n=0}^{\infty} a_n z^n,\ |z| < 1.$$

The complex-valued functional L is continuous over $\mathcal{B}(\mathcal{D})$ and $\operatorname{Re} L$ is not constant there. Further

$$\begin{aligned} \operatorname{Re} L(f) = \operatorname{Re} \sum_{n=0}^{\infty} a_n(f) e^{-i\theta_n} r^n &\leq \sum_{n=0}^{\infty} |a_n(f)| r^n \\ &\leq \sum_{n=0}^{\infty} |a_n(f_r)| r^n = L(f_r) = \operatorname{Re} L(f_r). \end{aligned}$$

Since

$$f \in \mathcal{B}(\mathcal{D}) \Longleftrightarrow f = w \circ F$$

where $w \in \mathcal{B}(\mathbb{D})$ and F is a conformal map of $\mathcal{D}$ onto $\mathbb{D}$, we may therefore think of L as a continuous linear functional over $\mathcal{B}(\mathbb{D})$ whose real part is nonconstant there and maximized by a function $w_r \in \mathcal{B}(\mathbb{D})$ with

$$f_r(z) = w_r\big(F(z)\big), \qquad z \in \mathcal{D}.$$

It is well-known [5, Chapter 4] that such a function w_r must be a finite Blaschke product. We may therefore state that any function f for which the sup is attained in (9) is such that $f \circ F^{-1}$ is a finite Blaschke product.

References

1. F. G. Avkhadiev and K.-J. Wirths, *Schwarz – Pick type inequalities*, Front. Math., Birkhäuser, Basel, 2009.
2. H. Bohr, *A theorem concerning power series*, Proc. London Math. Soc. **13** (1914), 1 – 5.
3. E. Bombieri, *Sopra un teorema di H. Bohr e G. Ricci sulle funzioni maggioranti delle serie di potenze*, Boll. Un. Mat. Ital. (3) **17** (1962), 276 – 282.
4. E. Bombieri and J. Bourgain, *A remark on Bohr's inequality*, Int. Math. Res. Not. **80** (2004), 4307 – 4330.
5. D. J. Hallenbeck and T. H. MacGregor, *Linear problems and convexity techniques in geometric function theory*, Monogr. Stud. Math., vol. 22, Pitman, Boston, MA, 1984.
6. G. Kresin and V. Maz'ya, *Sharp Bohr type real part estimates*, Comput. Methods Funct. Theory **7** (2007), no. 1, 151 – 165.
7. ______, *Sharp real-part theorems: A unified approach*, Lecture Notes in Math., vol. 1903, Springer, Berlin, 2007.
8. G. Ricci, *Complementi a un teorema di H. Bohr riguardante le serie di potenze*, Rev. Un. Mat. Argentina **17** (1955), 185 – 195.
9. St. Ruscheweyh, *Two remarks on bounded analytic functions*, Serdica **11** (1985), no. 2, 200 – 202.

Centre de recherches mathématiques, Université de Montréal, C.P. 6128, succ. Centre-ville, Montréal, QC H3C 3J7, Canada
E-mail address: fournier@dms.umontreal.ca

Mathematisches Institut, Universität Würzburg, 97074 Würzburg, Germany
E-mail address: ruscheweyh@mathematik.uni-wuerzburg.de

Centre de Recherches Mathématiques
CRM Proceedings and Lecture Notes
Volume **51**, 2010

Completeness of the System $\{f(\lambda_n z)\}$ in $L_a^2[\Omega]$

André Boivin and Changzhong Zhu

ABSTRACT. Given an analytic function f defined on the Riemann surface of the logarithm and represented by a generalized power series with complex exponents $\{\tau_k\}$, a sequence of complex numbers $\{\lambda_n\}$, and an unbounded domain Ω on the complex z plane, we study the completeness of the system $\{f(\lambda_n z)\}$ in $L_a^2[\Omega]$ (mean square approximation).

1. Introduction

For an entire function $f(z)$ and a sequence $\{\lambda_n\}$ of complex numbers, the completeness in a domain Ω of the system $\{f(\lambda_n z)\}$ in L^p-norms has been studied by many authors under various conditions on f, $\{\lambda_n\}$, p and Ω. See, for example, [1; 9; 10; 12, Chapter 4; 14–16].

In particular, M. M. Dzhrbasian studied the completeness of $\{f(\lambda_n z)\}$ in L^2 when Ω is an unbounded domain which does not contain the origin and is located outside an angle with vertex at the origin (see [6, Section 5]). In [3], we also considered this problem. We gave some sufficient conditions under which the system $\{f(\lambda_n z)\}$ is complete. Our conditions are different, though similar, to Dzhrbasian's. These results are recalled in Section 2. Besides, in [3, Section 3], we also studied a similar question where this time, the entire function f is replaced by one analytic on the Riemann surface of the logarithm. This generalized a result obtained by X. Shen in [22]. We used in [3] the classical Ritt order and Ritt type to characterize the growth of the function. This seems not suitable in general, as we pointed out in a later paper [5]. Hence, we introduced a modified Ritt order and modified Ritt type in [5]. After recalling the definitions, we will use them in this paper. Corresponding to this change, we modify related arguments used in (the second part of) [3]. Besides, comparing with [3], in this paper we will impose weaker conditions on the sequence of exponents $\{\tau_k\}$. In particular, we do not assume that the limit $\lim_{k\to\infty} k/|\tau_k|$ exists.

The paper is organized as follows. In the next two sections, we review some of the definitions, notations, examples and results from [3–5]. In Sections 4 and 5,

---

2000 *Mathematics Subject Classification.* 30E10, 30B60.

Key words and phrases. mean square approximation, completeness, analytic functions, unbounded domain.

Research supported by NSERC (Canada).

This is the final form of the paper.

we present three new results: An estimate of the coefficients in the (generalized) Dirichlet series of an entire function in terms of the modified Ritt order and type; a uniqueness theorem for analytic functions defined on the Riemann surface of the logarithm; and finally conditions for the completeness of the system $\{z^{\tau_k}\}$ in $L_a^2[\Omega]$, where $\{\tau_k\}$ is a sequence of complex numbers. In Section 6, we give our main result on the completeness of the system $f(\lambda_n z)\}$ in $L_a^2[\Omega]$. The proofs of a few lemmas are gathered in the last section.

2. Dzhrbasian's theorem

Let Ω be a domain in the complex z-plane. Denote by $L_a^2[\Omega]$ the space of functions g analytic on Ω which satisfy

$$\iint_\Omega |g(z)|^2 \,\mathrm{d}x\,\mathrm{d}y < \infty \qquad (z = x + \mathrm{i}y).$$

Endowed with the inner product

$$(g,h) = \iint_\Omega g(z)\overline{h(z)}\,\mathrm{d}x\,\mathrm{d}y,$$

and associated norm $\|g\| = (g,g)^{1/2}$, $L_a^2[\Omega]$ becomes a Hilbert space. A sequence $\{h_n\}$ ($h_n \in L_a^2[\Omega]$, $n = 1, 2, \dots$) is complete in $L_a^2[\Omega]$ if for any $g \in L_a^2[\Omega]$ and any $\varepsilon > 0$, there is a finite linear combination h of elements of the sequence $\{h_n\}$, such that $\|g - h\| < \varepsilon$.

For a function $f(z)$ and a sequence $\{\lambda_n\}$ of complex numbers, we studied in [3] the completeness of $\{f(\lambda_n z)\}$ in $L_a^2[\Omega]$.

For $f(z)$, we assumed that it is an entire function with power series expansion

$$f(z) = \sum_{k=0}^{\infty} a_k z^k, \qquad a_k \neq 0 \ (k = 0, 1, 2, \dots). \tag{2.1}$$

A simple example of such an entire function is $f(z) = \mathrm{e}^z$. Hence, our results provided sufficient conditions for the completeness of the system $\{\mathrm{e}^{\lambda_n z}\}$ in $L_a^2[\Omega]$.

For $\{\lambda_n\}$, we assumed it to be a sequence of complex numbers with

$$\lambda_n \neq 0. \tag{2.2}$$

For other conditions on $\{\lambda_n\}$, see Theorem 2.1 below.

For Ω, we assumed it is an unbounded simply-connected domain satisfying the following two conditions:

Condition Ω(I). *For $r > 0$, let $\sigma(r)$ denote the Lebesgue measure of the intersection of the circle $|z| = r$ and Ω; we assume that there exists $r_0 > 0$ such that for $r > r_0$,*

$$\sigma(r) \le \exp\bigl(-p(r)\bigr),$$

where $p(r)$ has the form

$$p(r) = \alpha' r^{s'}, \tag{2.3}$$

for two positive constants α' and s'.

Condition Ω(II). *The complement of Ω consists of m unbounded simply connected domains G_i ($i = 1, 2, \dots, m$), each G_i contains an angle domain Δ_i with measure π/α_i, $\alpha_i > \frac{1}{2}$* (see Figure 1).

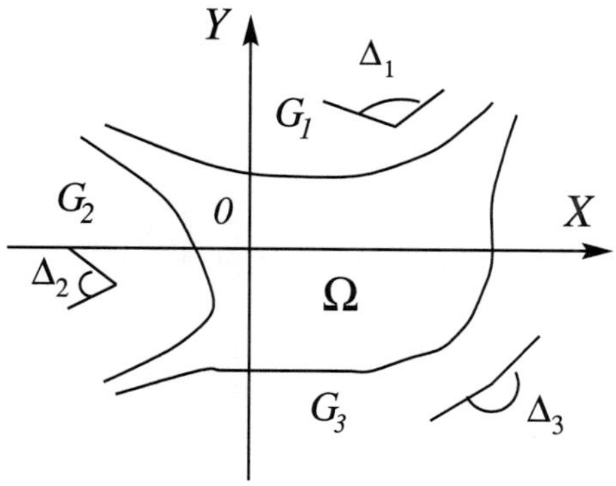

FIGURE 1. Domain Ω

Remark 1. Since α' and s are positive, it follows from condition Ω(I) that there exists a constant C, $0 < C < \infty$, such that

$$\iint_\Omega d\sigma \leq \int_0^{r_0} 2\pi r \, dr + \int_{r_0}^\infty \sigma(r) \, dr \leq C + \int_{r_0}^\infty e^{-\alpha' r^{s'}} \, dr < +\infty.$$

To study the completeness of $\{f(\lambda_n z)\}$ in $L^2_a[\Omega]$, we needed a result on the completeness of the sequence $\{1, z, z^2, z^3, \dots\}$ in $L^2_a[\Omega]$, which is a special case of a result of M. M. Dzhrbasian (see [6], or [19, Theorem 10.1]):

Let us define ϑ by

$$\vartheta = \max\{\alpha_1, \dots, \alpha_m\}, \tag{2.4}$$

where the α_i are the constants appearing in condition Ω(II).

Theorem (M. M. Dzhrbasian). *Suppose that Ω is a domain satisfying conditions Ω(I) and Ω(II) and that s' and ϑ are defined by* (2.3) *and* (2.4), *respectively. If*

$$\int^\infty \frac{1}{r^{1+\vartheta-s'}} \, dr = +\infty,$$

where $\int^\infty$ means that the lower limit of the integral is a sufficiently large number, then the system $\{1, z, z^2, z^3, \dots\}$ is complete in $L^2_a[\Omega]$.

Remark 2. The "sufficient large number" in the integral condition above can be replaced by "strictly positive number." In fact, the integral condition can be replaced by the simpler condition $\vartheta \leq s'$.

Example 1. This example is taken from [3]. Let Ω be the unbounded domain containing the real axis and having the curves $y = \pm\frac{1}{8}e^{-x^2}$ $(-\infty < x < \infty)$ as its boundary. It is not hard to see that Ω satisfies conditions Ω(I) and Ω(II) with $p(r) = \frac{1}{2}r^2$ (i.e., $s' = 2$ and $\alpha' = \frac{1}{2}$) and $m = 2$, $\alpha_1 = \alpha_2 = 1$. So we have $\vartheta = 1$, and

$$\int^\infty \frac{dr}{r^{1+\vartheta-s'}} = \int^\infty dr = +\infty.$$

Let us now recall some concepts from the theory of entire functions (see, for example, [12, Chapters 1 and 4]). Let ϕ be an entire function. The quantity

$$\rho = \limsup_{r\to\infty} \frac{\log\log M_\phi(r)}{\log r}$$

is called the order of ϕ, and if $0 < \rho < \infty$, the quantity

$$\sigma = \limsup_{r\to\infty} \frac{\log M_\phi(r)}{r^\rho}$$

is called the type of ϕ, where

$$M_\phi(r) = \max_{|z|=r} |\phi(z)|.$$

For a sequence $\{\lambda_n\}$ of complex numbers, we denote by $n(r)$ the number (counted according to multiplicities) of λ_n with $|\lambda_n| < r$.

Now we state the first main result from [3].

Theorem 2.1. *Assume that f, $\{\lambda_n\}$ and Ω satisfy conditions* (2.1), (2.2), Ω(I), *and* Ω(II) *stated previously in this section. Let ρ and σ be the order and type of the entire function $f(z)$, respectively, and assume that $0 < \rho < s'$ and $0 < \sigma < \infty$, where s' is the exponent appearing in* (2.3). *Suppose that either*

$$\limsup_{r\to\infty} \frac{n(r)}{r^{s'\rho\beta}} > \mathrm{e}\left(\frac{2}{s'\alpha'}\right)^{\rho\beta} (\rho\sigma)^{s'\beta}, \tag{2.5}$$

or

$$\liminf_{r\to\infty} \frac{n(r)}{r^{s'\rho\beta}} > \left(\frac{2}{s'\alpha'}\right)^{\rho\beta} (\rho\sigma)^{s'\beta}, \tag{2.5'}$$

where $\beta = 1/(s'-\rho) > 0$.

If

$$\int^{\infty} \frac{\mathrm{d}r}{r^{1+\vartheta-s'}} = +\infty$$

where ϑ is defined in (2.4), *then the system $\{f(\lambda_n z)\}$, $n = 1, 2, 3, \ldots$ is complete in $L^2_a[\Omega]$.*

Example 2 (Taken from [3]). Let Ω be the domain described in Example 1 and let $f(z) = \mathrm{e}^z$. Then f is an entire function with $\rho = \sigma = 1$. It thus follows that $\beta = 1/(s'-\rho) = 1$ and $s'\rho\beta = 2$. Consequently (2.5) and (2.5′) become

$$\limsup_{r\to\infty} \frac{n(r)}{r^2} > 2\mathrm{e}, \tag{2.5*}$$

and

$$\liminf_{r\to\infty} \frac{n(r)}{r^2} > 2, \tag{2.5$'^*$}$$

respectively. Hence by Theorem 2.1, if $\{\lambda_n\}$ satisfies (2) and (2.5)* or (2.5′)* (for example $\lambda_n = \sqrt{n/3\mathrm{e}}\,\mathrm{i}$), then the system $\{\mathrm{e}^{\lambda_n z}\}_{n=1}^{\infty}$, is complete in $L^2_a[\Omega]$.

3. The Riemann surface of the logarithm

For the remaining of this paper, we assume that f is an analytic function defined on the Riemann surface of the logarithm and is represented by a generalized power series

$$f(z) = \sum_{k=1}^{\infty} d_k z^{\tau_k}, \qquad z = r\mathrm{e}^{\mathrm{i}\theta}\ (r > 0, |\theta| < \infty), \tag{3.1}$$

or, which is the same, we assume that $F(s) = f(\mathrm{e}^{-s})$ is an entire function represented by the (generalized) Dirichlet series

$$F(s) = \sum_{k=1}^{\infty} d_k \mathrm{e}^{-\tau_k s}, \qquad s = u + \mathrm{i}v \tag{3.2}$$

where $\{\tau_k\}$ is a sequence of complex numbers.

In the sections that follow, we will study the completeness of the system $\{f(\lambda_n z)\}$ in $L_a^2[\Omega]$. X. Shen studied this problem for the case when $\{\tau_k\}$ is a sequence of real numbers (see [22]). We consider the case when the τ_k are complex.

We now make some assumptions on the exponential sequence $\{\tau_k\}$, the domain Ω and the sequence $\{\lambda_n\}$.

For the sequence of complex numbers $\{\tau_k\}$, we will assume some or all of the following (for (II)(3.5)), see [21]; for (III)′ and (III)″, see [23]):

(I) $0 < |\tau_1| < |\tau_2| < \cdots < |\tau_k| < \cdots$;

(II)

$$\limsup_{k\to\infty} \frac{k}{|\tau_k|} = D^* \qquad (0 < D^* < \infty), \tag{3.3}$$

$$\liminf_{k\to\infty} \frac{k}{|\tau_k|} = D_* \qquad (D_* > 0), \tag{3.4}$$

and

$$\sup_{0<\xi<1} \limsup_{r\to\infty} \frac{n_\tau(r) - n_\tau(r\xi)}{r - r\xi} = \tau < +\infty, \tag{3.5}$$

where $n_\tau(r)$ denotes the number of the elements of the sequence $\{\tau_k\}$ with $|\tau_k| < r$;

(III)′ there is a $K > 0$ such that for sufficiently large x and each $t \geq 1$, the number of τ_k with $x < |\tau_k| \leq x + t$ is less than Kt;

(III)″ for any fixed $\delta > 0$, the inequality $\big||\tau_l| - |\tau_k|\big| > \mathrm{e}^{-|\tau_k|\delta}$ holds for sufficiently large k and any $l \neq k$;

(IV) there is an α with $0 < \alpha < \pi/2$ such that

$$|\arg(\tau_k)| < \alpha. \tag{3.6}$$

Assume that the domain Ω satisfies conditions Ω(I) and Ω(II) given in Section 2. Moreover assume that one of the angle domains defined in Ω(II), say Δ_1, is given by

$$|\arg(z) - \pi| < \frac{\pi}{2\gamma} \tag{3.7}$$

where the constant $(\alpha_1 =)\gamma > \frac{1}{2}$. Note that this implies that Ω does not contain the origin 0. Moreover, we assume that for any $z \in \Omega$,

$$|z| \geq r_\Omega > 0, \qquad r_\Omega < r_0. \tag{3.8}$$

For $\{\lambda_n\}$ $(n = 1, 2, \dots)$, assume that it is a sequence of complex numbers with

$$|\lambda_n| \geq r_\lambda > 0, \qquad |\arg(\lambda_n)| < \alpha_\lambda, \tag{3.9}$$

where $0 < \alpha_\lambda < \infty$. Since α_λ can be greater than π, we think of λ_n as being located on the Riemann surface of the logarithm.

If the sequence of exponents $\{\tau_k\}$, $(k = 1, 2, \dots)$ consists only of strictly positive (thus real) terms tending to ∞, that is, if F is represented by an *ordinary* Dirichlet

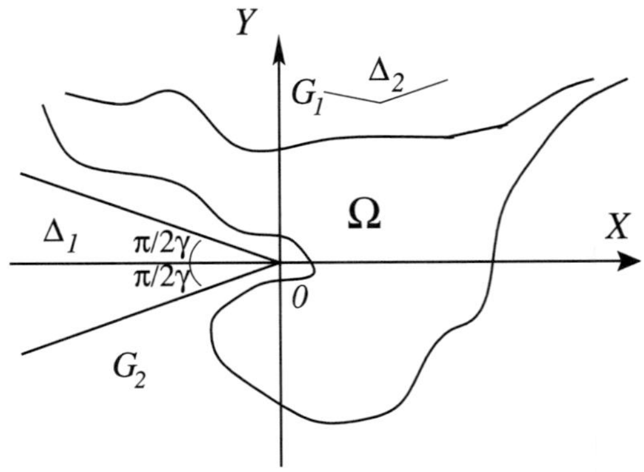

FIGURE 2

series, the classical (R)-order ρ^* and (R)-type σ^* of F (the Ritt order and Ritt type of F) are defined respectively by (see [17], [22] or [11])

$$\rho^* = \limsup_{u \to -\infty} \frac{\log \log M_F^*(u)}{-u} \tag{3.10}$$

and when $0 < \rho^* < \infty$,

$$\sigma^* = \limsup_{u \to -\infty} \frac{\log M_F^*(u)}{\mathrm{e}^{-u\rho^*}}, \tag{3.11}$$

where

$$M_F^*(u) = \sup_{|v|<\infty} |F(u + \mathrm{i}v)|.$$

It is known (see, for example, [22] and [11, Chapter 2]) that, in this case, if

$$\limsup_{k \to \infty} \frac{\log k}{\tau_k} < \infty,$$

then

$$\rho^* = \limsup_{k \to \infty} \frac{\tau_k \log \tau_k}{\log |1/d_k|}, \tag{3.12}$$

and when $0 < \rho^* < \infty$ and

$$\lim_{k \to \infty} \frac{\log k}{\tau_k} = 0,$$

then

$$\sigma^* = \limsup_{k \to \infty} \left(\frac{\tau_k}{\mathrm{e}\rho^*} |d_k|^{\rho^*/\tau_k} \right). \tag{3.13}$$

We would like to derive similar relations when the entire function F is represented by a generalized Dirichlet series (3.2). But if $\{\tau_k\}$ contains non-real element(s), the definitions of order and type given above seem no longer suitable. For example, if $F(s) = \mathrm{e}^{-s\tau}$ with $|\tau| > 0$ and $\arg \tau = \alpha \neq 0$, we see that for any $u < 0$,

$$M_F^*(u) = \sup_{|v|<\infty} \mathrm{e}^{-|\tau|(u \cos \alpha - v \sin \alpha)} = +\infty.$$

For this reason, we modified in [5] the above definitions as follows:

Define, for $u < 0$,

$$M_F(u) = \sup_{|v| \leq -u} |F(u+iv)|, \tag{3.14}$$

then define the (mR)-order ρ and (mR)-type σ of F (the *modified* Ritt order and *modified* Ritt type of F) by

$$\rho = \limsup_{u \to -\infty} \frac{\log \log M_F(u)}{-u} \tag{3.15}$$

and if $0 < \rho < \infty$,

$$\sigma = \limsup_{u \to -\infty} \frac{\log M_F(u)}{e^{-u\rho}} \tag{3.16}$$

respectively. In other words, we consider the growth of $|F(s)|$ only in the angle domain $|\arg s - \pi| \leq \pi/4$.

Example 3. For $F(s) = e^{-s\tau}$ with $|\tau| > 0$ and $\arg \tau = \alpha$, $0 < |\alpha| < \pi/2$, we have for $u < 0$,

$$M_F(u) = \sup_{|v| \leq -u} e^{-|\tau|(u\cos\alpha - v\sin\alpha)} = e^{-|\tau| u(\cos\alpha + |\sin\alpha|)},$$

so the (mR)-order of F is 0.

Example 4. Let τ be a complex number with $|\tau| > 0$ and $\arg \tau = \alpha = \pi/4$. Consider the entire function

$$F(s) = e^{e^{-s\tau}} - 1 = \sum_{n=1}^{\infty} \frac{1}{n!} e^{-s(n\tau)} \qquad (s = u + iv). \tag{3.17}$$

For the function $G(s) = e^{e^{-s\tau}}$, we have

$$\begin{aligned} |G(s)| = |G(u+iv)| &= \exp\big[e^{-|\tau|(u\cos\alpha - v\sin\alpha)} \cdot \cos(-|\tau|(u\sin\alpha + v\cos\alpha))\big] \\ &= \exp\left[\exp\left(-\frac{1}{\sqrt{2}}|\tau|(u-v)\right) \cdot \cos\left(\frac{1}{\sqrt{2}}|\tau|(u+v)\right)\right]. \end{aligned}$$

So, for $u < 0$,

$$M_G(u) = \sup_{|v| \leq -u} |G(u+iv)| = |G(u-iu)| = \exp\big(e^{-\sqrt{2}u|\tau|}\big).$$

Thus the (mR)-order of G and the (mR)-type of G are $\rho = \sqrt{2}|\tau|$ and $\sigma = 1$ respectively. From the inequalities $|G(s)| - 1 \leq |F(s)| \leq |G(s)| + 1$, it immediately follows that the (mR)-order and (mR)-type of F are also $\rho = \sqrt{2}|\tau|$ and $\sigma = 1$.

4. An estimate of $|d_k|$ and a uniqueness theorem

Now we give an estimate for the upper bound of $|d_k|$, where d_k $(k = 1, 2, \dots)$ are the coefficients of the generalized Dirichlet series (3.2) which represents an entire function $F(s)$.

First we need the following estimate. Its proof will be given in Section 7.

Lemma 4.1. *Under the assumptions* (I), (II)(3.3) *and* (3.4), (III)′ *and* (III)″, *denoting by* $n(t)$ *the number of* τ_i *with* $|\tau_i| \leq t$, *if* (a) *there is a number* p *with* $0 < p < 1$, *for sufficiently large* n,

$$\frac{n}{n(3|\tau_n|)} < p \cdot \frac{D_*}{D^*}; \tag{4.1}$$

or (b) *for sufficiently large* n, $n(3|\tau_n|) = n$, *then we have*

$$\limsup_{k\to\infty} \frac{\log|T_k(\tau_k)|^{-1}}{|\tau_k|} \le H, \tag{4.2}$$

where

$$T_k(z) = \prod_{\substack{i=1\\ i\neq k}}^{\infty} \left(1 - \frac{z^2}{\tau_i^2}\right), \tag{4.3}$$

and, for case (a),

$$H = \big(L + 3\pi - 3\log(1-p)\big)D^*; \tag{4.4}$$

for case (b),

$$H = (L + 3\pi)D^* - 2K. \tag{4.5}$$

with L *satisfying* $LD^* > 5K$.

Theorem 4.1. *Assume that the exponential sequence* $\{\tau_k\}$ *satisfies conditions* (I), (II)(3.3) *and* (3.4), (III)$'$, (III)$''$ *and* (IV) *given in Section* 3, *and the condition* (a) *or* (b) *given in Lemma* 4.1, *and that* ρ *and* σ *are the* (mR)*-order and* (mR)*-type of* $F(s)$, *respectively. If* $0 < \rho < \infty$, *then, given* $\varepsilon > 0$, *for* k *sufficiently large,*

$$|d_k| < C_1 \mathrm{e}^{(\pi D^* + H' + 2\varepsilon)\cdot \mathrm{Re}(\tau_k)} \left[\frac{\mathrm{e}\rho(\sigma+\varepsilon)}{\mathrm{Re}(\tau_k)}\right]^{\mathrm{Re}(\tau_k)/\rho};$$

If $\rho = 0$, *then, given* $\varepsilon > 0$, *for* k *sufficiently large,*

$$|d_k| < C_1 \mathrm{e}^{(\pi D^* + H' + 2\varepsilon)\cdot \mathrm{Re}(\tau_k)} \left[\frac{\mathrm{e}\varepsilon}{\mathrm{Re}(\tau_k)}\right]^{\mathrm{Re}(\tau_k)/\varepsilon},$$

where C_1 *is a constant*

$$C_1 = (\pi D^* + \varepsilon)\int_0^{\infty} g(t)\mathrm{e}^{-(\pi D^* + \varepsilon)t}\,\mathrm{d}t,$$

with

$$g(r) = \prod_{k=1}^{\infty}\left(1 + \frac{r^2}{|\tau_k|^2}\right), \tag{4.6}$$

and $H' = H/\cos\alpha$ *with* H *given by* (4.4) *or* (4.5) *corresponding to the condition* (a) *or* (b).

The proof is basically the same as that of Theorem 3 in [5] except for the following two points: (1) We can use D^* instead of D. Indeed in [5] it is assumed that the limit $\lim_{k\to\infty} k/|\tau_k| = D < \infty$ exists, but here by [11, Lemma 2.2] and the assumption (II)(3.3), we have

$$\limsup_{r\to\infty} \frac{\log g(r)}{r} \le \pi D^*. \tag{4.7}$$

Hence we can still use [5, Lemma 3.2]) if we use D^* instead of D. (2) We need to use the above Lemma 4.1 to estimate $1/|T_k(-\tau_k)|$ rather than using [5, Lemma 3.1].

We now prove a uniqueness theorem which is a modification of Lemma 1 in [22].

Let $\{\lambda_n\}$ be a sequence of complex numbers satisfying (3.9) given in Section 3, γ' be a fixed number with $\pi\gamma'/2 > \alpha_\lambda$, let $\Phi(z)$ be a function analytic in a domain

$$\mathbb{D} = \left\{ z : |z| \geq r_\lambda, |\arg(z)| \leq \frac{\pi\gamma'}{2} \right\}$$

on the Riemann surface of the logarithm, and $\{\lambda_n\}$ $(n = 1, 2, \dots)$ be its zeroes, i.e., $\Phi(\lambda_n) = 0$. Denote by $n_\lambda(r)$ the number of the elements of the sequence $\{\lambda_n\}$ with $|\lambda_n| < r$. Define, for $r > r_\lambda$,

$$M_\Phi(r, \gamma') = \sup_{|\theta| \leq \pi\gamma'/2} |\Phi(re^{i\theta})|. \tag{4.8}$$

Let $B = b\cos b$ be the maximum of the function $x\cos x$ in $(0, \pi/2)$.

Theorem 4.2. *If for some $\rho > 0$, $\sigma < \infty$,*

$$\limsup_{r\to\infty} \frac{\log M_\Phi(r, \gamma')}{r^\rho} \leq \sigma, \tag{4.9}$$

and one of the following two conditions holds

$$\liminf_{r\to\infty} \frac{n_\lambda(r)}{r^\rho} > \frac{\sigma\alpha_\lambda\rho}{\pi B}, \qquad \text{if } \alpha_\lambda > \frac{b}{\rho}; \tag{4.10}$$

$$\liminf_{r\to\infty} \frac{n_\lambda(r)}{r^\rho} > \frac{\sigma}{\pi\cos(\alpha_\lambda\rho)}, \qquad \text{if } \alpha_\lambda < \frac{\pi}{2\rho}, \tag{4.11}$$

then $\Phi(z) \equiv 0$ for $z \in \mathbb{D}$.

PROOF. Consider the function $G(z) = \Phi(z^{\gamma'})$, where $z^{\gamma'}$ is the branch with $z^{\gamma'} > 0$ for $z = x > 0$. Now $G(z)$ is an analytic function on the domain

$$\mathbb{D}_1 = \left\{ z : |z| \geq r_\lambda^{1/\gamma'}, |\arg(z)| \leq \frac{\pi}{2} \right\},$$

and $G(\lambda_n^{1/\gamma'}) = 0$, i.e., $b_n = \lambda_n^{1/\gamma'}$ $(n = 1, 2, \dots)$ are the zeroes of $G(z)$. Clearly, $|\arg(b_n)| < \pi/2$. We claim that $G(z) = G(re^{i\theta}) \equiv 0$ for $z \in \mathbb{D}_1$, and hence, $\Phi(z) \equiv 0$ for $z \in \mathbb{D}$. If $G(z) \not\equiv 0$ for $z \in \mathbb{D}_1$, then by Carleman formula (see, for example, [13]), for $r_\lambda^{1/\gamma'} < \lambda < R$, we have

$$\begin{aligned}\sum_{\lambda<|b_n|<R} \left(\frac{1}{|b_n|} - \frac{|b_n|}{R^2}\right) \cos\frac{\theta_n}{\gamma'} &\leq \frac{1}{\pi R}\int_{-\pi/2}^{\pi/2} \log|G(Re^{i\theta})| \cos\theta \, d\theta \\ &+ \frac{1}{2\pi}\int_\lambda^R \left(\frac{1}{y^2} - \frac{1}{R^2}\right) \log|G(iy)G(-iy)| \, dy + O(1),\end{aligned} \tag{4.12}$$

where $\theta_n = \arg(\lambda_n)$, and $O(1)$ is with respect to $R \to \infty$ for fixed λ. Since for $|\theta| \leq \pi/2$, we have that

$$\begin{aligned}|G(Re^{i\theta})| = |\Phi(R^{\gamma'}e^{i\gamma'\theta})| &\leq \sup_{|\theta|\leq\pi/2} |\Phi(R^{\gamma'}e^{i\gamma'\theta})| \\ &= \sup_{|\phi|\leq\pi\gamma'/2} |\Phi(R^{\gamma'}e^{i\phi})| = M_\Phi(R^{\gamma'}, \gamma'),\end{aligned}$$

it follows that, by (4.9), given $\varepsilon > 0$, and for R sufficiently large, we have

$$\log|G(Re^{i\theta})| \leq (\sigma + \varepsilon)R^{\gamma'\rho}.$$

The remaining of the proof is to estimate the upper bound of right hand side and the lower bound of left hand side of the inequality (4.12). By the same arguments as in the proof of Lemma 1 in [22, pp. 105–107], we then obtain

$$\liminf_{r\to\infty} \frac{n_\lambda(r)}{r^\rho} \le \frac{\sigma\alpha_\lambda\rho}{\pi B} \qquad \text{if } \alpha_\lambda > \frac{b}{\rho};$$

and

$$\liminf_{r\to\infty} \frac{n_\lambda(r)}{r^\rho} \le \frac{\sigma}{\pi\cos(\alpha_\lambda\rho)} \qquad \text{if } \alpha_\lambda < \frac{\pi}{2\rho},$$

a contradiction with (4.10) or (4.11). □

5. Completeness of the system $\{z^{\tau_k}\}$

The next theorem is on the completeness of the system $\{z^{\tau_k}\}$ in $L^2_a[\Omega]$. We studied this problem in [4], but there we assume a stronger condition for the sequence of exponents $\{\tau_k\}$, namely, the limit

$$\lim_{k\to\infty} \frac{k}{|\tau_k|} = D \qquad (0 < D < \infty)$$

exists. Instead of this, in this paper we assume the weaker condition (II) given in Section 3.

Theorem 5.1. *Assume conditions* Ω(I), Ω(II) *for the domain* Ω, *and* (I), (II) *and* (IV) *for the sequence* $\{\tau_k\}$ *given in Section* 3. *Furthermore assume that*

$$D_* - (2+\tau)(1-\cos\alpha) - \left(1 - \frac{1}{2\gamma}\right) > 0. \tag{5.1}$$

Denote

$$h = \max_{(x,y)\in\mathbb{D}} g(x,y), \tag{5.2}$$

where

$$g(x,y) = \frac{2x}{\sqrt{(2+\tau+y)^2\sin^2\alpha + x^2}}\left[D_* - (2+\tau+y)(1-\cos\alpha) - 2x - \left(1-\frac{1}{2\gamma}\right)\right],$$

and

$$\mathbb{D} = \left\{(x,y) : x > 0, y > 0, D_* - (2+\tau+y)(1-\cos\alpha) - 2x - \left(1-\frac{1}{2\gamma}\right) > 0\right\}.$$

Let

$$\eta = \max\left\{\alpha_1, \dots, \alpha_m, \frac{1}{h} + \varepsilon_0\right\} \tag{5.3}$$

with ε_0 *some positive number.*

If

$$\int^\infty \frac{\mathrm{d}r}{r^{1+\eta-s}} = +\infty, \tag{5.4}$$

then the system $\{z^{\tau_k}\}$ $(k = 1, 2, \dots)$ *is complete in* $L^2_a[\Omega]$.

PROOF. We only need to prove that if $f \in L_a^2[\Omega]$ and

$$(f(z), z^{\tau_k}) = 0, \qquad k = 1, 2, \ldots, \tag{5.5}$$

then $f(z) \equiv 0$. So we assume that (5.5) holds. By Lemma 2 in [21], condition (II)(3.5) implies that for any h with $0 < h < \frac{1}{2+\tau}$, we can find a number $q > 0$ and a sequence $\{\nu_k\}$ of positive numbers with $\nu_{k+1} - \nu_k \geq q$ such that the sequence $\{\mu_k\} = \{\tau_k\} \cup \{\nu_k\}$ satisfies

$$\lim_{k\to\infty} \frac{k}{|\mu_k|} = \frac{1}{h}.$$

By (IV), clearly

$$|\arg(\mu_k)| \leq \alpha < \frac{\pi}{2}.$$

And we have (see [21])

$$D_\nu^* = \limsup_{k\to\infty} \frac{k}{\nu_k} = \frac{1}{h} - \liminf_{k\to\infty} \frac{k}{|\tau_k|} = \frac{1}{h} - D_*. \tag{5.6}$$

Let

$$T(z) = \prod_{k=1}^{\infty} \left(1 - \frac{z^2}{\mu_k^2}\right),$$

and

$$I(s) = -\frac{1}{2\pi} \int_{-\infty}^{\infty} \frac{\mathrm{e}^{-isy}}{T(\mathrm{i}y)} \mathrm{d}y, \qquad s = u + \mathrm{i}v.$$

Denote

$$Q' = \left\{\xi = \xi_1 + \mathrm{i}\xi_2 : |\xi_2| < \pi\left(1 - \frac{1}{2\gamma}\right)\right\},$$

$$Q_\gamma = \left\{s = u + iv : |v| < \pi\left[\frac{1}{h}\cos\alpha - \left(1 - \frac{1}{2\gamma}\right)\right]\right\},$$

and for sufficiently small $\delta > 0$ and $\delta_1 > 0$, denote

$$S_\delta = \left\{s = u + \mathrm{i}v : |v| \leq \pi\left[\frac{1}{h}\cos\alpha - \delta\right]\right\},$$

$$Q_\gamma^\delta = \left\{s = u + \mathrm{i}v : |v| \leq \pi\left[\frac{1}{h}\cos\alpha - \delta - \left(1 - \frac{1}{2\gamma}\right)\right]\right\}$$

and

$$Q = \left\{s = u + \mathrm{i}v : |v| < \pi\left[D_* - (2 + \tau + \delta_1)(1 - \cos\alpha) - 2\delta - \left(1 - \frac{1}{2\gamma}\right)\right]\right\}.$$

Let us take

$$h = \frac{1}{2 + \tau + \delta_1}.$$

Then, by (5.1) and (5.6), we will have for δ and δ_1 sufficiently small,

$$\begin{aligned}
&\left[\frac{1}{h}\cos\alpha - \delta - \left(1 - \frac{1}{2\gamma}\right)\right] - (D_\nu^* + \delta) \\
&\qquad = \left[\frac{1}{h}\cos\alpha - 2\delta - \left(1 - \frac{1}{2\gamma}\right)\right] - \left(\frac{1}{h} - D_*\right) \\
&\qquad = D_* - (2 + \tau + \delta_1)(1 - \cos\alpha) - 2\delta - \left(1 - \frac{1}{2\gamma}\right) > 0.
\end{aligned}$$

Thus, the strip Q is located inside the strip Q^δ_γ, and the distance from the boundary of Q to the boundary of Q^δ_γ is greater than $\pi(D^*_\nu+\delta)$.

Let $z=e^\xi$, $\xi=\xi_1+i\xi_2$ and Ω' be the image of Ω in the ξ plane. By condition Ω(II), Ω' must be located inside Q'. If $s\in Q_\gamma$ and $\xi\in\Omega'$, $s-\xi$ must be inside S_δ, so the function $I(s-\xi)$ is analytic (see [4, Lemma 2.2]). If $s\in Q^\delta_\gamma$ and $\xi\in\Omega'$, $I(s-\xi)$ must be uniformly bounded. Fix $f(z)\in L^2_a[\Omega]$, and define, for $s\in Q^\delta_\gamma$, the function

$$G(s)=\iint_{\Omega'}\overline{f(e^\xi)}|e^\xi|^2 I(s-\xi)\,d\xi_1\,d\xi_2,\qquad \xi=\xi_1+i\xi_2. \tag{5.7}$$

As mentioned above, $G(s)$ is analytic in Q_γ (hence in Q^δ_γ) and uniformly bounded in Q^δ_γ.

By [4, Lemma 2.4], if for $s\in Q^\delta_\gamma$ the above $G(s)\equiv 0$, then

$$\iint_\Omega \overline{f(z)}z^n\,dz=0,\qquad n=1,2,\dots. \tag{5.8}$$

By Dzhrbasian's theorem, if (5.4) holds, the system $\{z^n\}$ $(n=0,1,2,\dots)$ is complete in $L^2_a[\Omega]$. Thus, by the Hahn–Banach theorem, from (5.8) it follows that $f(z)\equiv 0$ for $z\in\Omega$. So, we only need to prove that $G(s)\equiv 0$ for $s\in Q^\delta_\gamma$ whenever f satisfies (5.5). We will use $p(r)$ to denote $\alpha' r^s$ below.

We will now make use of the sequence $\{\nu_k\}$. Let

$$l(z)=\prod_{k=1}^\infty\left(1-\frac{z^2}{\nu_k^2}\right)=\sum_{n=0}^\infty \frac{l_n}{n!}z^n,$$

and $\gamma(z)$ be the Borel transform of $l(z)$, that is

$$\gamma(z)=\sum_{n=0}^\infty \frac{l_n}{z^{n+1}}.$$

We know that $l(z)$ is an entire function of exponential type πD^*_ν, and $\gamma(z)$ is analytic outside the vertical line segment with centre at the origin and length $2\pi D^*_\nu$. For sufficiently small $\delta>0$, define the convolution operator

$$L[y(s)]=\frac{1}{2\pi i}\int_{|\xi-s|=\pi(D^*_\nu+\delta)}\gamma(\xi-s)\cdot y(\xi)\,d\xi,$$

where the function $y(s)$ is analytic in Q^δ_γ.

Since the series representing $\gamma(\xi-s)$ is convergent uniformly on $|\xi-s|=\pi(D^*_\nu+\delta)$, we can interchange the order of the integration and the summation as follows:

$$\begin{aligned} L[y(s)]&=\frac{1}{2\pi i}\int_{|\xi-s|=\pi(D^*_\nu+\delta)}\left(\sum_{n=0}^\infty\frac{l_n}{(\xi-s)^{n+1}}\right)\cdot y(\xi)\,d\xi \\ &=\sum_{n=0}^\infty\frac{l_n}{2\pi i}\int_{|\xi-s|=\pi(D^*_\nu+\delta)}\frac{y(\xi)}{(\xi-s)^{n+1}}\,d\xi=\sum_{n=0}^\infty\frac{l_n}{n!}\cdot y^{(n)}(s). \end{aligned} \tag{5.9}$$

Note that $G(s)$ is analytic and uniformly bounded in Q^δ_γ, the strip Q is located inside the strip Q^δ_γ and the distance from the boundary of Q to the boundary of Q^δ_γ is greater than $\pi(D^*_\nu+\delta)$. So, for $s\in Q$, we can define the function

$$\Phi(s)=L[G(s)].$$

By (5.7), for $s \in Q_\gamma^\delta$, we have

$$G(s) = \iint_{\Omega'} \overline{f(\mathrm{e}^\xi)}|\mathrm{e}^\xi|^2 \Big[I(s-\xi) - \sum_{|\mu_k|<t_n} \frac{\mathrm{e}^{-\mu_k(s-\xi)}}{T'(\mu_k)} \Big] \,\mathrm{d}\xi_1 \,\mathrm{d}\xi_2 + \iint_{\Omega'} \overline{f(\mathrm{e}^\xi)}|\mathrm{e}^\xi|^2 \sum_{|\mu_k|<t_n} \frac{\mathrm{e}^{-\mu_k(s-\xi)}}{T'(\mu_k)} \,\mathrm{d}\xi_1 \,\mathrm{d}\xi_2$$

$$= G_{1,t_n}(s) + G_{2,t_n}(s),$$

where the sequence $\{t_n\}$ satisfies $n \geq t_n \geq (1-\lambda)n$ while λ is a sufficiently small positive number, $G_{1,t_n}(s)$ and $G_{2,t_n}(s)$ denote the above first and second integration, respectively.

Note that, using condition (5.5), we have

$$\begin{aligned} G_{2,t_n}(s) &= \sum_{|\mu_k|<t_n} \frac{1}{T'(\mu_k)} \Big[\iint_{\Omega'} \overline{f(\mathrm{e}^\xi)}|\mathrm{e}^\xi|^2 e^{\mu_k \xi} \,\mathrm{d}\xi_1 \,\mathrm{d}\xi_2 \Big] \mathrm{e}^{-\mu_k s} \\ &= \sum_{|\mu_k|<t_n} \frac{1}{T'(\mu_k)} \Big[\iint_{\Omega} \overline{f(z)} z^{\mu_k} \,\mathrm{d}x \,\mathrm{d}y \Big] \mathrm{e}^{-\mu_k s} \\ &= \sum_{\nu_k<t_n} \frac{1}{T'(\nu_k)} \Big[\iint_{\Omega} \overline{f(z)} z^{\nu_k} \,\mathrm{d}x \,\mathrm{d}y \Big] \mathrm{e}^{-\nu_k s}. \end{aligned}$$

We claim that for $n \in \mathbb{N}$, and for $s \in Q$,

$$L[G_{2,t_n}(s)] = 0. \tag{5.10}$$

Since $G_{2,t_n}(s)$ is a linear combination of $e^{-\nu_k s}$, it is enough to show that $L[\mathrm{e}^{-\nu_k s}] = 0$. Indeed, by (5.9), we have for any $k \in \mathbb{N}$,

$$L[\mathrm{e}^{-\nu_k s}] = [\sum_{i=0}^{\infty} \frac{l_i}{i!}(-\nu_k)^i] \cdot \mathrm{e}^{-\nu_k s} = l(-\nu_k) \cdot \mathrm{e}^{-\nu_k s} = 0.$$

As in [4, pp. 13–14], we have that for $\mathrm{Re}(s) = u > 0$, and $s \in Q_\gamma^\delta$,

$$|G_{1,t_n}(s)| \leq \inf_{n\in\mathbb{N}} C^n \frac{\big(\int_0^\infty r^{2n} \mathrm{e}^{-p(r)} \,\mathrm{d}r\big)^{1/2}}{|\mathrm{e}^s|^{(1-\lambda)n \sin(\pi\mu)}}, \tag{5.11}$$

where C is a constant independent of s and n, and μ is a small positive number satisfying

$$\tan(\pi\mu) < \frac{\delta}{(1/h)\sin\alpha}.$$

For $s \in Q$, noting (5.10), we have

$$\begin{aligned} |\Phi(s)| &= |L[G(s)]| = |L[G_{1,t_n}(s)]| \\ &\leq C_1(\delta) \cdot \max\big\{ |G_{1,t_n}(\xi)| : \xi \in Q_\gamma^\delta, |\mathrm{Re}(\xi) - \mathrm{Re}(s)| \leq \pi(D_\nu^* + \delta) \big\}, \end{aligned}$$

where $C_1(\delta)$ is a constant only depending on δ. By (5.11), for $s \in Q$, $\mathrm{Re}(s) \geq 0$, we have

$$|\Phi(s)| \leq \inf_{n\in\mathbb{N}} C_2^n \frac{\big(\int_0^\infty r^{2n} \mathrm{e}^{-p(r)} \,\mathrm{d}r\big)^{1/2}}{|\mathrm{e}^s / e^{\pi(D_\nu^*+\delta)}|^{(1-\lambda)n \sin(\pi\mu)}} \leq \inf_{n\in\mathbb{N}} C_3^n \frac{\big(\int_0^\infty r^{2n} \mathrm{e}^{-p(r)} \,\mathrm{d}r\big)^{1/2}}{|\mathrm{e}^s|^{(1-\lambda)n \sin(\pi\mu)}},$$

where C_2 and C_3 are constants independent of s and n.

Let

$$M_n = \int_0^\infty r^n e^{-p(r)}\, dr,$$

and

$$H(\bar{r}) = \sup_{n\in\mathbb{N}} \frac{\bar{r}^n}{\sqrt{M_{2n}}},$$

where

$$\bar{r} = c|e^s|^{(1-\lambda)\sin(\pi\mu)}$$

with c a constant independent of s and n. By [4, Lemma 2.6], we have for $s \in Q$ and $\mathrm{Re}(s) > 0$ sufficiently large,

$$|\Phi(s)| \le \frac{1}{H(\bar{r})} \le \exp\big[-q\cdot p\big(c|e^s|^{(1-\lambda)\sin(\pi\mu)}\big)\big],$$

where $q > 0$ is a constant.

Now we transform Q (with respect to s) into the upper half-plane $\mathrm{Im}(w) \ge 0$: (i) by $w_1 = e^s$, Q is transformed into an angle domain $|\arg(w_1)| \le \pi l$ with

$$l = D_* - (2+\tau+\delta_1)(1-\cos\alpha) - 2\delta - \left(1 - \frac{1}{2\gamma}\right); \tag{5.12}$$

(ii) by $w_2 = w_1^{1/(2l)}$, the above angle domain is then transformed into the right half-plane $\mathrm{Re}(w_2) \ge 0$; (iii) by $w = \mathrm{i}w_2$, the right half-plane is transformed into the upper half-plane $\mathrm{Im}(w) \ge 0$. The remaining of the proof is the same as that in [4, pp. 15−17] except that the quantity l is now given by (5.12) rather than $l = D\cos\alpha - \delta - 1 + 1/(2\gamma)$ given in [4, (23)], and correspondingly the quantity h is also given by the expression (5.2) rather than that expressed in [4, (18)]. Thus, by the assumption (5.4), we have $\Phi(s) \equiv 0$ for $s \in Q$. Hence $G(s) \equiv 0$ for $s \in Q$, and $G(s) \equiv 0$ for $s \in Q_\gamma^\delta$ since $G(s)$ is analytic in Q_γ^δ and $Q \subset Q_\gamma^\delta$. The proof is complete. □

6. Completeness of the system $\{f(\lambda_n z)\}$

We now present the main result of this paper:

Theorem 6.1. *Assume that:*

(i) *the functions $f(z)$ and $F(s)$ are given by* (3.1) *and* (3.2), *respectively, their complex coefficients $d_k \ne 0$ $(k = 1, 2, \dots)$, and the sequence $\{\tau_k\}$ of complex exponents satisfies conditions* (I), (II), (III)′, (III)″, (IV) *and the condition* (a) *or* (b) *given in Lemma* 4.1;

(ii) *the unbounded domain Ω satisfies conditions* Ω(I) *and* Ω(II);

(iii) *the complex sequence $\{\lambda_n\}$ satisfies condition* (3.9).

Moreover assume that

(iv) *the entire function $F(s)$ has (mR)-order ρ and (mR)-type σ, with $\rho < s\cos\alpha$; and either*

$$\liminf_{r\to\infty} \frac{n_\lambda(r)}{r^{s\rho\beta}} > \frac{\alpha_\lambda}{\pi B}\left(\frac{2}{s\alpha'}\right)^{\rho\beta} (\rho\sigma)^{s\beta}, \qquad \textit{if } \alpha_\lambda > \frac{b}{s\rho\beta}; \tag{6.1}$$

or

$$\liminf_{r\to\infty} \frac{n_\lambda(r)}{r^{s\rho\beta}} > \frac{1}{\pi s\rho\beta\cos(\alpha_\lambda s\rho\beta)}\left(\frac{2}{s\alpha'}\right)^{\rho\beta} (\rho\sigma)^{s\beta}, \qquad \textit{if } \alpha_\lambda < \frac{\pi}{2s\rho\beta}, \tag{6.2}$$

where $B = b\cos b$ *is the maximum of the function* $x\cos x$ *in* $(0, \pi/2)$ *as in Theorem* 4.2, *and* $\beta = 1/(s-\rho) > 0$.

If

$$D_* - (2+\tau)(1-\cos\alpha) - \left(1 - \frac{1}{2\gamma}\right) > 0,$$

and

$$\int^{\infty} \frac{\mathrm{d}r}{r^{1+\eta-s}} = +\infty,$$

where η *is given as in Theorem* 5.1, *then the system* $\{f(\lambda_n z)\}$ *is complete in* $L_a^2[\Omega]$.

PROOF. Consider the function $f(wz)$ with $z \in \Omega$ and w on the Riemann surface of the logarithm. Clearly, for fixed z, $f(wz)$ is an analytic function with respect to w on the Riemann surface of $\log w$. Now we restrict w to the domain

$$\mathbb{D} = \left\{z : |z| \geq r_\lambda, |\arg(z)| \leq \frac{\pi\gamma'}{2}\right\},$$

where γ' is a fixed number satisfying $\pi\gamma'/2 > \alpha_\lambda$. It is clear that $\lambda_n \in \mathbb{D}$ for $n = 1, 2, \ldots$.

Since $F(s) = f(\mathrm{e}^{-s})$ has (mR)-order ρ and (mR)-type σ, for any $A' > \sigma$, and for $|u|$ sufficiently large with $u < 0$, say $u < -u_1$ with

$$u_1 > \alpha_\Omega + \frac{\pi\gamma'}{2},$$

we have

$$\log M_F(u) \leq A'\mathrm{e}^{-u\rho}.$$

Hence, noting that $z = r\mathrm{e}^{\mathrm{i}\theta} = \mathrm{e}^{-s} = \mathrm{e}^{-(u+\mathrm{i}v)}$, for $|z| = r$ sufficiently large, say $r > r_1$, and

$$|\theta| \leq \alpha_\Omega + \frac{\pi\gamma'}{2},$$

we have

$$|f(z)| = |f(r\mathrm{e}^{\mathrm{i}\theta})| \leq \mathrm{e}^{A'r^\rho}.$$

Thus, for $|wz| \geq r_1$ and $|\arg(wz)| \leq \alpha_\Omega + \pi\gamma'/2$,

$$|f(wz)| < e^{A'|w|^\rho|z|^\rho}.$$

For fixed $w \in \mathbb{D}$, letting $\Omega_1 = \Omega \cap \{z : |wz| < r_1\}$ and $\Omega_2 = \Omega \cap \{z : |wz| \geq r_1\}$, we have (noting that $z = x + \mathrm{i}y$, and $r_\Omega < r_0$ by (3.8))

$$\begin{aligned}(6.3)\qquad \iint_{\Omega_2} |f(wz)|^2 \,\mathrm{d}x\,\mathrm{d}y &\leq \iint_{\Omega_2} \mathrm{e}^{2A'|w|^\rho|z|^\rho}\,\mathrm{d}x\,\mathrm{d}y \leq \iint_{\Omega} \mathrm{e}^{2A'|w|^\rho|z|^\rho}\,\mathrm{d}x\,\mathrm{d}y \\ &\leq \int_{r_\Omega}^{r_0} 2\pi r\mathrm{e}^{2A'|w|^\rho r^\rho}\,\mathrm{d}r + \int_{r_0}^{\infty} \mathrm{e}^{2A'|w|^\rho r^\rho}\sigma(r)\,\mathrm{d}r \\ &\leq c_2\mathrm{e}^{c_3|w|^\rho} + \int_0^\infty \mathrm{e}^{2A'|w|^\rho r^\rho}\mathrm{e}^{-\alpha' r^s}\,\mathrm{d}r,\end{aligned}$$

where $c_2 > 0$ and $c_3 > 0$ are constants independent of w. It is clear that, for $w \in \mathbb{D}$, since $|w| > r_\lambda$, for $|wz| < r_1$, and $z \in \Omega$ (so $|z| > r_\Omega$), we have $r_\lambda < |w| < r_1/r_\Omega$. And since $|\arg(wz)| < \alpha_\Omega + \pi\gamma'/2$, we must have

$$(6.4)\qquad \iint_{\Omega_1} |f(wz)|^2 \,\mathrm{d}x\,\mathrm{d}y \leq c_1,$$

where c_1 is a constant independent of w. Thus, by (6.4) and (6.3), we have

$$\iint_\Omega |f(wz)|^2 \,\mathrm{d}x\,\mathrm{d}y \le c_1 + c_2 e^{c_3|w|^\rho} + \int_0^\infty \mathrm{e}^{2A'|w|^\rho r^\rho} \mathrm{e}^{-\alpha' r^s}\,\mathrm{d}r.$$

Hence, for any μ with $0 < \mu < \alpha'$, we have

$$\iint_\Omega |f(wz)|^2 \,\mathrm{d}x\,\mathrm{d}y \le c_1 + c_2 e^{c_3|w|^\rho} + c_4 \sup_{r\ge 0} \cdot \exp[-\mu r^s + 2A'|w|^\rho r^\rho],$$

where c_4 is a constant independent of w. As in [3, p. 282], we have

$$\iint_\Omega |f(wz)|^2 \,\mathrm{d}x\,\mathrm{d}y < c_1 + c_2 e^{c_3|w|^\rho} + c_4 \exp\left[2\left(\frac{2}{s\mu}\right)^{\rho\beta} \cdot \frac{1}{s\rho\beta} \cdot (A'\rho)^{s\beta} \cdot |w|^{s\rho\beta}\right]. \tag{6.5}$$

where $\beta = 1/(s-\rho)$. Hence for any fixed $w \in \mathbb{D}$, $f(wz) \in L_a^2[\Omega]$, and $f(\lambda_n z) \in L_a^2[\Omega]$ since $\lambda_n \in \mathbb{D}$ for $n = 1, 2, \dots$. To prove the theorem we only need to prove that for any $h(z) \in L_a^2[\Omega]$, if

$$\big(f(\lambda_n z), h(z)\big) = \iint_\Omega f(\lambda_n z)\overline{h(z)}\,\mathrm{d}x\,\mathrm{d}y = 0, \qquad n = 1, 2, \dots, \tag{6.6}$$

then $h(z) \equiv 0$ for $z \in \Omega$. So we assume that (6.6) holds. Consider the function

$$\Phi(w) = \big(f(wz), h(z)\big) = \iint_\Omega f(wz)\overline{h(z)}\,\mathrm{d}x\,\mathrm{d}y, \qquad w \in \mathbb{D},$$

where $h(z)$ satisfies (6.6). By (6.6), we see that $\Phi(\lambda_n) = 0$ $(n = 1, 2, \dots)$. We need to prove that $\Phi(w) \equiv 0$ for $w \in \mathbb{D}$. By (6.5), as in [3, p. 283], we have, for $w \in \mathbb{D}$,

$$|\Phi(w)| \le c_5 \left[c_6 + \mathrm{e}^{c_7|w|^\rho} + c_8 \exp\left[\left(\frac{2}{s\mu}\right)^{\rho\beta} \cdot \frac{1}{s\rho\beta}(A'\rho)^{s\beta}|w|^{s\rho\beta}\right]\right], \tag{6.7}$$

where c_5, c_6, c_7, c_8 are positive constants independent of w. Thus, by Appendix A in [3], $\Phi(w)$ is analytic in $\mathbb{D}$. By (6.7), letting $\mu \to \alpha'$ and $A' \to \sigma$, we have

$$\limsup_{|w|\to\infty} \frac{\log M_\Phi(|w|, \gamma')}{|w|^{s\rho\beta}} \le \left(\frac{2}{s\alpha'}\right)^{\rho\beta} \cdot \frac{1}{s\rho\beta}(\rho\sigma)^{s\rho}.$$

Thus, by Theorem 4.2 and either condition (6.1) or condition (6.2), we must have $\Phi(w) \equiv 0$ for $w \in \mathbb{D}$. Then we have for $w \in \mathbb{D}$,

$$\Phi(w) = \iint_\Omega \left[\sum_{k=1}^\infty d_k (wz)^{\tau_k}\right]\overline{h(z)}\,\mathrm{d}x\,\mathrm{d}y = \sum_{k=1}^\infty d_k \left[\iint_\Omega z^{\tau_k}\overline{h(z)}\,\mathrm{d}x\,\mathrm{d}y\right] w^{\tau_k} \equiv 0, \tag{6.8}$$

and, since $d_k \ne 0$ $(k = 1, 2, \dots)$, we get

$$\iint_\Omega z^{\tau_k}\overline{h(z)}\,\mathrm{d}x\,\mathrm{d}y = 0, \qquad k = 1, 2, \dots.$$

Note that Theorem 4.1 is used in the justification for interchanging the integration and summation in (6.8) (see Appendix B in [3]), except that here we need to use the condition $\rho < s\cos\alpha$. The remaining of the proof is the same as in [3]: by Theorem 5.1, using the completeness of the system $\{z^{\tau_k}\}$ $(k = 1, 2, \dots)$ in $L_a^2[\Omega]$, we get $h(z) \equiv 0$. The proof is complete. □

7. Proof of Lemma 4.1

First we need a few more lemmas:

Lemma 7.1. *Under conditions* (I) *and* (II)(3.3), $T_n(z)$ *is an entire function of exponential type* πD^*.

PROOF. By (II)(3.3), given $\varepsilon > 0$, there exists an $I > 0$ such that for all $i > I$, $|\tau_i| > i/(D^* + \varepsilon)$. Thus, for any $R > 0$, if $|z| < R$, we have

$$\left|\frac{z^2}{\tau_i^2}\right| < \frac{R^2(D^* + \varepsilon)^2}{i^2}.$$

Hence the infinite product in (4.3) converges uniformly in any bounded domain of the complex plane, and $T_n(z)$ is an entire function. For $r > 0$, let (see (4.6))

$$g(r) = \prod_{i=1}^{\infty}\left(1 + \frac{r^2}{|\tau_i|^2}\right).$$

Since for $|z| = r$,

$$|T_n(z)| \le \prod_{\substack{i=1\\ i\ne n}}\left(1 + \frac{r^2}{|\tau_i|^2}\right) \le g(r),$$

and by (4.7),

$$\limsup_{r\to\infty}\frac{\log g(r)}{r} \le \pi D^*,$$

hence

$$\limsup_{r\to\infty}\frac{\log|M_{T_n}(r)|}{r} \le \pi D^*,$$

where

$$M_{T_n}(r) = \sup_{|z|=r}|T_n(z)|. \qquad \square$$

The following two estimates can be found in [13, pp. 76–78]:

Lemma 7.2. *Let* $z_1, \dots, z_n$ *be any* n *complex numbers. Given* H *with* $0 < H < 1$. *If*

$$P(z) = \prod_{k=1}^{n}(z - z_k),$$

then the inequality

$$|P(z)| \ge \left(\frac{H}{\mathrm{e}}\right)^n$$

holds outside exceptional disks with the sum of diameters not exceeding $10H$.

Lemma 7.3. *If an analytic function* $f(z)$ *has no zeros in a disk* $\{z : |z| \le R\}$ *and if* $|f(0)| = 1$, *then as* $|z| = r < R$,

$$\log|f(z)| \ge -\frac{2r}{R-r}\log M_f(R),$$

where

$$M_f(R) = \max_{|z|=R}|f(z)|.$$

Let $w_i = |\tau_i|$ $(i = 1, 2, \dots)$. For fixed n, denote

$$P_2 = \prod_{\substack{|w_i - w_n|<1 \\ i\neq n}} \left|1 - \frac{w_n}{w_i}\right|.$$

Using Lemma 7.2, we can prove

Lemma 7.4. *Under conditions* (I), (II)(3.3), (III)′ *and* (III)″, *we have, for n sufficiently large*

$$P_2 > e^{-2K(\delta+1)w_n}, \tag{7.1}$$

where we choose

$$\delta = \frac{LD^* - 5K}{2K}, \tag{7.2}$$

with L satisfying $LD^* > 5K$.

Proof. Consider the function

$$P(z) = \prod_{\substack{|w_i - w_n|<1 \\ i\neq n}} \frac{z - w_i}{w_i} = \prod_{\substack{|w_i - w_n|<1 \\ i\neq n}} (z - w_i) \prod_{\substack{|w_i - w_n|<1 \\ i\neq n}} \frac{1}{w_i}.$$

Suppose the numerator is a polynomial of degree q. By (III)′, we know that $q \leq 2K$. By (III)″, for n sufficiently large and $p \neq n$,

$$|w_n - w_p| > e^{-w_n\delta}. \tag{7.3}$$

Taking $H = (1/10)e^{-w_n\delta}$, by Lemma 7.2, the inequality

$$\left|\prod_{\substack{|w_i - w_n|<1 \\ i\neq n}} (z - w_i)\right| \geq \left(\frac{1}{10e} e^{-w_n\delta}\right)^q \tag{7.4}$$

holds outside exceptional disks with the sum of diameters not exceeding $e^{-w_n\delta}$. It is not hard to see that in every exceptional disk there is at least one w_i with $|w_n - w_i| < 1$ (if some disk does not contain such a w_i, then this disk should not be an exceptional disk since in this disk the inequality (7.4) holds for n sufficiently large). Thus, by (7.3), we know that w_n must be outside these exceptional disks. Hence

$$\left|\prod_{\substack{|w_i - w_n|<1 \\ i\neq n}} (w_n - w_i)\right| \geq \left(\frac{1}{10e} e^{-w_n\delta}\right)^q.$$

When n is sufficiently large, we have $\big(1/(10e)\big)e^{-w_n\delta} < 1$, hence, noting that $q \leq 2K$, for n sufficiently large,

$$\left|\prod_{\substack{|w_i - w_n|<1 \\ i\neq n}} (w_n - w_i)\right| \geq \left(\frac{1}{10e} e^{-w_n\delta}\right)^{2K}. \tag{7.5}$$

Obviously, we have

$$\prod_{\substack{|w_i - w_n|<1 \\ i\neq n}} \frac{1}{w_i} \geq \prod_{\substack{|w_i - w_n|<1 \\ i\neq n}} \frac{1}{w_n + 1} > \left(\frac{1}{2w_n}\right)^q \geq \left(\frac{1}{2w_n}\right)^{2K}. \tag{7.6}$$

Thus, by (7.5) and (7.6), for n sufficiently large, we have

$$P_2 > \left(\frac{1}{20\mathrm{e}}\right)^{2K} \left(\frac{\mathrm{e}^{-w_n\delta}}{w_n}\right)^{2K} > \mathrm{e}^{-2K(\delta+1)w_n}. \tag{7.7}$$

□

For $t > 0$, use $n(t)$ to denote the number of τ_i with $|\tau_i| \leq t$. For fixed n, let

$$l = n(3w_n)$$

and

$$P_4 = \prod_{w_i > w_l} \left|1 - \frac{w_n^2}{w_i^2}\right|.$$

Using Lemma 7.3, we can prove

Lemma 7.5. *Under conditions* (I) *and* (II)(3.3), *given* $\varepsilon' > 0$, *we have, for* n *sufficiently large,*

$$P_4 > \mathrm{e}^{-(3\pi D^* + \varepsilon')w_n}. \tag{7.8}$$

PROOF. Consider the function

$$Q(z) = \prod_{w_i > w_l} \left(1 - \frac{z^2}{w_i^2}\right).$$

By the proof of Lemma 7.1, we see that $Q(z)$ is an entire function of exponential type πD^*. Clearly $Q(0) = 1$, and $Q(z)$ has no zeros in $|z| < 3w_n$ (since when $w_i > w_l$ we have $w_i \geq w_{l+1}$, but $w_{l+1} > 3w_n$). Thus, by Lemma 7.3,

$$\log|Q(w_n)| > -\frac{2w_n}{3w_n - w_n} \log M_Q(3w_n) = -\log M_Q(3w_n).$$

When n is sufficiently large (since $Q(z)$ is of exponential type πD^*),

$$\log M_Q(3w_n) < (\pi D^* + \frac{\varepsilon'}{3})3w_n = (3\pi D^* + \varepsilon')w_n,$$

hence, noting that $P_4 = |Q(w_n)|$, we get (7.8). □

We now prove Lemma 4.1:

PROOF. As before, denote $w_i = |\tau_i|$ $(i = 1, 2, \dots)$, $n(t)$ the number of τ_i with $|\tau_i| \leq t$. For any fixed n, let $l = n(3w_n)$, denote w_{L_n} the nearest left w_i to w_n with

$|w_i - w_n| \geq 1$, and w_{R_n} the nearest right w_i to w_n with $|w_i - w_n| \geq 1$. We have

$$
\begin{aligned}
(7.9)\quad & |T_n(\tau_n)| \\
&= \left|\prod_{\substack{i\geq 1\\ i\neq n}}\left(1-\frac{\tau_n^2}{\tau_i^2}\right)\right| = \left|\prod_{\substack{i\geq 1, i\neq n\\ w_i\leq w_l}}\left(1-\frac{\tau_n^2}{\tau_i^2}\right)\right|\left|\prod_{w_i>w_l}\left(1-\frac{\tau_n^2}{\tau_i^2}\right)\right| \\
&\geq \prod_{\substack{i\geq 1, i\neq n\\ w_i\leq w_l}}\left|\left(1-\frac{w_n}{w_i}\right)\right|\left(1+\frac{w_n}{w_i}\right)\cdot\prod_{w_i>w_l}\left|1-\frac{w_n^2}{w_i^2}\right| \\
&\geq \prod_{\substack{i\geq 1, i\neq n\\ w_i\leq w_l}}\left|1-\frac{w_n}{w_i}\right|\cdot\prod_{w_i>w_l}\left|1-\frac{w_n^2}{w_i^2}\right| \\
&= \prod_{w_1\leq w_i\leq w_{L_n}}\frac{w_n-w_i}{w_i}\cdot\prod_{\substack{|w_i-w_n|<1\\ i\neq n}}\left|1-\frac{w_n}{w_i}\right|\cdot\prod_{w_{R_n}\leq w_i\leq w_l}\frac{w_i-w_n}{w_i}\cdot\prod_{w_i>w_l}\left|1-\frac{w_n^2}{w_i^2}\right| \\
&= P_1\cdot P_2\cdot P_3\cdot P_4,
\end{aligned}
$$

where P_1, P_2, P_3, P_4 denote the above four products in order.

We have estimated P_2 and P_4 in Lemmas 7.4 and 7.5, so we only need to estimate P_1 and P_3.

For P_1:

$$
\begin{aligned}
\log P_1 &= -\log(w_1 w_2\cdots w_{L_n}) + \log[(w_n - w_{L_n})(w_n - w_{L_n-1})\cdots(w_n - w_1)] \\
&=: P_{1,1} + P_{1,2}.
\end{aligned}
$$

$$
\begin{aligned}
P_{1,1} = &- L_n\log w_{L_n} + (L_n - 1)[\log w_{L_n} - \log w_{L_n-1}] \\
&+ (L_n - 2)[\log w_{L_n-1} - \log w_{L_n-2}] + \cdots + [\log w_2 - \log w_1] \\
= &-L_n\log w_{L_n} + \sum_{j=1}^{L_n-1} j(\log w_{j+1} - \log w_j).
\end{aligned}
$$

Since when $w_j \leq t < w_{j+1}$, $n(t) = j$. Thus, we have

$$
\begin{aligned}
P_{1,1} &= -L_n\log w_{L_n} + \sum_{j=1}^{L_n-1} j\int_{w_j}^{w_{j+1}}\frac{1}{t}dt \\
&= -L_n\log w_{L_n} + \sum_{j=1}^{L_n-1}\int_{w_j}^{w_{j+1}}\frac{n(t)}{t}\,\mathrm{d}t = -L_n\log w_{L_n} + \int_{w_1}^{w_{L_n}}\frac{n(t)}{t}\,\mathrm{d}t.
\end{aligned}
$$

$$
\begin{aligned}
P_{1,2} &= L_n\log(w_n - w_1) - (L_n - 1)[\log(w_n - w_1) - \log(w_n - w_2)] \\
&\quad - (L_n - 2)[\log(w_n - w_2) - \log(w_n - w_3)] - \cdots \\
&\quad - [\log(w_n - w_{L_n-1}) - \log(w_n - w_{L_n})] \\
&= L_n\log(w_n - w_1) - \sum_{j=1}^{L_n-1} j[\log(w_n - w_{L_n-j}) - \log(w_n - w_{L_n-j+1})] \\
&= L_n\log(w_n - w_1) - \sum_{j=1}^{L_n-1}\int_{w_n-w_{L_n-j+1}}^{w_n-w_{L_n-j}}\frac{j}{t}\,\mathrm{d}t.
\end{aligned}
$$

Let $n_1(t)$ be the number of w_i with $|w_i - w_n| \le t$ and $w_1 \le w_i \le w_{L_n}$. Since when $w_n - w_{L_n-j+1} \le t < w_n - w_{L_n-j}$, $n_1(t) = j$, we have

$$\begin{aligned} P_{1,2} &= L_n \log(w_n - w_1) - \sum_{j=1}^{L_n-1} \int_{w_n-w_{L_n-j+1}}^{w_n-w_{L_n-j}} \frac{n_1(t)}{t}\,\mathrm{d}t \\ &= L_n \log(w_n - w_1) - \int_{w_n-w_{L_n}}^{w_n-w_1} \frac{n_1(t)}{t}\,\mathrm{d}t. \end{aligned}$$

Now we have

$$\begin{aligned} \log P_1 &= P_{1,1} + P_{1,2} \\ &= -L_n \log w_{L_n} + \int_{w_1}^{w_{L_n}} \frac{n(t)}{t}\,\mathrm{d}t + L_n \log(w_n - w_1) - \int_{w_n-w_{L_n}}^{w_n-w_1} \frac{n_1(t)}{t}\,\mathrm{d}t \\ &> -L_n \log w_n + L_n \log(w_n - w_1) - \int_{w_n-w_{L_n}}^{w_n-w_1} \frac{n_1(t)}{t}\,\mathrm{d}t \\ &= L_n \log\left(1 - \frac{w_1}{w_n}\right) - \int_{w_n-w_{L_n}}^{w_n-w_1} \frac{n_1(t)}{t}\,\mathrm{d}t. \end{aligned}$$

Since $n_1(t) \le n(w_n) - n(w_n - t)$, we have

$$\begin{aligned} \log P_1 &> L_n \log\left(1 - \frac{w_1}{w_n}\right) - \int_{w_n-w_{L_n}}^{w_n-w_1} \frac{n(w_n) - n(w_n - t)}{t}\,\mathrm{d}t \\ &= L_n \log\left(1 - \frac{w_1}{w_n}\right) - \int_{w_1}^{w_{L_n}} \frac{n(w_n) - n(x)}{w_n - x}\,\mathrm{d}x. \end{aligned}$$

Given $\varepsilon' > 0$, for n sufficiently large,

$$\log\left(1 - \frac{w_1}{w_n}\right) > -\varepsilon'.$$

By (III)$'$,

$$n(w_n) - n(x) < K(w_n - x).$$

So, we have

$$\log P_1 > -\varepsilon' L_n - K(w_{L_n} - w_1) > -\varepsilon' L_n - K w_n.$$

By the definition of w_{L_n}, $L_n < n$, hence

$$-\varepsilon' L_n > -\varepsilon' n.$$

By (II)(3.3), for n sufficiently large, $n < (D^* + \varepsilon')w_n$. Hence we have, for n sufficiently large,

$$\log P_1 > -\varepsilon'(D^* + \varepsilon')w_n - K w_n = -[\varepsilon'(D^* + \varepsilon') + K]w_n. \tag{7.10}$$

For P_3:

First, consider the case when the condition (a) holds. Assume that w_{R_n}, $w_{R_n+1}, \dots, w_l$ are all the w_i satisfying $w_{R_n} \le w_i \le w_l$, then

$$\begin{aligned} \log P_3 &= -\log(w_{R_n} w_{R_n+1} \cdots w_l) + \log[(w_{R_n} - w_n)(w_{R_n+1} - w_n) \cdots (w_l - w_n)] \\ &=: P_{3,1} + P_{3,2}. \end{aligned}$$

Denote $n_2(t)$ the number of w_i with $w_i \le t$ and $w_{R_n} \le w_i \le w_l$. Denote $n_3(t)$ the number of such w_i with $|w_i - w_n| \le t$. Suppose the total number of $w_{R_n}, w_{R_n+1}, \dots, w_l$ is m_n. It is not hard to see that

$$m_n = n(3w_n) - R_n + 1 = l - R_n + 1.$$

Similar to that in $\log P_1$, we can get

$$P_{3,1} = -m_n \log w_l + \int_{w_{R_n}}^{w_l} \frac{n_2(t)}{t}\, \mathrm{d}t.$$

and

$$\begin{aligned} P_{3,2} &= m_n \log(w_l - w_n) - \sum_{j=1}^{m_n - 1} j[\log(w_{R_n+j} - w_n) - \log(w_{R_n+j-1} - w_n)] \\ &= m_n \log(w_l - w_n) - \sum_{j=1}^{m_n-1} \int_{w_{R_n+j-1}-w_n}^{w_{R_n+j}-w_n} \frac{j}{t} dt. \end{aligned}$$

Since when $w_{R_n+j-1} - w_n \le t < w_{R_n+j} - w_n$, $n_3(t) = j$, we have

$$\begin{aligned} P_{3,2} &= m_n \log(w_l - w_n) - \sum_{j=1}^{m_n-1} \int_{w_{R_n+j-1}-w_n}^{w_{R_n+j}-w_n} \frac{n_3(t)}{t}\, \mathrm{d}t \\ &= m_n \log(w_l - w_n) - \int_{w_{R_n}-w_n}^{w_l - w_n} \frac{n_3(t)}{t}\, \mathrm{d}t. \end{aligned}$$

By (III)$'$, $n_3(t) < Kt$. Thus

$$\log P_3 = P_{3,1} + P_{3,2} > m_n \log\left(1 - \frac{w_n}{w_l}\right) - K(w_l - w_{R_n}).$$

Now we estimate the value of $\frac{w_n}{w_l}$ for n sufficiently large. By (4.1) in the condition (a), there is a sufficiently small positive number ε_0 with $\varepsilon_0 < D_*$ such that for n sufficiently large,

$$\frac{n}{n(3w_n)} < p \cdot \frac{D_* - \varepsilon_0}{D^* + \varepsilon_0}.$$

Thus, by (3.3) and (3.4), for n sufficiently large,

$$\frac{w_n}{w_l} = \frac{w_n}{n} \cdot \frac{l}{w_l} \cdot \frac{n}{l} < \frac{D^* + \varepsilon_0}{D_* - \varepsilon_0} \cdot \frac{n}{n(3w_n)} < p.$$

Hence

$$\log P_3 > m_n \log(1-p) - K(w_l - w_{R_n}).$$

Noting that $w_{R_n} > w_n$ hence $w_l - w_{R_n} < 2w_n$, and $R_n > 1$, we have, for n sufficiently large

$$\begin{aligned} \log P_3 &> m_n \log(1-p) - 2Kw_n = \log(1-p)[n(3w_n) - R_n + 1] - 2Kw_n \\ &> \log(1-p) n(3w_n) - 2Kw_n \ge \log(1-p)(D^* + \varepsilon')3w_n - 2Kw_n \\ &= -(c'D^* + c'\varepsilon' + 2K)w_n, \end{aligned} \tag{7.11}$$

where $c' = -3\log(1-p)$.

Combining (7.7), (7.8), (7.10) and (7.11), we have for n sufficiently large,

$$\log(P_1P_2P_3P_4) > -[K + \varepsilon'(D^* + \varepsilon') + 2K(\delta+1) + c'D^* + c'\varepsilon' + 2K + 3\pi D^* + \varepsilon']w_n.$$

Given $\varepsilon > 0$, take $\varepsilon' > 0$ such that $\varepsilon'(D^* + \varepsilon') + c'\varepsilon' + \varepsilon' < \varepsilon$. Let

$$H = 3K + 2K(\delta + 1) - 3\log(1-p)D^* + 3\pi D^*,$$

i.e., by (7.2),

$$H = \big(L + 3\pi - 3\log(1-p)/bigr)D^*,$$

then we have for n sufficiently large

$$|T_n(\tau_n)| > e^{-(H+\varepsilon)w_n},$$

hence (4.2) holds.

For the case when the condition (b) holds, since $l = n$, $w_l = w_n$ for n sufficiently large, and since, by its definition, $w_{R_n} \geq w_n + 1$, we have $w_{R_n} \geq w_l + 1 > w_l$. In this case, it is impossible to have a w_i satisfying $w_{R_n} \leq w_i \leq w_l$ so, for n sufficiently large, the factor P_3 should not appear in the product $P_1 \cdot P_2 \cdot P_3 \cdot P_4$ in (7.9), or we should set it to be $P_3 = 1$. Thus, by the above calculation, it is not hard to see that, in this case, the H should be changed to $H = (L + 3\pi)D^* - 2K$. The proof is complete. □

Acknowledgement. We thank the referee for careful reading of the paper.

References

1. R. P. Boas Jr., *Fundamental sets of entire functions*, Ann. of Math. (2) **47** (1946), 21–32.
2. ———, *Entire functions*, Academic Press, New York, 1954.
3. A. Boivin and C. Zhu, *On the L^2-completeness of some systems of analytic functions*, Complex Variables Theory Appl. **45** (2001), no. 3, 273–295.
4. ———, *On the completeness of the system $\{z^{\tau_n}\}$ in L^2*, J. Approx. Theory **118** (2002), no. 1, 1–19.
5. ———, *The growth of an entire function and its Dirichlet coefficients and exponents*, Complex Var. Theory Appl. **48** (2003), no. 5, 397–415.
6. M. M. Dzhrbashyan, *Metric theorems on completeness and on the representation of analytic functions*, Uspehi Matem. Nauk (N.S.) **5** (1950), no. 3(37), 194–198 (Russian).
7. D. Gaier, *Lectures on complex approximation*, Birkhäuser, Boston, MA, 1987.
8. G. M. Goluzin, *Geometric theory of functions of a complex variable*, Transl. Math. Monogr., vol. 26, Amer. Math. Soc., Providence, RI, 1969.
9. I. I. Ibragimov, *On the completeness of systems of analytic functions $\{F(a_k z)\}$*, Izvestiya Akad. Nauk SSSR. Ser. Mat. **13** (1949), 45–54 (Russian).
10. A. F. Leont′ev, *On the completeness of a system of analytic functions*, Mat. Sbornik N.S. **31(73)** (1952), 381–414 (Russian).
11. ———, *Entire functions: Series of exponentials*, "Nauka", Moscow, 1983 (Russian).
12. B. Ya. Levin, *Distribution of zeros of entire functions*, Revised edition, Transl. Math. Monogr., vol. 5, Amer. Math. Soc., Providence, RI, 1980.
13. ———, *Lectures on entire functions*, Transl. Math. Monogr., vol. 150, Amer. Math. Soc., Providence, RI, 1996.
14. I. F. Lohin, *On completeness of a system of functions of the form $\{F(\lambda_n z)\}$*, Doklady Akad. Nauk SSSR (N.S.) **81** (1951), 141–144 (Russian).
15. ———, *On completeness of the system of functions $\{f(\lambda_n z)\}$*, Mat. Sb. N.S. **35(77)** (1954), 215–222 (Russian).
16. A. I. Markushevich, *On the basis in the space of analytic functions*, Mat. Sb. N.S. **17(59)** (1945), 211–252 (Russian).
17. S. Mandelbrojt, *Séries adhérentes, régularisation des suites, applications*, Gauthier-Villars, Paris, 1952.
18. ———, *Dirichlet series: Principles and methods*, D. Reidel, Dordrecht, 1972.
19. S. N. Mergeljan, *On the completeness of systems of analytic functions*, Amer. Math. Soc. Transl. (2) **19** (1962), 109–166.
20. W. Rudin, *Real and complex analysis*, 3rd ed., McGraw-Hill, New York, 1987.

21. X.-c. Shen, *On the closure of* $\{z^{\tau_n}\log^j z\}$ *on unbounded curves in the complex plane*, Acta Math. Sinica **13** (1963), 170–192 (Chinese); English transl., Chinese Math. **4** (1963), 188–210.
22. ———, *Completeness of the sequence of functions* $\{f(\lambda_n z)\}$, Acta Math. Sinica **14** (1964), 103–118 (Chinese); English transl., Chinese Math. **5** (1964), 112–128.
23. C.-Y. Yu, *On generalized Dirichlet series*, Acta Math. Sinica **5** (1955), 295–311 (Chinese, with English summary).

DEPARTMENT OF MATHEMATICS, UNIVERSITY OF WESTERN ONTARIO, LONDON, ON N6A 5B7, CANADA
E-mail address, A. Boivin: `boivin@uwo.ca`
E-mail address, C. Zhu: `czhu28@uwo.ca`

Centre de Recherches Mathématiques
CRM Proceedings and Lecture Notes
Volume **51**, 2010

A Formula for the Logarithmic Derivative and Its Applications

Javad Mashreghi

ABSTRACT. We show how an explicit formula for the imaginary part of the logarithmic derivative of f, where f is in the Cartwright class of entire functions of exponential type leads to a new integral representation of the Hilbert transform of $\log|f|$ and also to a representation for the first moment of $|\hat{f}|^2$.

1. Introduction

An entire function $f(z)$ is said to be of exponential type if there are constants A and B such that $|f(z)| \le Be^{A|z|}$ for all $z \in \mathbb{C}$. In this note, we are interested in two special subclasses of entire functions of exponential type. Both classes are defined by putting a growth restriction on the modulus of the function on the real line. The Cartwright class Cart consists of entire functions of exponential type satisfying the boundedness condition

$$\int_{-\infty}^{\infty} \frac{\log^+|f(x)|}{1+x^2}\,\mathrm{d}x < \infty,$$

and the Paley – Wiener class PW contains entire functions of exponential type fulfilling $f \in L^2(\mathbb{R})$. The inequality $\log^+|f| \le \frac{1}{2}|f|^2$ shows that PW is contained in Cart.

In this paper we obtain an explicit formula for $\Im\big(f'(t)/f(t)\big)$, $f \in \text{Cart}$, in terms of nonreal zeros of f and its rate of growth on the imaginary axis. Then we provide two applications of this formula. First, we derive an integral representation of the Hilbert transform of $\log|f|$. Secondly, we calculate the first moment of $|\hat{f}|^2$, where $\hat{f}$ is the Fourier – Plancherel transform of f, for functions in the Paley – Wiener space.

2. Reminder on representation theorems

In this section we gather some well known representation theorems about entire functions of exponential type. These results can be found for example in [1–3].

2000 *Mathematics Subject Classification.* Primary 30D20; Secondary 42A38.
Key words and phrases. Entire functions, Fourier – Plancherel transform, Hilbert transform.
This work was supported by NSERC (Canada) and FQRNT (Québec).
This is the final form of the paper.

Let $f \in \text{Cart}$ and let $\{z_n\}$ denote the sequence of zeros of f in the upper half-plane. Since $\sum_n \Im z_n/|z_n|^2 < \infty$ and $\lim_{n\to\infty}|z_n| = \infty$, the Blaschke product

$$B_u(z) = \prod_n \left(\frac{1 - z/z_n}{1 - z/\bar{z}_n}\right)$$

formed with this sequence is a well defined meromorphic function. Let

$$\sigma_u[f] = \limsup_{y\to+\infty} \frac{\log|f(\mathrm{i}y)|}{y}, \qquad \sigma_l[f] = \limsup_{y\to-\infty} \frac{\log|f(\mathrm{i}y)|}{|y|}.$$

In what follows, for simplicity we will write σ_u and σ_l respectively for $\sigma_u[f]$ and $\sigma_l[f]$. The following theorem is a celebrated result of Cartwright.

Theorem 1 (Cartwright). *Let $f \in \text{Cart}$. Then, for $\Im z > 0$,*

$$f(z) = c\mathrm{e}^{-\mathrm{i}\sigma_u z} B_u(z) \exp\left(\frac{i}{\pi}\int_{-\infty}^{\infty}\left(\frac{1}{z-t} + \frac{t}{1+t^2}\right)\log|f(t)|\,\mathrm{d}t\right),$$

where c is a constant of modulus one.

Put $f^*(z) := \overline{f(\bar{z})}$. Then $f^* \in \text{Cart}$ and

$$\sigma_u[f^*] = \limsup_{y\to+\infty} \frac{\log|f^*(\mathrm{i}y)|}{y} = \limsup_{y\to+\infty} \frac{\log|f(-\mathrm{i}y)|}{y} = \sigma_l[f].$$

Moreover, the upper half-plane zeros of f^* are conjugates of the lower half plane zeros of f, say $\{\overline{w}_n\}_{n\geq1}$, and for the Blaschke product formed with this sequence we write

$$B_l(z) = \prod_n \left(\frac{1 - z/\overline{w}_n}{1 - z/w_n}\right).$$

Therefore, by Theorem 1,

$$f^*(z) = c'\mathrm{e}^{-\mathrm{i}\sigma_l z} B_l(z) \exp\left(\frac{i}{\pi}\int_{-\infty}^{\infty}\left(\frac{1}{z-t} + \frac{t}{1+t^2}\right)|\log f(t)|\,\mathrm{d}t\right) \tag{1}$$

for all $\Im z > 0$. We also need the following celebrated theorem of Paley – Wiener. We remind that l. i. m. stands for the limit in mean and implicitly implies that the sequence is convergent in L^2-norm.

Theorem 2 (Paley – Wiener). *Let $f \in \text{PW}$. Then*

$$f(z) = \int_{-\sigma_u}^{\sigma_l} \hat{f}(\lambda)\mathrm{e}^{\mathrm{i}\lambda z}\,\mathrm{d}\lambda,$$

where

$$\hat{f}(\lambda) = \operatorname*{l.i.m.}_{N\to\infty} \frac{1}{2\pi}\int_{-N}^{N} f(t)\mathrm{e}^{-\mathrm{i}\lambda t}\,\mathrm{d}t$$

is the Fourier – Plancherel transform of f on the real line. Furthermore, the supporting interval of $\hat{f}$ is precisely $[-\sigma_u, \sigma_l]$.

In particular, if $f(\mathbb{R}) \subset \mathbb{R}$, then $\overline{\hat{f}(\lambda)} = \hat{f}(-\lambda)$ and thus the supporting interval of $\hat{f}$ is symmetric with respect to the origin, i.e., $\sigma_u = \sigma_l$.

3. The logarithmic derivative

Let z_n be a point in the upper half-plane. Then the Blaschke factor

$$b_{z_n}(z) = \frac{1 - z/z_n}{1 - z/\bar{z}_n}$$

satisfies $|b_{z_n}(t)| = 1$ for all $t \in \mathbb{R}$. As a matter of fact, there exists a unique real function $\arg b_{z_n} \in \mathcal{C}^\infty(\mathbb{R})$ such that

$$b_{z_n}(t) = \mathrm{e}^{\mathrm{i} \arg b_{z_n}(t)} \qquad (t \in \mathbb{R}),$$

with $\arg b_{z_n}(0) = 0$. Hence, by taking the logarithmic derivative of b_{z_n}, we obtain

$$\frac{\mathrm{d}}{\mathrm{d}t}\Big(\arg b_{z_n}(t)\Big) = \frac{b'_{z_n}(t)}{\mathrm{i} b_{z_n}(t)} = \frac{2\Im z_n}{|t - z_n|^2},$$

and thus $\arg b_{z_n}$ is given by

$$\text{(2)} \qquad \arg b_{z_n}(t) = \int_0^t \frac{2\Im z_n}{|s - z_n|^2}\,\mathrm{d}s = 2\arctan\left(\frac{t - \Re z_n}{\Im z_n}\right) + 2\arctan\left(\frac{\Re z_n}{\Im z_n}\right).$$

Let $\{z_n\}$ be a sequence of complex numbers in the upper half-plane such that

$$\sum_n \frac{\Im z_n}{|z_n|^2} < \infty$$

and $\lim_{n\to\infty}|z_n| = \infty$. Let $B = \prod_n b_{z_n}$. Since the zeros of B do not accumulate at any finite point of the complex plane, the function B is a meromorphic Blaschke product. In particular, B is analytic at *every point* of the real line. Hence, for all $t \in \mathbb{R}$,

$$\text{(3)} \qquad \frac{B'(t)}{B(t)} = 2\mathrm{i} \sum_n \frac{\Im z_n}{|t - z_n|^2} \qquad (t \in \mathbb{R}),$$

and the series is uniformly convergent on compact subsets of $\mathbb{R}$.

Lemma 3. *Let $f \in$ Cart. Then, for all $t \in \mathbb{R}$,*

$$\Im\left(\frac{f'(t)}{f(t)}\right) = \frac{\sigma_l - \sigma_u}{2} + \sum_n \frac{\Im \zeta_n}{|t - \zeta_n|^2}$$

where $\{\zeta_n\}$ is the sequence of zeros of f in $\mathbb{C} \setminus \mathbb{R}$.

PROOF. Let $F = f/f^*$. Then, one one hand,

$$\frac{F'(t)}{F(t)} = \frac{f'(t)}{f(t)} - \overline{\left(\frac{f'(t)}{f(t)}\right)} = 2\mathrm{i}\Im\left(\frac{f'(t)}{f(t)}\right).$$

On the other hand, by Theorem 1 and (1),

$$F(z) = c\mathrm{e}^{\mathrm{i}(\sigma_l - \sigma_u)z} \frac{B_u(z)}{B_l(z)} \qquad (\Im z > 0).$$

By continuity, this relation holds for $\Im z \geq 0$. Hence, we also have

$$\frac{F'(t)}{F(t)} = \mathrm{i}(\sigma_l - \sigma_u) + \frac{B'_u(t)}{B_u(t)} - \frac{B'_l(t)}{B_l(t)} \qquad (t \in \mathbb{R}).$$

Thus, by (3),

$$\begin{aligned}\frac{F'(t)}{F(t)} &= \mathrm{i}(\sigma_l - \sigma_u) + 2\mathrm{i}\sum_n \frac{\Im z_n}{|t-z_n|^2} - 2\mathrm{i}\sum_n \frac{\Im \overline{w}_n}{|t-\overline{w}_n|^2}\\ &= \mathrm{i}(\sigma_l - \sigma_u) + 2\mathrm{i}\sum_n \frac{\Im z_n}{|t-z_n|^2} + 2\mathrm{i}\sum_n \frac{\Im w_n}{|t-w_n|^2}\\ &= 2\mathrm{i}\left(\frac{\sigma_l-\sigma_u}{2} + \sum_n \frac{\Im \zeta_n}{|t-\zeta_n|^2}\right).\end{aligned}$$

Therefore, comparing with the first formula, we obtain the result. □

Note that the real zeros of f do not cause discontinuity in $\Im\big(f'(t)/f(t)\big)$. Their effect appears in $\Re\big(f'(t)/f(t)\big)$. An immediate consequence of Lemma 3 and (2) is the following result.

Corollary 4. *Let $f \in$ Cart. Let $\{z_n\}$ and $\{w_n\}$ be respectively the sequence of upper and lower half plane zeros of f. Then, for all $t \in \mathbb{R}$,*

$$\int_0^t \Im\left(\frac{f'(s)}{f(s)}\right) \mathrm{d}s = \left(\frac{\sigma_l-\sigma_u}{2}\right)t + \frac{1}{2}\sum_n \arg b_{z_n}(t) - \frac{1}{2}\sum_n \arg b_{\overline{w}_n}(t),$$

where $\arg b$ *is given by* (2).

4. An integral formula for $\widetilde{\log|f|}$ and the first moment of $|\hat{f}|^2$

Let $\{x_n\}$ be a sequence of real numbers such that $\lim_{|n|\to\infty}|x_n| = \infty$ and $x_k < x_l$ if $k < l$. Let $\{m_n\}$ be a sequence of nonnegative integers. The counting function $\nu_{\{x_n\}}$ of the sequence $\{x_n\}$ is defined to be constant between x_{n-1} and x_n and at each point x_n jumps up by m_n units. The value of $\nu_{\{x_n\}}(t)$ at x_n is not important. For one-sided or finite sequences, $\nu_{\{x_n\}}$ is defined similarly and it is adjusted such that its value between $-\infty$ and the first point of the sequence is zero.

Let $f \in$ Cart. In [4], we showed that

$$\widetilde{\log|f|}(t) = -\pi\nu_{\{x_n\}}(t) + \left(\frac{\sigma_u+\sigma_l}{2}\right)t - \frac{1}{2}\sum_n \arg b_{z_n}(t) - \frac{1}{2}\sum_n \arg b_{\overline{w}_n}(t),$$

where $\sim$ stands for the Hilbert transform. This formula has been used to obtain a partial characterization of the argument of outer functions on the real line [5]. By a standard technique, one can shift all zeros of f in the lower half plane to the upper half-plane without changing $|f|$ on the real line. Furthermore, one can multiply f by $\mathrm{e}^{-\mathrm{i}\sigma_l z}$, to get a new function with the same absolute value on the real line, but instead $\sigma_l = 0$. Therefore, to find $\widetilde{\log|f|}$, without loss of generality we can assume that f has no zeros in the lower half plane and besides $\sigma_l = 0$. Therefore, by Corollary 4, we find the following formula for the Hilbert transform of $\log|f|$.

Theorem 5. *Let $f \in$ Cart. Suppose that f has no zeros in the lower half-plane and that $\sigma_l = 0$. Let ν denote the counting function of the sequence of real zeros of f. Then*

$$\widetilde{\log|f|}(t) = -\pi\nu(t) - \int_0^t \Im\left(\frac{f'(s)}{f(s)}\right) \mathrm{d}s.$$

Another consequence of Lemma 3 is an explicit formula for the first moment of $|\hat{f}|^2$, in terms of σ_u, σ_l and nonreal zeros of f, for functions in the Paley–Wiener space PW.

Theorem 6. *Let $f \in \mathrm{PW}$. Then*

$$\int_{-\sigma_u}^{\sigma_l} \lambda |\hat{f}(\lambda)|^2 \,\mathrm{d}\lambda = \frac{1}{2\pi} \int_{-\infty}^{\infty} \left(\frac{\sigma_l - \sigma_u}{2} + \sum_n \frac{\Im \zeta_n}{|t-\zeta_n|^2} \right) |f(t)|^2 \,\mathrm{d}t,$$

where $\{\zeta_n\}$ is the sequence of nonreal zeros of f.

PROOF. Since $\lambda \hat{f}(\lambda) \in L^1(\mathbb{R})$, the Fourier–Plancherel transform of $f'(t)$ is $\mathrm{i}\lambda\hat{f}(\lambda)$. Thus, by the Parseval's identity,

$$\int_{-\infty}^{\infty} f'(t)\overline{f(t)} \,\mathrm{d}t = 2\pi \int_{-\sigma_u}^{\sigma_l} \mathrm{i}\lambda \hat{f}(\lambda)\overline{\hat{f}(\lambda)} \,\mathrm{d}\lambda = 2\pi \int_{-\sigma_u}^{\sigma_l} \mathrm{i}\lambda |\hat{f}(\lambda)|^2 \,\mathrm{d}\lambda.$$

The right-hand side is purely imaginary. Therefore, by Lemma 3,

$$\begin{aligned} 2\pi \int_{-\sigma_u}^{\sigma_l} \lambda |\hat{f}(\lambda)|^2 \,\mathrm{d}\lambda &= \int_{-\infty}^{\infty} \Im\Big(f'(t)\,\overline{f(t)} \Big) \,\mathrm{d}t = \int_{-\infty}^{\infty} \Im\left(\frac{f'(t)}{f(t)} |f(t)|^2 \right) \mathrm{d}t \\ &= \int_{-\infty}^{\infty} \Im\left(\frac{f'(t)}{f(t)} \right) |f(t)|^2 \,\mathrm{d}t \\ &= \int_{-\infty}^{\infty} \left(\frac{\sigma_l - \sigma_u}{2} + \sum_n \frac{\Im \zeta_n}{|t-\zeta_n|^2} \right) |f(t)|^2 \,\mathrm{d}t. \qquad \square \end{aligned}$$

Acknowledgements. The author deeply thanks the anonymous referee for his/her valuable remarks and suggestions.

References

1. R. P. Boas Jr., *Entire functions*, Academic Press, New York, 1954.
2. P. Koosis, *The logarithmic integral.* I, Cambridge Stud. Adv. Math., vol. 12, Cambridge Univ. Press, Cambridge, 1988.
3. B. Ya. Levin, *Distribution of zeros of entire functions*, Revised edition, Transl. Math. Monogr., vol. 5, Amer. Math. Soc., Providence, RI, 1980.
4. J. Mashreghi, *Hilbert transform of* $\log|f|$, Proc. Amer. Math. Soc. **130** (2002), no. 3, 683–688.
5. J. Mashreghi and M. R. Pouryayevali, *Argument of outer functions on the real line*, Illinois J. Math. **51** (2007), no. 2, 499–511.

DÉPARTEMENT DE MATHÉMATIQUES ET DE STATISTIQUE, UNIVERSITÉ LAVAL, QUÉBEC, QC G1V 0A6, CANADA

E-mail address: `javad.mashreghi@mat.ulaval.ca`

Centre de Recherches Mathématiques
CRM Proceedings and Lecture Notes
Volume **51**, 2010

Composition Operators on the Minimal Möbius Invariant Space

Hasi Wulan and Chengji Xiong

ABSTRACT. Two sufficient and necessary conditions are given for ϕ to ensure the composition operator C_ϕ to be compact on the minimal Möbius invariant space. Meanwhile, our results show that some known results about the compactness of C_ϕ on the Besov spaces are still valid for the minimal Möbius invariant space.

1. Introduction

Throughout this paper $\mathbb{D}$ will denote the open unit disc in the complex plane $\mathbb{C}$. The set of all conformal automorphisms of $\mathbb{D}$ forms a group, called Möbius group and denoted by $\mathrm{Aut}(\mathbb{D})$. It is well-known that each element of $\mathrm{Aut}(\mathbb{D})$ is a fractional transformation φ of the following form

$$\varphi(z) = \mathrm{e}^{\mathrm{i}\theta}\sigma_a(z), \qquad \sigma_a(z) = \frac{a-z}{1-\bar{a}z},$$

where θ is real and $a \in \mathbb{D}$. Denote by dA the normalized area measure:

$$\mathrm{dA}(z) = \frac{1}{\pi}\,\mathrm{d}x\,\mathrm{d}y, \qquad z = x + \mathrm{i}y.$$

Let X be a linear space of analytic functions on $\mathbb{D}$ which is complete in a norm or seminorm $\|\cdot\|_X$. X is called *Möbius invariant* if for each function f in X and each element φ in $\mathrm{Aut}(\mathbb{D})$, the composition function $f\circ\varphi$ also lies in X and satisfies that $\|f\circ\varphi\|_X = \|f\|_X$; see [2]. For example, the space H^∞ of bounded analytic functions f on $\mathbb{D}$ with the norm $\|f\|_\infty = \sup\{|f(z)| : z \in \mathbb{D}\}$ is Möbius invariant. BMOA, the space of analytic functions f on $\mathbb{D}$ for which

$$\sup\left\{\frac{1}{2\pi}\int_0^{2\pi} |f(\mathrm{e}^{\mathrm{i}\theta})|^2 \frac{1-|a|^2}{|1-\bar{a}\mathrm{e}^{\mathrm{i}\theta}|^2}\,\mathrm{d}\theta - |f(a)|^2 : a \in \mathbb{D}\right\} < \infty,$$

is Möbius invariant. Actually, some other spaces of analytic functions on $\mathbb{D}$ such as Q_p and Q_K spaces are Möbius invariant, too. See [2–4]. However, the Hardy

2000 *Mathematics Subject Classification.* 30D45, 47B38.

Key words and phrases. composition operator, minimal Möbius invariant space, Besov spaces.

The authors are partially supported by NSF of China (No. 10671115), RFDP of China (No. 20060560002) and NSF of Guangdong Province of China (No. 06105648).

This is the final form of the paper.

spaces H^p are not Möbius invariant. Now we return to our primary interest, the Besov spaces.

For $1 < p < \infty$ the space B_p consists of all analytic functions f on $\mathbb{D}$ for which

$$\int_{\mathbb{D}} |f'(z)|^p (1-|z|^2)^{p-2}\, \mathrm{dA}(z) < \infty. \tag{1.1}$$

For $p = \infty$ the requirement is that the quantity

$$\sup_{z\in\mathbb{D}} (1-|z|^2)|f'(z)| \tag{1.2}$$

be finite. When $1 < p < \infty$ the space B_p is called the Besov space and $B_\infty = \mathcal{B}$ is called the Bloch space. The seminorm $\|\cdot\|_{B_p}$ on B_p is the pth root of the left of (1.1) if $1 < p < \infty$ and the quantity (1.2) if $p = \infty$. The space B_2 is known as the Dirichlet space and usually denoted by $\mathcal{D}$. It is immediately clear that the Besov spaces are Möbius invariant. Unlike B_p spaces for $p > 1$, we define the B_1 by other way since (1.1) does not converge when $p = 1$ for any non-constant function.

Arazy, Fisher and Peetre [2] defined B_1 as a set of those analytic functions f on $\mathbb{D}$ which have a representation as

$$f(z) = \sum_{k=1}^{\infty} c_k \sigma_{a_k}(z), \qquad a_k \in \mathbb{D} \text{ and } \sum_{k=1}^{\infty} |c_k| < \infty. \tag{1.3}$$

Since a function f could conceivably have several such representation, the norm of B_1 can be defined by

$$\|f\| = \inf\left\{ \sum_{k=1}^{\infty} |c_k| : (1.3) \text{ holds} \right\}.$$

By [2] we know that the space B_1 is the minimal Möbius invariant space since it is contained in any Möbius invariant space X. Also, we say that the Bloch space $\mathcal{B}$ is the maximal Möbius invariant space; see [7].

We know that for $1 < p < \infty$ a function f belongs to B_p if and only if the seminorm

$$\|f\|_{B_p}^p \approx \int_{\mathbb{D}} |f''(z)|^p (1-|z|^2)^{2p-2}\, \mathrm{dA}(z) < \infty. \tag{1.4}$$

Arazy, Fisher and Peetre showed in [2] that there exist constants c_1 and c_2 such that

$$c_1\|f\| \le |f(0)| + |f'(0)| + \int_{\mathbb{D}} |f''(z)|\, \mathrm{dA}(z) \le c_1\|f\|. \tag{1.5}$$

Hence, (1.4) and (1.5) do permit us to pass the case $p = 1$ and the connect the space B_1 with the Besov spaces B_p. We define now the seminorm of B_1 as

$$\|f\|_{B_1} := \int_{\mathbb{D}} |f''(z)|\, \mathrm{dA}(z) < \infty. \tag{1.6}$$

Modulo constants, B_1 is a Banach space under the norm defined in (1.6).

2. Composition operators on B_1

Let ϕ be a holomorphic mapping from $\mathbb{D}$ into itself and $f \in H(\mathbb{D})$, the set of all analytic functions on $\mathbb{D}$. Then ϕ induces a composition operator $C_\phi\colon f \to f \circ \phi$ on $H(\mathbb{D})$. Tjani [9] gave the following result.

Theorem A. *Let ϕ be a holomorphic mapping from $\mathbb{D}$ into itself and $1 < p \leq q \leq \infty$. Then the following are equivalent*:

(a) *$C_\phi\colon B_p \to B_q$ is a compact operator.*
(b) *$\|C_\phi \sigma_a\|_{B_q} \to 0$ as $|a| \to 1$.*

It is natural to ask what condition for ϕ ensures the composition operator C_ϕ to be compact for the critical case $p = 1$. This paper mainly answers this question.

Theorem 1. *Let ϕ be a holomorphic mapping from $\mathbb{D}$ into itself. Then C_ϕ is compact on B_1 if and only if*

$$\lim_{|a|\to 1} \|C_\phi \sigma_a\|_{B_1} = 0. \tag{2.1}$$

Proof. Necessity. Assume that C_ϕ is compact on B_1. We have that $\|C_\phi \sigma_a\|_{B_1} = \|C_\phi(\sigma_a - a)\|_{B_1} \to 0$ as $|a| \to 1$ since $\sigma_a(z) - a \to 0$ uniformly on compacts of $\mathbb{D}$. Thus (2.1) holds.

Sufficiency. We first show that (2.1) implies that

$$\lim_{r\to 1} \sup_{a\in\mathbb{D}} \int_{|\phi(z)|>r} \left|\big(\sigma_a \circ \phi(z)\big)''\right| \mathrm{dA}(z) = 0. \tag{2.2}$$

In fact, by (1.5) we see that (2.1) gives that

$$\lim_{|a|\to 1} \int_{\mathbb{D}} \left|\big(\sigma_a \circ \phi(z)\big)''\right| \mathrm{dA}(z) = 0. \tag{2.3}$$

Hence

$$\int_{\mathbb{D}} \left|\big(\sigma_a \circ \phi(z)\big)''\right| \mathrm{dA}(z) \leq 1$$

for some $a \in \mathbb{D}$. It follows that $\sigma_a \circ \phi \in B_1$ for some $a \in \mathbb{D}$. Since $B_1 \subset B_2 = \mathcal{D}$ and σ_a is analytic on $\overline{\mathbb{D}}$, we have $\phi = \sigma_a \circ \sigma_a \circ \phi \in B_1$. Note that by (2.3) we know that for given $\epsilon > 0$ there exists a $\delta > 0$ such that

$$\sup_{\delta<|a|<1} \int_{|\phi(z)|>r} \left|\big(\sigma_a \circ \phi(z)\big)''\right| \mathrm{dA}(z) < \epsilon$$

for all $r \in (0, 1)$. Letting $r \to 1^-$ gives

$$\begin{aligned}
&\sup_{|a|\leq\delta} \int_{|\phi(z)|>r} \left|\big(\sigma_a \circ \phi(z)\big)''\right| \mathrm{dA}(z) \\
&\quad \leq \sup_{|a|\leq\delta} \int_{|\phi(z)|>r} |\sigma_a''(\phi(z))||\phi'(z)|^2 \,\mathrm{dA}(z) + \sup_{|a|\leq\delta} \int_{|\phi(z)|>r} |\sigma_a'(\phi(z))||\phi''(z)| \,\mathrm{dA}(z) \\
&\quad \leq C\bigg(\int_{|\phi(z)|>r} |\phi'(z)|^2 \,\mathrm{dA}(z) + \int_{|\phi(z)|>r} |\phi''(z)| \,\mathrm{dA}(z)\bigg) < \epsilon,
\end{aligned}$$

where

$$C = \max\bigg\{ \sup_{\substack{|a|\leq\delta \\ z\in\mathbb{D}}} |\sigma_a'(z)|, \sup_{\substack{|a|\leq\delta \\ z\in\mathbb{D}}} |\sigma_a''(z)| \bigg\}.$$

Note that $\phi \in B_1$ and $\phi \in \mathcal{D}$ have been used in the above estimate. Thus we show that (2.2) holds.

Now we prove the compactness of C_ϕ. For any bounded sequence $\{f_n\} \subset B_1$, without loss of generality, we assume that f_n converges to zero uniformly on any compact subset of $\mathbb{D}$ and $\|f_n\|_{B_1} \leq 1$. To end our proof it suffices to show that $\|f_n \circ \phi\|_{B_1} \to 0$ as $n \to \infty$ since $|f_n \circ \phi(0)|+|(f_n \circ \phi)'(0)| \to 0$ as $n \to \infty$. We write

$$f_n(z) = \sum_{k=1}^{\infty} c_{n,k}\sigma_{a_{n,k}}(z), \qquad a_{n,k} \in \mathbb{D}$$

with

$$\|f_n\|_{B_1} \leq \sum_{k=1}^{\infty}|c_{n,k}| \leq 2, \qquad n = 1, 2, \dots .$$

Using (1.5), it suffices to prove

$$\int_{\mathbb{D}} |(f_n(\phi(z))''| \,\mathrm{dA}(z) \to 0, \qquad n \to \infty.$$

By (2.2) for given $\epsilon > 0$, there exists an r, $0 < r < 1$ such that for all $a \in \mathbb{D}$

$$\sup_{a\in\mathbb{D}} \int_{|\phi(z)|>r} |(\sigma_a \circ \phi(z))''| \,\mathrm{dA}(z) \leq \frac{\epsilon}{2}.$$

Hence

$$\begin{aligned}
&\int_{\mathbb{D}} |(f_n(\phi(z))''| \,\mathrm{dA}(z) \\
&\quad = \int_{|\phi(z)|\leq r} |(f_n(\phi(z))''| \,\mathrm{dA}(z) + \int_{|\phi(z)|>r} |(f_n(\phi(z))''| \,\mathrm{dA}(z) \\
&\quad \leq \int_{|\phi(z)|\leq r} |(f_n(\phi(z))''| \,\mathrm{dA}(z) + \sum_{k=1}^{\infty}|c_{n,k}| \int_{|\phi(z)|>r} \left|(\sigma_{a_{n,k}}(\phi(z)))''\right| \,\mathrm{dA}(z) \\
&\quad \leq \int_{|\phi(z)|\leq r} |(f_n(\phi(z))''| \,\mathrm{dA}(z) + \epsilon.
\end{aligned}$$

Notice that

$$\int_{|\phi(z)|\leq r} |(f_n(\phi(z))''| \,\mathrm{dA}(z) \to 0$$

as $n \to \infty$. We obtain $\|f_n \circ \phi\|_{B_1} \to 0$. The proof is completed. □

Arazy, Fisher and Peetre [2] obtained following theorem.

Theorem B. *The composition operator C_ϕ is bounded on B_1 if and only if*

$$\sup_{a\in\mathbb{D}} \int_{\mathbb{D}} |\sigma_a''(\phi(z))||\phi'(z)|^2 \,dm(z) < \infty$$

and

$$\sup_{a\in\mathbb{D}} \int_{\mathbb{D}} |\sigma_a'(\phi(z))||\phi''(z)| \,dm(z) < \infty.$$

Now we show a similar result for compact operators.

Theorem 2. *Let ϕ be a holomorphic mapping from $\mathbb{D}$ into itself. Then C_ϕ is compact on B_1 if and only if the following two expressions are true*:

$$\lim_{|a|\to 1}\int_{\mathbb{D}}|\sigma_a''(\phi(z))||\phi'(z)|^2\,dA(z)=0 \tag{2.4}$$

and

$$\lim_{|a|\to 1}\int_{\mathbb{D}}|\sigma_a'(\phi(z))\phi''(z)|\,dA(z)=0. \tag{2.5}$$

PROOF. Since

$$\|C_\phi\sigma_a\|_{B_1}\approx\int_{\mathbb{D}}\left|\sigma_a''(\phi(z))\left(\phi'(z)\right)^2+\sigma_a'(\phi(z))\phi''(z)\right|\,dA(z),$$

it is easy to see that (2.4) and (2.5) imply

$$\lim_{|a|\to 1}\|C_\phi\sigma_a\|_{B_1}=0.$$

By Theorem 1, C_ϕ is compact on B_1.

Conversely, assume that C_ϕ is compact on B_1. By Theorem 1 we have

$$\lim_{|a|\to 1}\|C_\phi\sigma_a\|_{B_1}=\lim_{|a|\to 1}\int_{\mathbb{D}}\left|\left(\sigma_a\circ\phi(z)\right)''\right|\,dA(z)=0.$$

Since σ_a'' is zero-free and analytic in $\mathbb{D}$, we can find a function f_a analytic in $\mathbb{D}$ with $f_a(0)=0$ such that $(f_a')^2=\sigma_a''$. So $f_a\in B_2$ and $\|f_a\|_{B_2}$ is bounded. By the estimate

$$\|C_\phi\sigma_a\|_{B_2}\le C\|C_\phi\sigma_a\|_{B_1}$$

and the assumption, we know that C_ϕ is compact on B_2 by Theorem A. A direct computation gives $|\sigma_a''(w)|\to 0$ in $\mathbb{D}$ as $|a|\to 1$. Hence f_a tends to 0 uniformly on any compact subset of $\mathbb{D}$. Thus $\|C_\phi f_a\|_{B_2}\to 0$ and

$$\lim_{|a|\to 1}\|C_\phi f_a\|_{B_2}=\lim_{|a|\to 1}\int_{\mathbb{D}}|\sigma_a''(\phi(z))||\phi'(z)|^2\,dA(z)=0.$$

On the other hand, since

$$(\sigma_a\circ\varphi)''=\sigma_a''(\varphi)(\phi')^2+\sigma_a'(\varphi)\phi'',$$

we have

$$\int_{\mathbb{D}}|\sigma_a'(\phi(z))\phi''(z)|\,dA(z)\le\int_{\mathbb{D}}|(\sigma_a\circ\varphi)''(z)|\,dA(z)+\int_{\mathbb{D}}|\sigma_a''(\phi(z))||\phi'(z)|^2\,dA(z).$$

By (2.4) and the assumption we obtain (2.5). We complete the proof. □

3. Composition operators between B_1 and B_p

Theorem 3. *Let φ be a holomorphic mapping of $\mathbb{D}$ into itself and $1<p\le\infty$. Then C_ϕ is compact from B_1 to B_p if and only if*

$$\lim_{|a|\to 1}\|C_\phi\sigma_a\|_{B_p}=0. \tag{3.1}$$

PROOF. Suppose C_ϕ is compact from B_1 to B_p. Choose $\sigma_a(z)-a\in B_1$ which converges to 0 uniformly on any compact subset of $\mathbb{D}$. Thus

$$\lim_{|a|\to 1}\|C_\phi\sigma_a\|_{B_p}=\lim_{|a|\to 1}\|C_\phi(\sigma_a-a)\|_{B_p}=0.$$

Conversely, for the case $1 < p < \infty$ we consider a bounded sequence $\{f_n\} \subset B_1$ converges to zero uniformly on any compact subset of $\mathbb{D}$ and $\|f_n\|_{B_1} \leq 1$. To end our proof it suffices to show that $\|f_n \circ \phi\|_{B_p} \to 0$ as $n \to \infty$. Let

$$f_n(z) = \sum_{k=1}^{\infty} c_{n,k}\sigma_{a_{n,k}}(z), \qquad a_{n,k} \in \mathbb{D}$$

with

$$\|f_n\|_{B_1} \leq \sum_{k=1}^{\infty} |c_{n,k}| \leq 2, \qquad n = 1, 2, \ldots .$$

Similar to the proof of Theorem 1 one can prove that (3.1) implies that

$$\lim_{r\to 1} \sup_{a\in\mathbb{D}} \int_{|\phi(z)|>r} \left|(\sigma_a \circ \phi(z))''\right|^p (1 - |z|^2)^{2p-2}\, dA(z) = 0. \tag{3.2}$$

Thus, for given $\epsilon > 0$, there exists an r, $0 < r < 1$ such that for all $a \in \mathbb{D}$

$$\sup_{a\in\mathbb{D}} \int_{|\phi(z)|>r} \left|(\sigma_a \circ \phi(z))''\right|^p (1 - |z|^2)^{2p-2}\, dA(z) \leq \frac{\epsilon}{2^{p+1}}.$$

Therefore, by Hölder's inequality

$$\begin{aligned}
&\int_{|\phi(z)|>r} \left|(f_n(\phi(z)))''\right|^p (1 - |z|^2)^{2p-2}\, dA(z) \\
&= \int_{|\phi(z)|>r} \left|\sum_{k=1}^{\infty} c_{n,k}\left(\sigma_{a_{n,k}}(\phi(z))\right)''\right|^p (1 - |z|^2)^{2p-2}\, dA(z) \\
&\leq \left(\sum_{k=1}^{\infty} |c_{n,k}|\right)^{p-1} \int_{|\phi(z)|>r} \sum_{k=1}^{\infty} |c_{n,k}| \left|\left(\sigma_{a_{n,k}}(\phi(z))\right)''\right|^p (1 - |z|^2)^{2p-2}\, dA(z) \\
&\leq \left(\sum_{k=1}^{\infty} |c_{n,k}|\right)^{p} \sup_{a\in\mathbb{D}} \int_{|\phi(z)|>r} \left|\left(\sigma_a(\phi(z))\right)''\right|^p (1 - |z|^2)^{2p-2}\, dA(z) \leq \epsilon/2.
\end{aligned}$$

On the other hand, we have

$$\int_{|\phi(z)|\leq r} \left|(f_n(\phi(z)))''\right|^p (1 - |z|^2)^{2p-2}\, dA(z) < \epsilon/2$$

as $n \to \infty$. Hence

$$\lim_{|a|\to 1} \int_{\mathbb{D}} \left|(f_n(\phi(z)))''\right|^p (1 - |z|^2)^{2p-2}\, dA(z) = 0.$$

Thus, C_ϕ is compact from B_1 to B_p.

For the case $p = \infty$, if (3.1) holds, then

$$\lim_{|a|\to 1} \|C_\phi \sigma_a\|_{\mathcal{B}} = 0. \tag{3.3}$$

By Theorem 1, C_ϕ is compact from $\mathcal{B}$ to $\mathcal{B}$. Since $B_1 \subset \mathcal{B}$, C_ϕ is compact from B_1 to $\mathcal{B}$. The proof is completed. □

Combining our theorems with Tjani's result, we are able to build the following new theorem.

Theorem 4. *Let ϕ be a holomorphic mapping from $\mathbb{D}$ into itself and $1 \leq p \leq q \leq \infty$. Then the following are equivalent*:

(a) *$C_\phi \colon B_p \to B_q$ is a compact operator.*

(b) $\|C_\phi \sigma_a\|_{B_q} \to 0$ *as* $|a| \to 1$.

References

1. J. M. Anderson, J. Clunie, and Ch. Pommerenke, *On Bloch functions and normal functions*, J. Reine Angew. Math. **270** (1974), 12–37.
2. J. Arazy, S. D. Fisher, and J. Peetre, *Möbius invariant function spaces*, J. Reine Angew. Math. **363** (1985), 110–145.
3. R. Aulaskari, J. Xiao, and R. H. Zhao, *On subspaces and subsets of BMOA and UBC*, Analysis **15** (1995), no. 2, 101–121.
4. M. Essén and H. Wulan, *On analytic and meromorphic functions and spaces of Q_K-type*, Illinois J. Math. **46** (2002), no. 4, 1233–1258.
5. K. Madigan and A. Matheson, *Compact composition operators on the Bloch space*, Trans. Amer. Math. Soc. **347** (1995), no. 7, 2679–2687.
6. J. Peetre, *New thoughts on Besov spaces*, Duke Univ. Math. Ser., vol. 1, Mathematics Department, Duke University, Durham, NC, 1976.
7. L. A. Rubel and R. M. Timoney, *An extremal property of the Bloch space*, Proc. Amer. Math. Soc. **75** (1979), no. 1, 45–49.
8. J. H. Shapiro, *Composition operators and classical function theory*, Universitext Tracts Math., Springer, New York, 1993.
9. M. Tjani, *Compact composition operators on Besov spaces*, Trans. Amer. Math. Soc. **355** (2003), no. 11, 4683–4698.
10. N. Zorboska, *Composition operators on weighted Dirichlet spaces*, Proc. Amer. Math. Soc. **126** (1998), no. 7, 2013–2023.

Department of Mathematics, Shantou University, Shantou, Guangdong 515063, People's Republic of China

E-mail address, H. Wulan: `wulan@stu.edu.cn`

E-mail address, C. Xiong: `chengji_xiong@yahoo.com`

Centre de Recherches Mathématiques
CRM Proceedings and Lecture Notes
Volume **51**, 2010

Whether Regularity is Local for the Generalized Dirichlet Problem

Paul M. Gauthier

ABSTRACT. We give an example which shows that a regular boundary point for the classical Dirichlet problem need not be regular for the generalized Dirichlet problem.

Let G be a bounded open set in $\mathbb{R}^n$. The classical Dirichlet problem for G is the problem of the existence, for every continuous function φ on ∂G, of a harmonic function u in G having φ as boundary values. To solve the Dirichlet Problem, Lejeune Dirichlet introduced a variational method, which asserts that a solution u can be attained as a minimizer of the Dirichlet energy in a certain function space. Bernard Riemann named this method the Dirichlet Principle and assumed that such a minimizer exists. However, Karl Weierstrass, in 1870, provided a counterexample to the existence in general of a minimizer. In 1899, David Hilbert gave a rigourous solution to the Dirichlet problem by justifying the Dirichlet principle, under certain conditions, thereby foreshadowing the introduction of Hilbert space.

The Dirichlet problem is attacked by somehow providing a candidate u_φ for a solution. Let us call such a candidate a generalized solution. Once we have a generalized solution u_φ, there remains the problem of showing that u_φ has the desired boundary values φ. For any continuous function φ on ∂G, the Perron method provides a generalized solution, which we denote by u_φ^G and call the Perron solution.

A boundary point $p \in \partial G$ is said to be a *regular point* for the (classical) Dirichlet problem for G, if for each continuous function φ on ∂G, the Perron solution u_φ^G has the desired boundary behavior at p. That is,

$$\lim_{x \to p} u_\varphi^G(x) = \varphi(p). \tag{1}$$

Thus, it is a tautology to say that the classical solution to the Dirichlet problem exists for a bounded open set G if and only if each boundary point is regular.

Of course, if φ is not continuous, then there is no solution to the classical Dirichlet problem for the boundary data φ. However, the Dirichlet problem can be

2000 *Mathematics Subject Classification.* 31B20, 31B05, 31A25, 31A05.
Key words and phrases. Dirichlet problem.
Research supported by NSERC (Canada).
This is the final form of the paper.

generalized as follows. For many functions φ defined on ∂G, but not necessarily continuous, Perron's method still produces a harmonic function u_φ^G, which it is natural to call a Perron solution to the Dirichlet problem for the boundary function φ. In some sense, the Perron solution is the harmonic function which makes the best attempt at being a classical solution. If a classical solution exists, then the Perron solution exists and coincides with the classical solution.

Marcel Brelot has shown that the Perron solution u_φ^G exists if and only if φ is integrable with respect to harmonic measure μ_a^G, for some (equivalently, for every) $a \in G$. Moreover, we have the integral representation

$$u_\varphi^G(a) = \int_{\partial G} \varphi \, \mathrm{d}\mu_a^G, \qquad a \in G.$$

Norbert Wiener showed that a boundary point $p \in \partial G$ is regular for the (classical) Dirichlet problem if and only if the complement of G is not *thin* at p. For example, if the complement of G contains a cone with vertex at p, then p is a regular point.

Of course this implies that regularity is a local condition. The regularity or non-regularity of a boundary point $p \in \partial G$ depends only on the nature of the open set G *near* the point p.

Our definition of regularity (which is the usual one) is for the *classical* Dirichlet problem. Now that we have introduced the more general Perron solution to the Dirichlet problem, it is very tempting to think that regularity is local also in terms of Perron solutions. Namely, one might think that if $p \in \partial G$ is regular for the (classical) Dirichlet problem, then (1) holds whenever u_φ^G makes sense and φ is continuous at p.

The purpose of this note is to provide a counterexample. This example was formulated in a conversation with Aurel Cornea over 30 years ago.

Example 1. There exists a bounded simply connected domain D in $\mathbb{R}^2$, having the point $(0,0)$ as a regular boundary point for the (classical) Dirichlet problem and containing the interval $\{(x,0) : 0 < x < 1\}$, and there exists a function φ on ∂D integrable with respect to harmonic measure (so the Perron solution u_φ^D exists) and a neighborhood U of $(0,0)$ in $\mathbf{R^2}$ such that: $\varphi(x,y) = 0$ for $(x,y) \in U \cap \partial D$, but

$$\limsup_{x \searrow 0} u_\varphi^D(x,0) = +\infty.$$

PROOF. It is easy to give an example of an open set G having the required properties, except that it is not connected. We shall then add (carefully chosen) connecting channels between components of G to obtain the desired domain D.

For $n = 0,1,2,\ldots$, let R_n be the open rectangle

$$R_n = \{(x,y) \in \mathbb{R}^2 : 2^{-n-1} < x < 2^{-n}, |y| < 1\}$$

and denote by

$$H_n = \{(x,y) : 2^{-n-1} < x < 2^{-n}, y = \pm 1\}$$

the upper and lower boundary segments of R_n. Let $(x_n, 0)$ be the mid-point of the rectangle R_n. Thus, $x_n = (2^{-n-1} + 2^{-n})/2)$.

Set $G = \bigcup_m R_n$. We now define the function φ on ∂G. First of all, we put $\varphi(x,y) = 0$ on all vertical boundary segments $\{(2^{-n}, y) : |y| \le 1\}$, $n = 0,1,2,\ldots$, and $\{(0,y) : |y| \le 1\}$. On horizontal boundary segments $\{(x, \pm 1) :$

$2^{-n-1} < x < 2^{-n}\}$, $n = 0, 1, 2, \ldots$, we set $\varphi(x, \pm 1) = \lambda_n$, where $\lambda_n > 0$ is chosen so large that the value of u_φ^G, at the mid-point $(x_n, 0)$ of the rectangle R_n, is greater than n:

(2) $$u_\varphi^G(x_n, 0) > n.$$

The open set G and the boundary function φ have all of the required properties with the exception that G is not connected.

We shall now construct a domain D from G. For each $n = 1, 2, \ldots$, let S_n be a segment

$$S_n = \{(2^{-n}, y) : |y| < \epsilon_n\},$$

for some $0 < \epsilon_n < 1$ to be chosen later. Set

$$D_n = \bigcup_{k=0}^{n} R_k \cup \bigcup_{k=1}^{n} S_k$$

and

$$D = \bigcup_{k=0}^{\infty} R_k \cup \bigcup_{k=1}^{\infty} S_k.$$

By abuse of notation, we denote by u_φ^n the Perron solution of the Dirichlet problem on D_n, with boundary values φ restricted to ∂D_n. This makes sense, since $\partial D_n \subset \partial G$. Let $\mu_{a,b}^D$ and $\mu_{a,b}^n$ denote harmonic measure for the domains D and D_n at a point (a, b) respectively in D or D_n.

From the maximum principle,

$$\mu_{x,y}^1(H_1 \cup S_2) < \mu_{x,y}^{R_0}(S_1), \qquad (x, y) \in R_0.$$

Thus, we may choose ϵ_1 so small, that

(3) $$\lambda_1 \cdot \mu_{x_0,0}^1(H_1 \cup S_2) < \frac{1}{2^1}.$$

We note that on D_1, by the maximum principle,

$$\mu_{x,y}^D(H_1) < \mu_{x,y}^1(H_1 \cup S_2),$$

and so, by (3),

(4) $$\lambda_1 \cdot \mu_{x_0,0}^D(H_1) < \frac{1}{2^1}.$$

Suppose, for $j = 1, 2, \ldots, n-1$, we have defined ϵ_j such that

$$\lambda_j \cdot \mu_{x_0,0}^D(H_j) < \frac{1}{2^j}.$$

We may choose ϵ_n so small, that

(5) $$\lambda_n \cdot \mu_{x_0,0}^n(H_n \cup S_{n+1}) < \frac{1}{2^n}.$$

We note that on D_n, by the maximum principle,

$$\mu_{x,y}^D(H_n) < \mu_{x,y}^n(H_n \cup S_{n+1}),$$

and so, by (5),

(6) $$\lambda_n \cdot \mu_{x_0,0}^D(H_n) < \frac{1}{2^n}.$$

Thus, by induction, (6) holds for all $n = 1, 2, \ldots$, and so

$$\int_{\partial D} \varphi \, d\mu^D_{x_0,0} = \sum_{n=0}^{\infty} \lambda_n \cdot \mu^D_{x_0,0}(H_n) < \lambda_0 \cdot \mu^D_{x_0,0}(H_0) + 1 < +\infty.$$

Hence, φ is integrable with respect to harmonic measure for D. Therefore, the Perron solution u^D_φ, by which we mean the Perron solution for the restriction of φ to ∂D, exists.

It follows from Theorem 6.3.6 in [1] that $u^D_\varphi > u^G_\varphi$ on R_n and so by the maximum principle and (2) it follows that

$$u^D_\varphi(x_n, 0) > n, \qquad n = 0, 1, 2, \ldots .$$

Thus, the Perron solution u^D_φ has all of the required properties. □

References

1. D. H. Armitage and S. J. Gardiner, *Classical potential theory*, Springer Monogr. Math., Springer, London, 2001.

Département de mathématiques et de statistique, Université de Montréal, CP 6128, succ. Centre-ville, Montréal, QC H3C 3J7, Canada

E-mail address: gauthier@dms.umontreal.ca

Titles in This Series

51 **Javad Mashreghi, Thomas Ransford, and Kristian Seip, Editors,** Hilbert spaces of analytic functions, 2010

50 **P. Robert Kotiuga, Editor,** A celebration of the mathematical legacy of Raoul Bott, 2010

49 **Miguel Abreu, François Lalonde, and Leonid Polterovich, Editors,** New perspectives and challenges in symplectic field theory, 2009

48 **David Avis, David Bremner, and Antoine Deza, Editors,** Polyhedral computation, 2009

47 **John Harnad and Pavel Winternitz, Editors,** Groups and symmetries: From Neolithic Scots to John McKay, 2009

46 **Jean-Marie De Koninck, Andrew Granville, and Florian Luca, Editors,** Anatomy of integers, 2008

45 **Panos M. Pardalos and Pierre Hansen, Editors,** Data mining and mathematical programming, 2008

44 **Stanley Alama, Lia Bronsard, and Peter J. Sternberg, Editors,** Singularities in PDE and the calculus of variationa, 2007

43 **Andrew Granville, Melvyn B. Nathanson, and Jozsef Solymosi, Editors,** Additive combinatorics, 2007

42 **Donald A. Dawson, Vojkan Jakšić, and Boris Vainberg, Editors,** Probability and mathematical physics: a volume in honor of Stanislav Molchanov, 2007

41 **André Bandrauk, Michel C. Delfour, and Claude Le Bris, Editors,** High-dimensional partial differential equations in science and engineering, 2007

40 **V. Apostolov, A. Dancer, N. Hitchin, and M. Wang, Editors,** Perspectives in Riemannian geometry, 2006

39 **P. Winternitz, D. Gomez-Ullate, A. Iserles, D. Levi, P. J. Olver, R. Quispel, and P. Tempesta, Editors,** Group theory and numerical analysis, 2005

38 **Jacques Hurtubise and Eyal Markman, Editors,** Algebraic structures and moduli spaces, 2004

37 **P. Tempesta, P. Winternitz, J. Harnad, W. Miller, Jr., G. Pogosyan, and M. Rodriguez, Editors,** Superintegrability in classical and quantum systems, 2004

36 **Hershy Kisilevsky and Eyal Z. Goren, Editors,** Number theory, 2004

35 **H. E. A. Eddy Campbell and David L. Wehlau, Editors,** Invariant theory in all characteristics, 2004

34 **P. Winternitz, J. Harnad, C. S. Lam, and J. Patera, Editors,** Symmetry in physics, 2004

33 **André D. Bandrauk, Michel C. Delfour, and Claude Le Bris, Editors,** Quantum control: Mathematical and numerical challenges, 2003

32 **Vadim B. Kuznetsov, Editor,** The Kowalevski property, 2002

31 **John Harnad and Alexander Its, Editors,** Isomonodromic deformations and applications in physics, 2002

30 **John McKay and Abdellah Sebbar, Editors,** Proceedings on moonshine and related topics, 2001

29 **Alan Coley, Decio Levi, Robert Milson, Colin Rogers, and Pavel Winternitz, Editors,** Bäcklund and Darboux transformations. The geometry of solitons, 2001

28 **J. C. Taylor, Editor,** Topics in probability and Lie groups: Boundary theory, 2001

27 **I. M. Sigal and C. Sulem, Editors,** Nonlinear dynamics and renormalization group, 2001

26 **J. Harnad, G. Sabidussi, and P. Winternitz, Editors,** Integrable systems: From classical to quantum, 2000

25 **Decio Levi and Orlando Ragnisco, Editors,** SIDE III—Symmetries and integrability of difference equations, 2000

24 **B. Brent Gordon, James D. Lewis, Stefan Müller-Stach, Shuji Saito, and Noriko Yui, Editors,** The arithmetic and geometry of algebraic cycles, 2000